TOURISM AND SUSTAINABILITY

Increasingly it is argued that the growth of tourism offers a means for Third World countries to escape the confines of 'underdevelopment', and that new forms of tourism allow this transition to be achieved sustainably and equitably. Building upon this fundamental precept, this book explores and challenges the notion of sustainability and its relationship to contemporary tourism in the Third World. It takes examples from throughout the Third World, and in particular draws upon primary research from Central America and the Caribbean, regions of prime importance in terms of new forms of tourism.

Taking a broad geographical and conceptual perspective, the authors contend that a clear understanding of the tourism process and its relationship to development can only be achieved by an interdisciplinary approach touching on environmentalism, socio-cultural studies, economics and development studies.

In the first part of the book the emergence of the concepts of sustainability and its application to development in general and contemporary tourism specifically, are critically examined. Tracing the inception of sustainability within environmentalism – and its extension into the realms of socio-cultural and economic thinking, policy and practice – it is argued that sustainability has emerged as a hegemonic discourse. The meaning of sustainability is competed over by a variety of interests: international agencies, national governments, new social and environmental movements, tour operators, tourists, and so on. The relationship between this and the growth of new tourism (*'ego'* 'alternative' or 'green' tourism) is then discussed.

Developing this conceptual framework, the second part of the book explores the relationship between a number of critical themes. There are chapters on 'tourists' and their relationship to new social movements, on tour operators, on the policies adopted by national governments, and on the impact of international and regional consortia.

The conclusion of the book draws the various strands together and offers a number of alternative ways for the development of tourism in the Third World to progress.

Martin Mowforth is a Research Fellow in Human Geography at the University of Plymouth. **Ian Munt** is an Associate Lecturer at the Open University.

TOURISM AND SUSTAINABILITY

New tourism in the Third World

Martin Mowforth and Ian Munt

London and New York

First published 1998
by Routledge
11 New Fetter Lane, London EC4P 4EE

Simultaneously published in the USA and Canada
by Routledge
29 West 35th Street, New York, NY 10001

Reprinted 1998, 2000

Routledge is an imprint of the Taylor & Francis Group

© 1998 Martin Mowforth and Ian Munt

Typeset in Garamond by Keystroke, Jacaranda Lodge, Wolverhampton
Printed and bound in Great Britain by Biddles Ltd, Guildford and King's Lynn

British Library Cataloguing in Publication Data
A catalogue record for this book is available from the British Library,

Library of Congress Cataloging in Publication Data
A catalogue record for this book has been requested

ISBN 0–415–13763–2 (hbk)
0–415–13764–0 (pbk)

For Herbie and Feathers

CONTENTS

vii

CONTENTS

CONTENTS

FIGURES

TABLES

BOXES

ACKNOWLEDGEMENTS

We acknowledge permission granted by the following persons and organisations to use extracts from their work: Professor Tej Vir Singh, Editor of *Tourism Recreation Research*, Lucknow; Evelyne Hong, an anthropologist working with the Consumers' Association of Penang and the Third World Network; the *Annual Review of Anthropology*; Elsevier Science Ltd, Kidlington, Oxford, for extracts from Stephen Britton's 'The political economy of tourism in the Third World'; Pion Ltd. London for extracts from Britton's 'Tourism, capital and place: towards a critical geography of tourism'; the WWF–UK for an extract from *Beyond the Green Horizon*; the Ecumenical Coalition on Third World Tourism for an extract of Paul Gonsalves' article in *Contours; Envío* (Managua) for an extract from Pierre Galand's letter of resignation to the World Bank's NGO Forum (1994); Aitken & Stone Ltd for permission to reproduce an extract from Martha Gellhorn's work (1990); the World Travel and Tourism Council for two extracts of articles by the WTTC's President, Geoffrey H. Lipman; Professor Colin Price of University College of North Wales for an extract of his paper presented to a Conference on Values and the Environment at the University of Surrey in 1993; the Programme for Belize for extracts from their 1989 newsletter; the Catalyst Collective Ltd for material from Oliver Tickell's article 'After The Summit' (1992); the Association of Independent Tour Operators; Rick Holland currently of the Department of Responsible Tourism of the Institute of Central American Studies, San José, Costa Rica for extracts from his writings for the San José Audubon Society in 1992; the *Tico Times* (San José) for extracts from Jeff Marshall's letters (1994); Tirso Maldonado of the Fundación Neotrópica (San José) for extracts from an article published in 1992; the *Ecologist* for an extract from Anita Pleumarom's article (1994); Dr Hugh Somerville of British Airways for an extract from BA's *Annual Environment Report 1995*; *Amandala* (Belize City) for extracts from articles; the IUCN for permission to reproduce the table of protected area categories (Table 6.2); Peter Stone of Green Horizons Travel for extracts of material from the company's information pack (1995); Sue Curtin for access to transcripts of interviews with environmental managers of numerous tour companies (1996); the *Guardian* for extracts from articles by Miles Warde (1992), Ros Coward (1996), R. Edwards

(1992) and Pass Notes 223 (1993); *New Internationalist* for extracts from articles by A. Carothers (1993) and A. Imam (1994); Biff Products for permission to reproduce Figure 4.1; Butterworth Heinemann for an extract from J. Krippendorf's *The Holidaymakers*; the *Financial Times* for extracts from Richard Gordon's article (1994); Zed Books for extracts from Wolfgang Sachs' work and from essays by Vandana Shiva; Blackwell for allowing us to adapt Gregory's table on Modernism and Fordism (1994); *Cultural Survival* for extracts from Daltabuit and Pi-Sunyer's work (1990); the *Independent* for extracts from articles by Isabel Wolff (1991), Steve McClarence (1995) and Martin Wright (1996); the Ecotourism Society of the USA for extracts from their International Membership Directory; *Wanderlust* magazine for extracts from an article by Paul Morrison; Dr Susan Stonich of the University of California for extracts from the article in the *Journal of Sustainable Tourism* (1995) of which she was co-author.

We especially wish to thank *Race and Class* and *Theory, Culture and Society* for allowing us to modify their respective articles 'Eco-tourism or ego-tourism?' (Munt, 1994a) and 'The Other postmodern tourism: travel, culture and the new middle class' (Munt, 1994b), the latter being reproduced also by permission of Sage Publications Ltd. We would also like to acknowledge the inspiration provided by the material reproduced from the Open University course (D215) 'The Shape of the World'.

For permission to use material communicated over the internet, we acknowledge Fred Morris of Ecopaz (Brazil), Dr Hilton da Silva (Brazil), and Regina McGoff of the Centre for Global Education, Augsburg College, Minneapolis.

The authors and publisher also wish to thank the following for permission to reproduce copyright material. Every effort has been made to contact copyright holders and we apologise for any inadvertent omissions: Cassel Academic, Earthscan Publications Ltd, Explore Worldwide Limited, The Institute of Race Relations, The Regency Corporation Ltd.

We owe much to the work of the following organisations which have produced reports of special relevance and significance for this field of study: the Catholic Institute for International Relations (CIIR), London; the International Institute for Environment and Development (IIED), London, especially the work of Jules Pretty on local participation (and acknowledgement is made here of the quotations from Jon Tinker in the IIED's 'UNCED: A User's Guide' no. 1 written by Koy Thomson); the Panos Institute, London; the Latin America Bureau (LAB), London; Survival, London, for their reports on tourism and tribal peoples; and of course to Tourism Concern, London, without whose reports, assistance and advice our analysis would be considerably weaker. Particular mention should be made of the forbearance of Tricia Barnett and Sue Wheat of Tourism Concern in the face of our many enquiries.

The work of Tim Absolom, Richard Freeman, Brian Rogers and Ian Stokes in the Cartography Unit of the Department of Geographical Sciences at the University of Plymouth deserves special mention for its professionalism and high quality and for the patience with which they accepted the regular changes we required to the maps, graphs, tables and drawings they created for us. David Griffiths and Tony Smith of the Photography Unit also deserve the same credit for their assistance. We also thank Norna Beadle for the specially commissioned cartoon (Figure 3.1) and all the changes that we required for this too. Help and encouragement have been given in abundance by our numerous editors at Routledge, Tristan Palmer, Matthew Smith, Sarah Lloyd, Casey Mein, Val Rose and Bill MacKeith.

We reserve our warmest thanks for Alison Stancliffe who not only supplied the material for the case studies of Bali but also commented on the draft manuscript throughout its production. We have found her comments and suggestions to be of enormous help. We could not have had an internal reviewer who was more thorough, prompt, intellectually rigorous and challenging or who was more aware of developments in the field than Alison. Many changes (we hope improvements) were made as a result of the extremely helpful comments made by our two invisible and unknown external reviewers, one of whom liked the material and one of whom hated it – the best combination.

Needless to say – but we most certainly should say it – June and Nicky bore the brunt of the late nights, extended telephone conversations, tiredness and lack of attention to other things caused by the writing of the book. They deserve more than gratitude for their patience, encouragement, opinions and help.

1

INTRODUCTION

In recent years the image of the Third World in western minds has emerged from that of cataclysmic crisis – of famine and starvation, deprivation and war – to represent the opportunity for an exciting 'new style' holiday. Offering the attraction of environmental beauty and ecological diversity, travel to many Third World countries has been promoted, especially among the middle classes, as an opportunity for exciting, 'off-the-beaten-track' holidays and as a means of preserving fragile, exotic and threatened landscapes and providing a culturally enhancing encounter. At the same time, some Third World governments have seized upon this new-found interest and have promoted tourism as an opportunity to earn much-needed foreign exchange – another attempt to break from the confines of 'under-development'.

There is something hugely contradictory in viewing the Third World through an analysis of tourism. At one extreme there is the association of the Third World with fundamentalist terrorism or overpopulation, poverty and disease. At the other there are also more subtle manipulations by the tourism industry of friendly Third World peoples, natural and pristine environments and ecological and cultural diversity. Both extremes, we will argue, form an important element in understanding contemporary Third World tourism. Importantly, and central to the line of argument developed in the following chapters, tourism is a way of representing the world to ourselves and to others. It cannot be understood as just a means of having some enjoyment and a break from the routine of every day, an entirely innocent affair with some unfortunate incidental impacts. Rather, a deeper understanding of tourism is needed to appreciate fully its content and expression as well as its potential impact.

PURPOSE AND LIMITS OF THE BOOK

This book has three focal points. The first is tourism, the second is sustainability, and the last is the Third World. It is the overlap between these themes that is explored here.

It is important to note that this is not an attempt to define and assess what is and what is not 'sustainable Third World tourism'. This would lead, of necessity, into prescribing what we consider the parameters of sustainability to be in the first place. Rather, we shall be discussing the way in which sustainability is reflected through (or enacted in) current developments in Third World tourism. Most particularly, the discussion focuses upon new forms of tourism that have arisen, in part as a response to the perceived unsustainability of much tourism development to date.

The focus is relatively tight in two important respects. There are now many studies of tourism, especially tourism in the Third World, that catalogue and discuss its growth and impacts. In particular, studies have tended to highlight the economic, environmental and socio-cultural impacts of conventional package tourism.[1] Rather than adding to this body of work, we focus particularly on the under-researched 'new' forms of tourism promoted in the First World and patronised almost exclusively by North Americans, Europeans and Australasians. Although proportionately small relative to all forms of Third World tourism, they are significant in terms of both the claims that are made about them and the rate at which they are growing.

Second, we do not seek to add to the growing number of accounts which attempt to identify sustainable tourism development (in terms of the environment, economy and culture) or prescribe good practice methods and tools for achieving this goal. Many of these have emerged from First World sources and necessarily involve judging the sustainability or otherwise of varying types of Third World tourism development. Instead, we explore the way in which the claim of sustainability is used and applied to new forms of tourism (for example, the way in which ecotourism – a new form of tourism – is premised upon the notion of sustainability). One way of capturing this important difference in our approach is to state that this is a book about *sustainability* and *Third World tourism*, rather than *sustainable Third World tourism*. The former approach, adopted here, signals the need for a critical analysis of the issues, while the latter implies the need to define and prescribe models of good practice.

TOURISM AS A MULTIDISCIPLINARY SUBJECT

Rather than commencing a study of Third World tourism with the environmental, economic and socio-cultural impacts *of* tourism (worthy though these are as research considerations in themselves), the starting-point here involves seeking to understand how socio-cultural, economic and political processes operate *on* and *through* tourism. In other words, it is necessary to take a step back in the analysis of tourism. This stems in part from the weaknesses present in the tourism literature. As Britton observes:

> Although over-simplifying, we could characterise the 'geography of tourism' as being primarily concerned with: the description of travel flows; microscale spatial structure and land use of tourist places and facilities; economic, social, cultural, and environmental impacts of tourist activity;

impacts of tourism in third world countries; geographic patterns of recreation and leisure pastimes; and the planning implications of all these topics. . . . These are vital elements of the study of travel and tourism. But these sections are dealt within descriptive and weakly theorised ways.

(1991: 451)

This problem is of fundamental importance as it has led to the absence of an adequate theoretical foundation for understanding the dynamics of tourism and the social activities it involves.

There are, therefore, two identifiable groups of research. The first is concerned primarily with auditing, categorising, listing and grouping the outputs or consequences of tourism; the second approach is concerned primarily with conceptualising the forces which impact on tourism and, through an analysis of these forces, providing a broader context for understanding tourism. The crucial difference in the latter approach is that tourism is seen as a focal lens through which broader considerations can be taken into account, and it confirms the multidisciplinary foundation upon which tourism research is built as the only way in which tourism can be comprehended. As a personal activity, tourism is practised by a diverse range of the population; as an industry, it is multi-sectoral; and as a means of economic and cultural exchange, it has many facets and forms. Any comprehensive analysis of the field must therefore be multidisciplinary.

Accordingly, our discussion draws on economics, environmental theory, social theory, politics, geography, and international relations, for example. Inevitably, this breadth of analysis will mean that a number of relevant aspects are not examined in depth and do not necessarily cover the complexity of the matters under discussion. We hope, however, that the discussion will serve as a pointer to further research and study.

At the same time as using the concepts and paradigms of a range of academic and intellectual fields in order better to understand tourism, the study of tourism helps us to illuminate more general economic, political, social, geographical and environmental processes. We try not to see tourism as a discrete field of study. Rather, it is an activity which helps us to understand the world and the ways in which humans interact with the planet and with each other in a range of senses.

KEY THEMES AND KEY WORDS

We have attempted to draw attention to the principal components of the discussion under the banners of the key themes and key words. The relationships between the analysis and the key themes and key words is summarised diagrammatically in Figure 1.1.

The three themes seek to underpin the discussions throughout the book. The first is the uneven and unequal development which underlines the relationship especially between the First World and Third World (but also at inter- and intra-

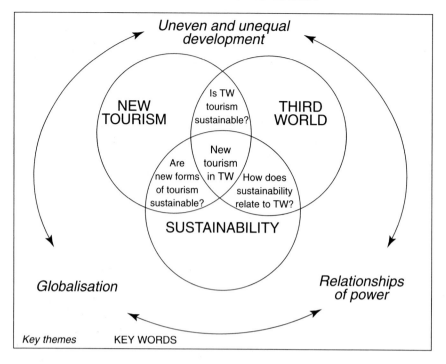

Figure 1.1 Key themes and key words

regional and intra-national levels), and through which all forms of tourism should be understood. At its most basic level this theme is reflected in the fact that it is people from the First World who make up the significant bulk of international tourists and it is they who have the resources to make relatively expensive journeys for pleasure, a clear example of inequality. Equally, processes of uneven development are reflected through the growing élite and newly wealthy classes in some Third World countries who are now able to participate in tourism. And of course tourism development is also highly uneven geographically. Areas that are in fashion today may fall out of fashion tomorrow.

The second and interrelated theme demonstrates why a critical understanding of tourism can be expanded through an analysis that places relationships of power at the heart of the enquiry. These relationships range from the power of First World countries in contrast to Third World countries, to the power wielded by international donors (from the World Bank to the European Union), the power of local élites in contrast to the local populations in tourist destination communities, and the power invested in tourists themselves.

The final theme is *globalisation*, a notion that at its core attempts to capture the idea that we are living in a shrinking world, a world in which places and countries are increasingly interdependent.

In addition to these three themes which inform the analysis of Third World tourism, there are a number of key words which will recur throughout the discussion. The idea of a key word is used here because it alerts us to the fact that words are open to debate and differing interpretations and that, consequently, there are problems in attaching uncontested meanings to words. As Williams (1988) suggests, the meanings of key words are closely related to the problems they seek to discuss. Take the idea of *sustainability* for example, a key word in this book. The very many meanings and interpretations of this idea are often a direct reflection of the problem that has been identified and both the explicit and implicit connections that people are making (Williams, 1988: 15). For environmentalists, to take just one example, the problem is one of degradation of the world's natural resources that activities such as tourism have caused. And the meaning attached to sustainability in this case is, at its core, ecological: the need to preserve and protect the natural environment. To industrialists, by contrast, sustainability may represent the opportunity to reduce costs and increase profit margins.

Similarly, *tourism* can also be regarded as a key word. The meanings attached to it are many and varied, and 'tourism' and 'tourist' have in some quarters become derided and ridiculed. How often are people heard to describe their holidays as being 'well away from tourists and the main tourism areas'? In other quarters, Third World tourism is regarded negatively, and the word symbolises a range of problems: environmental degradation, the distortion of national economies, the corruption of traditional cultures, and, on a trivial level for the individual, unsanitary conditions and food poisoning. The emergence of new forms of tourism (prefixed with *sustainable* and *eco-*, for example) is testimony to the identification of the problem and the attempt to signal that these new forms aim to overcome the problem and to be something that plain old 'tourism' is not.

For the purposes of this discussion, the term *new tourism* has been adopted. While this is a rather broad term, it helps to indicate that a variety of tourisms have emerged and that in some important respects these seek to distinguish themselves from what is referred to as mainstream or conventional mass tourism – the type of tourism that has attracted the most academic attention and the wrath of many commentators. It will be argued later that the 'new' in tourism also helps us to trace the relationships with *new* types of consumer (known as the *new* middle classes), *new* types of political movements (known as *new* socio-environmental movements) and *new* forms of economic organisation (known as post-Fordism).

Geography and global inequality

The history of development studies has thrown up a variety of terms that attempt to represent and categorise countries according to their wealth and social well-being. In particular, the terminology attached to countries lower down the 'human development index' (a widely adopted index ranking countries on a number of criteria) has been keenly disputed – should they be described as 'poorer', 'lower income', 'developing', 'under-developed', 'the South', 'Third World'? All such terms possess their advocates and detractors and no one term will suit all audiences. The problem of definition is made more acute when rapidly changing economic factors mean that old categories no longer hold good.

Principally, in this book, we refer to *Third World* (the last of our key words) and *First World*. Other terminology is used only where direct quotes from other sources necessitate this or where the meaning conveyed by the alternative term is distinctly different from these two terms (for example, in a differentiation between those relatively wealthy countries with significant new middle classes with a taste for new forms of tourism and those countries where the middle classes are currently demanding conventional forms of tourism, especially those in South Asia and the Pacific Rim). Occasionally, we also refer to the West and the Rest, where 'the West' represents the First World capitalist economies promoting western materialistic lifestyles; and the 'Rest' seeks to emphasise the inherent inequality in global development.

'Third World' also helps to reflect on and convey the way in which we are using this term in relationship to the notion of development. For example, the word 'developing' is avoided, because it implies that there is an end state to the process of development and that all countries will eventually reach a 'developed' state. By contrast, there are strong grounds for arguing that the process of development is one which actually causes under-development elsewhere. The term 'Third World', in other words, helps to emphasise the ways in which power, resources and development are unequally and unevenly shared globally. This is not to say, however, that the Third World is easily defined as a neat geographical entity coterminous with nation states. Inequality is of course not only manifest on a global scale; it occurs within and between countries and in relation to a variety of characteristics, particularly sex, ethnicity and class. Indeed, in Chapters 5–9 we examine a number of examples of inequalities at various different scales.

Although strict geographical divisions along national borders are increasingly meaningless, as a generalisation in talking of the First World at the end of the twentieth century we are referring to the power vested in the nation states and institutions of North America, Europe, Australasia and Japan. The Third World refers to those nation states and institutions that make up Latin America, the Caribbean, Africa and parts of Asia, although it is necessary to acknowledge the increasing wealth of some countries in South Asia and the Pacific Rim.

TOURISM AND GEOGRAPHICAL IMAGINATION

Tourism is one of the principal avenues through which our 'world-views' are shaped. This not only results from our holidays but also from the way destinations are represented through travel reviews, travel programmes and documentaries, travel brochures and guides, advertising and popular experience and exchange.

It is important to note at the outset that we are dealing with the nature of representation. Some geographers have adopted the term 'geographical imagination' as shorthand for these processes: the 'way we understand the geographical world, and the way in which we represent it, to ourselves and to others' (Massey, 1995: 41). It is also shorthand for emphasising that activities, issues, places and so on, are subject to competing interpretations. Benidorm, Bangkok and the Himalayas, for example, conjure up very different sorts of representations in our minds. As David Harvey suggests, the 'eye is never neutral and many a battle is fought over the "proper" way to see' (1989a: 1).

It is necessary to acknowledge this and work through the complexities rather than taking things for granted. Much of the argument presented in this book concerns the way in which tourism is a contested activity and how the field of tourism must be concerned with the nature of representation and interpretation. This involves the way in which we represent both our own activities (how we define ourselves as, for example, tourists, travellers, visitors, and what each of these categorisations entails)[2] and the places in which we holiday (for example, built-up beach resorts or remote regions).

The role of interpretation and representation is the key here. We all have our individual geographical imaginations and these are formed from a variety of factors – sex, age, class, ethnicity, culture, the media and many others. As authors, both white males, writing about Third World tourism from the 'privilege' of our comfortable First World lifestyles, our interpretation of the issues may differ markedly from the interpretation of others who are experiencing very different circumstances. Similarly, tourists interpret and represent their experiences in ways that may be fundamentally opposed to the experience of those being visited; and these interpretations and representations will differ between different types of tourists. Even the World Bank and International Monetary Fund have a particular geographical imagination of the Third World. Their representation of tourism and sustainability may also differ sharply from those of local communities in the countries where the policies of these supranational institutions are applied.

These differing geographical imaginations emphasise that representations of the world are socially constructed and that there is an array of factors that contribute to our understanding of the world; Benidorm, Bangkok and the Himalayas are, so to speak, the social products of tourism. Take, for example, the following extract discussing the impact of tourism in the mountainous regions of South Asia. This helps to emphasise that First World views of tropical environments or mountainous regions as those of unrestrained beauty and natural bounty, to be revered and mythologised, may not be held by all. It also suggests that what is seen as worthy of tourism changes over time.

Mountains for the local mountaineers have obtained an aura of mysticism acquired from the Western ideal. . . . The typical attitude of the moun-taineer toward the mountains once was 'if only they were flat, I could plough them', has changed through working and guiding Western Alpinists. This Western attitude is reinforced when the élite local porters and climbers are taken to Western countries for climbing workshops and tours. Gradually other notions that Westerners have about mountains such as conservation . . . seep into the vocabulary and thoughts of the local mountain populations as interaction with the tourists becomes more frequent and intense.

(Allan, 1988: 14)

This example of once divergent imaginations helps to emphasise two other points. First, some individuals, companies, institutions and countries will be better able to diffuse their particular imaginations to others. When we think of Bali, Goa or Hawaii, for example, the images and representations that are called forth are less likely to be of local people struggling to maintain identity in the wake of mass tourism development and more likely to be of palm-fringed beaches and crystal blue waters (often the products of travel brochures, travel reviews and holiday programmes). In short, some imaginations are more powerful than others (Allen and Massey, 1995).

Second, and related to this exercise of power, differing and highly divergent geographical imaginations imply that there is a high level of contest as domi-nant forms attempt to impose themselves on subordinate imaginations. The core of the argument in this book is about these contested views as they are expressed through tourism, ranging from the struggles between multinational tourism corporations and environmentalists, and the divergent understandings and definitions of sustainability, to the social struggle between so-called 'travellers' and tourists.

Finally, geographical imagination also makes a very distinctive contribution to our understanding of globalisation and its impacts (Allen and Massey, 1995). There is a sense that we are living in a smaller, more compressed and inter-connected world, and tourism is often invoked in this process of globalisation. On the one hand, places are drawn into the sphere of global tourism and the feeling of a smaller world encourages consumption of further places. On the other hand, some places deemed unattractive to tourism are marginalised from the processes of global interdependence. The relationship is, rather, a complex and symbiotic one. Tourism is both cause and consequence within globalisation.

LAYOUT OF THE BOOK

The book aims to present many of the issues and debates associated with different aspects of new tourism as questions. To some of these questions we give our own interpretations; to others we do not respond, choosing instead to leave the issue open for you to examine further through the references provided.

Given the multidisciplinary approach to the subject, there is clearly a myriad of non-tourism sources that inform the debates and issues within and surrounding tourism, while recent years have seen a large and growing number of works written on tourism itself. Inevitably, therefore, the references provide but a sample of the breadth of writing on this subject. (It should also be noted that for some authors we have for practical reasons cited a number of works, to enable you to access the necessary resources; it does not necessarily indicate that each piece of work develops a unique argument.)

The book is in two parts. The first, commencing with Chapter 2, attempts to broaden the conceptualisation of sustainability and trace its relationship to the processes of economic, cultural and political globalisation. This discussion also seeks to suggest how and why new forms of tourism have emerged and how they are related to sustainability. Central to this chapter is a consideration of power and the way in which sustainability is used to convey the interests of different groups which contest its meaning; these include international agencies, national governments, new social movements, tour operators and tourists.

Chapter 3 builds upon this notion that power must lie at the heart of tourism analysis. It looks at the existing critiques of tourism and argues that the majority of these are set within the context of mass international tourism. Relatively little thought has been given to the role of new forms of tourism; forms, that is, which seek to distinguish and define themselves in contrast to mass tourism. It is suggested that a critique of these forms of tourism is required and a number of characteristics of such a critique are set out.

Chapter 4 looks specifically at the recent emergence of the new forms of tourism and the terminology and definitions associated with them. Principles of 'sustainable' tourism are examined critically along with the tools generally used to define, assess and measure sustainability in tourism. We briefly question the future role of sustainability in tourism.

Chapters 2 to 4 therefore provide a conceptual framework within which the remainder of the book is developed. The second part of the book extends the relationships between sustainability, development and new forms of tourism as they apply to the major players: the tourists themselves, the socio-environmental movement, the operators in the industry, the communities at the destination end, and the national governments and international consortia which make decisions affecting tourism developments.

Chapter 5 examines the motives, the role and characteristics of those tourists (mostly from the First World) who take part in the new forms of tourism. Building upon the earlier discussion regarding the relationship between the new middle classes and new tourism, we suggest why new forms of tourism are of importance to the new middle classes of the First World and how they protect and enhance this activity.[3]

Many of the 'new' tourists are also represented by new social organisations (or what are referred to in this book as new socio-environmental organisations) which mobilise around issues such as ecology, the environment and human rights. Chapter 6 seeks to explore how new forms of tourism can be positioned in a discussion of environmentalism. A number of forms of environmentalism are presented and the ways in which sustainability and new tourism are related to them are assessed. The relationship between First World and Third World socio-environmental organisations represents a struggle for power and control over problem interpretation and policy. The charge that global environmental organisations often intervene in ways which can be seen as eco-colonialist is also considered.

Chapter 7 analyses the way in which the tourism industry, in all its guises and facets, absorbs and adapts itself to the notion of sustainability. Different players in the industry react in different ways, but they all attempt to define sustainability in a way that reflects their particular interest; and we look at the ways in which this process of self-adjustment and subsumption of the notion of sustainability by different sectors of the industry takes place.

Chapter 8 defines what is implied by the term 'host populations' and examines the importance of local participation in the activity of tourism. A spectrum of meanings of the word 'participation' exists and a number of case studies are presented and assessed against this range. Examples of the displacement of local populations by tourism developments are given, and it is clear that area protection measures (such as the creation of national parks) are often implicated in these displacements and subsequent resettlements. The techniques for measuring and improving local participation in planning, decision-making and implementing tourism schemes are assessed, and most of these appear to stem from First World professionals. The formation of local élites in exploiting the techniques and benefits of both tourism research and tourism activity is examined.

Chapter 9 looks at the role of governments and supranational institutions in tourism. The importance of political analysis in tourism studies and of analysis of the external influences on national governments is noted. With new forms of tourism emerging, many Third World countries have identified the opportunity of developing tourism around notions of ecology and environment and the adoption of discourses of sustainability by governments is described.

In the final chapter, we draw several general conclusions from the discussion, and indulge in some crystal ball gazing about the future development of new tourism and its relationships with the economic, social, cultural, political and geographical aspects of development. We also speculate on the future importance and absorption into 'reality' of the notion of sustainability.

2

GLOBALISATION AND SUSTAINABILITY

There are two essential strands to Chapter 2 which are subsequently followed through in the rest of the book. First, it is necessary to establish the relationship between the Third World and tourism. At first sight this may seem an obvious link, but it is important to establish the dynamic nature of the relationship and frame it within broader, global changes. It is necessary to demonstrate the way in which tourism has expanded in the Third World, and how, so to speak, the frontiers of tourism have increasingly been pushed back. In order to do this we refer to the concept of globalisation (the first of our key words), a notion that attempts to capture the way in which the world has shrunk in relative terms. The analysis starts therefore from changes in the First World and we review the ways in which globalisation is reflected in economics, culture and politics.

We also need to look at the underlying factors which have resulted in some profound changes to tourism and at how new forms of tourism (in contrast to mass or mainstream tourism) have emerged in Third World destinations. This is not to argue that mass tourism to Third World countries has been magically displaced, but that alternatives to such tourism are now well established; subsequent chapters will demonstrate what forms such new tourisms take and discuss the extent to which these are manifest in the Third World. It is the claims made by many of these new tourisms to be sustainable and appropriate alternatives to mass tourism that this book seeks to explore.

The second important strand is the notion of sustainability and sustainable development. We trace the relationship of these ideas to Third World tourism. An important question posed by many observers is whether and in what ways development in Third World countries, including tourism development, is (or is not) sustainable. This is especially reflected in the debates over Third World tourism where it is often argued that existing forms of mass tourism development are unsustainable in terms of the negative impacts on the environment, the way in which it corrupts and 'bastardises' local cultures and the manner in which any potential economic benefits are frittered away as a result of the First World ownership of much of the tourism industry globally. It is from this negative premise that much new tourism takes its cue, in an attempt to redress the impacts of tourism and establish forms of (new) tourism that are environmentally, economically and culturally sustainable.

But our interpretation and discussion of sustainability is broader. This chapter takes a step back and examines the ways in which sustainability is reflected through the wider processes in economics, culture and politics in the First World; and it is argued that these changes are closely associated with globalisation. It starts from the important question of what is being sustained, by whom and for whom; do all interest groups have the same intentions or aspirations in terms of sustainability? In other words, who decides what sustainability means and entails, and who dictates how it should be achieved and evaluated?

In order to begin to answer these questions, the chapter concludes by looking at different notions of power which are useful in our understanding of how and why power works; again, subsequent chapters explore the usefulness of such concepts in analysing Third World tourism.

This chapter (together with Chapters 3 and 4) introduces a range of concepts and debates that are built upon in the second part of the book. They are important concepts in that they provide a way of refocusing the analysis and arguing that the emergence and development of Third World tourism can only be understood within a much broader (or global) frame of analysis.

TOURISM IN A SHRINKING WORLD

Globalisation is a concept that is increasingly invoked in the analysis of tourism. With the seemingly limitless spread of tourism to the four corners of the world, the embracing of virtually any form of activity and the general ubiquity of tourists and tourism, the temptation to invoke globalisation in discussions of tourism has been great. For the most part, references to globalisation in tourism analysis have followed suit and tended towards casual and uncritical statements.

So what is globalisation and how can it most usefully be deployed in a critique of tourism? Essentially globalisation is a concept that seeks to encapsulate processes operating on a global scale. It refers to the ever-tightening network of connections which cut across national boundaries, integrating communities in new 'space–time combinations' (Hall, 1992a: 299) and resulting increasingly in the feeling that the world is a single interdependent whole, a shrinking world where local differences are steadily eroded and subsumed in a homogeneous mass or single social order. As Giddens puts it, 'Our lives, in other words, are increasingly influenced by activities and events happening well away from the social context in which we carry on our day-to-day activities' (1989: 520). Clearly then, globalisation is much more than an abstract concept and represents a fact of our everyday lives that the world, in some crucial respects, has shrunk.

Globalisation has been an especially appealing organising concept for geographers because it emphasises the way in which economic and social relationships have been stretched and interwoven across the globe. This is not to suggest, however, that globalisation is a new phenomenon. These processes of interconnection have been taking place for hundreds of years as part of an ongoing transition in the development of global capitalism. The qualitative difference today is the pace at which the process of globalisation is happening, with a current extraordinarily intensified phase of global transformation and change. So how can we understand contemporary changes? McGrew captures the richness of the concept and the impact of this acceleration by expanding on two distinct dimensions which he terms scope (or *stretching*) and intensity (or *deepening*):

> On the one hand it defines a process or set of processes which embrace most of the globe or which operate worldwide: the concept therefore has a spatial connotation. Politics and other social activities are becoming 'stretched' across the globe. On the other hand it also implies an intensification in the levels of interaction, interconnectedness, or interdependence between the states and societies which constitute the world community. Accordingly, alongside the 'stretching' goes a 'deepening' of the impact of global processes on national and local communities.
>
> (1992: 107)

The intensity or deepening of which McGrew speaks is reflected in the way that Third World countries and communities are increasingly drawn into tourism. A simple collage of newspaper travel pages (Figure 2.1) is testimony to the integration of once peripheral areas into this process.

Globalisation, then, provides an organising concept through which we are able to explore the extent and impacts of global change in terms of economics, culture and politics (Allen and Massey, 1995). Economic globalisation conveys the manner in which economic relationships and flows have been stretched across the globe. In the context of tourism, many point to the phenomenal growth of the industry in a global sense (it is now reputed to be the largest single industry) and the rapidity with which new places are continually drawn into the tourism process. Take for example an average travel agent and consider the range of destinations on offer. Not only has the number of holiday destinations increased but the distances between destinations and markets has increased markedly, and we will be examining how new tourism practices have helped to accelerate this process. This also suggests that globalisation is about capitalising on the revolutions in telecommunications, finance and transport, all of which have been instrumental in the 'globalisation' of tourism. In addition, as Box 2.1, taken from the *Financial Times*, emphasises, tourism for an increasing number of Third World countries is big business. Interestingly this suggests that it is not just capital and commodities that can be transported and transferred easily across the world, but tourists too. It is necessary therefore to consider how changes in contemporary global capitalism have impacted upon the development of tourism, a point we take up later.

Cultural globalisation, many would claim, signals the emergence of a single global culture reflected in global consumerism, most usually based on US lifestyles. As one of the most prominent tourism commentators, Dean MacCannell,

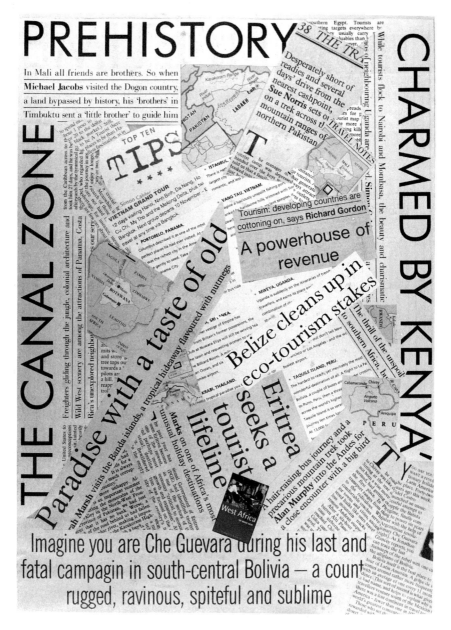

Figure 2.1 Tourism and the shrinking world

Box 2.1 *A powerhouse of revenue*

Tourism: developing countries are cottoning on

At a recent tourism conference . . . Stephen Dorell, the UK heritage secretary, told a group of tourism leaders that Britain needs to regain its share of the growing global tourism market. At that moment, a London bus, emblazoned with a sign inviting Londoners to 'Visit Korea in 1994', thundered overhead on Vauxhall Bridge.

The problem for Britain, and other traditional tourist destinations, is that the rest of the world has cottoned on to tourism. As the biggest growth industry, employer and source of revenue around the world, many developing countries have realised a quicker way to buy into first world affluence is by boosting their tourism potential rather than by selling tractors, bananas and rice.

Global tourism, according to the World Travel and Tourism Council, will double in size between 1990 and 2005. The market has been growing by 5 per cent a year in real terms since 1970. . . . The sheer size of the global industry has awakened many multinational companies to the possibilities of global brands and market dominance. As airlines form international networks and alliances, so too, travel agents, hotel brands and car hire firms are banding together . . .

The only areas not targeted by the global brands are the Middle East and Asia, where international arrivals in East Asia and the Pacific grew four times faster than the world average of 69m visitors. While arrivals were up by 12.6 per cent, revenue grew by 15.2 per cent to US$52.6bn. The World Tourism Organisation forecast 101m arrivals in East Asia and the Pacific by 2000 and 190m by 2010 . . .

Vietnam is the latest fashionable destination for tourists. There has been huge growth in tourism to Vietnam. . . . Foreign investment in Vietnam in the first quarter of this year jumped by 58 per cent compared to the same period last year. Between 1988 and 1990, most projects involving foreign money were in the hotel and oil sectors. The total amount of foreign investment in 1994 is expected to reach US$3.5bn.

The emergence and acceptability of Vietnam was confirmed recently when British Airways announced that it is negotiating to operate two flights per week from London to Ho Chi Minh City.

Robert Burns, chairman of the World Travel and Tourism Council, believes Shanghai will emerge in 10 years as the most important Asian city. A new airport, which could handle 150 landings an hour, is being built. Hotels in Shanghai are operating at near capacity and room rates are rocketing.

As Mr Burns pointed out, Japan now has a policy, the result of a balance of trade problem, that 20 per cent of its population should travel abroad by 2010. If China ever had just two per cent of its population travelling overseas, the rest of the world would be inundated with Chinese tourists.

(Gordon, 1994)

argues, in this 'giant, or global, socio-economic system, a "New World Order" ... there is a pretence that all the sub-groups and communities are actually phased together in some kind of relational equilibrium' (1992: 169). Burns and Holden (1995) also imply that the new world order has a direct relationship to the emergence of global consumerism, exemplified by 'McDonald's franchises' or what Ritzer refers to as the McDonaldisation of society (1993).

Much commentary on tourism, and particularly new forms of tourism, is dedicated to bemoaning the armies of mass tourists who, it is claimed, voraciously consume places and cultures, transforming them into Disney-like extravaganzas where cultural inauthenticity is actively promoted. It is a cultural order, pundits tell us, in which sophistication, difference and authenticity are increasingly denied and cultural homogenisation is the norm. The task before us is to try to discern some of the most prominent changes in the cultural sphere and judge whether this supposed homogenisation is a reasonable reflection of reality.

Finally, political globalisation focuses attention on what some argue is the erosion of the sovereignty of nation states, with territorial borders dismantled or melting away under the pressures of globalisation and the ascendency of trans-global politics and organisations. As will be argued in the second half of this book, such supranational organisations as the World Bank and International Monetary Fund have had far-reaching effects on many Third World countries and their development of tourism. Similarly, the globalisation of environmental issues, stressing the way in which our lives are inextricably linked and impact upon one another, has resulted in the emergence of vociferous debates over the environmental sustainability of tourism. Again such debates have had profound effects on the development of Third World tourism, and we shall be exploring them in the following sections.

This process of globalisation is clearly reflected in Box 2.2, which gives extracts from the final statement of the Asian Consultation on Tourism and makes some interesting observations and raises important issues. It locates the growth and expansion of tourism firmly within the 'complex nature of social change' and resonates some of the concepts that are discussed in this chapter. But it also sees the problem largely as one of 'mass tourism' and considers that solutions lie in the development of sustainable alternatives. The following chapters examine the scope of sustainable tourism and argue that the 'humani-sation' process which the Asian Consultation favours is much harder to achieve in practice, even given alternatives to mass tourism.

Uneven and unequal development

When, as a result of a coup d'état in the West African state of The Gambia, the Foreign Office advised British tourists to avoid the country (advice that was subsequently taken up by other European tourist-sending countries), the Gambian economy and tourism industry virtually collapsed. This is a clear

---

Box 2.2   Tourism as a global phenomenon

As a result of technological advancements and improvements in communications, tourism has become one of the fastest growing industries today. Global tourism is expected to continue to expand because people are beginning to discover more and more new destinations, and the travel industry is becoming more and more organised.

A critical attitude towards the negative effects of tourism on the culture of others and the environment has to lead to a creative endeavour to enter into dialogue, especially with those who are adversely affected by this process more than others. The complex nature of social change and the accompanying processes of modernisation, industrialisation, and economic development have thus to be the context in which tourism must be understood.

Today's mass tourism is not just the movement of people going from one country to another but the accompanying mass displacement of communities, its impact on traditional communities, and the involvement of large business corporations in this process . . .

The reality is one of many negative effects. These negative effects are the results of lack of proper attention paid to the conditions necessary for 'sustainable tourism'.

Tourism has to be humanised and not just driven by market forces and motivated solely by profit. The participation of the local people and attempts to incorporate their cultures and traditions in planning call for respect of their persons and communities. The autonomous growth of 'small business' organisations must be respected by large corporations.

(extracts from final statement of the Asian Consultation on Tourism, in *Contours* 7, 6: 36)

---

reflection of the way in which places have been drawn into one another and have become interdependent, but equally it is a stark illustration of the highly uneven and unequal nature of globalisation, which is highlighted in Figure 2.2. Clearly, The Gambia as a holiday destination is far more adversely affected by the cessation of tourism than are First World tourists. The latter only need consider alternative holiday destinations while The Gambia has little if no opportunity to consider alternative tourism markets. Tables 2.1–2.3 express the unequal nature of global tourism development in quantitative terms. These tables demonstrate the simple point that it is First World countries that are most visited, generate the most income from tourism and the largest number of tourists.

Globalisation also fails to acknowledge which places and peoples are included in this process and which are excluded. When Burns and Holden (1995) suggest that 'tourism is not so much about suntans as it is about being a major part of the globalisation of culture' (13), they cannot be implying that all people are implicated in this cultural process, owing among other factors to the glaring

Figure 2.2 Globalisation: uneven and unequal development
Source: 'El Fisgón', NACLA Report on the Americas xxxix,
4, January/February 1996:2

inequalities among people both within the First World and between First and Third Worlds. In other words globalisation is highly uneven too, and it is a concept that has become over-simplified in its application. It is important to remember, therefore, that most accounts of globalisation are by westerners (as this is) and are essentially about western globalisation as a result of the expansion of western capitalism: clearly an exercise in power. As Robins reflects, for 'all it has projected itself as transhistorical and transnational, as the transcendant and universalising force of modernisation and modernity, global capitalism has in reality been about westernisation – the export of western commodities, values, priorities, ways of life' (1991: 25; quoted in Hall, 1992a).

The notion that a 'new world order' has emerged, representing a triumph of western ideas, also preoccupies the work of Francis Fukuyama (Deputy Director, US State Department Policy Planning) who, following the end of the Cold War, proclaims we have reached the 'end of history' (1989, 1992):

Table 2.1 International tourist arrivals and receipts from First World countries

		1985	1986	1987	1988	1989	1990	1991	1992	1993	1994	1985–94 % mean annual change
Australia	Arrivals (000s)[1]	1,143	1,429	1,785	2,249	2,080	2,215	2,370	2,603	2,996	3,354	12.7
	% change	n.a.	25.0	24.9	26.0	-7.5	6.5	7.0	9.8	15.1	11.9	
	Receipts ($ m)	1,062	1,300	1,789	2,801	3,448	4,088	4,484	4,405	4,655	4,997	18.8
	% change	n.a.	22.4	37.6	56.6	23.1	18.6	9.7	-1.8	5.7	7.3	
Japan	Arrivals (000s)[1]	2,327	2,062	2,155	2,355	2,835	3,236	3,533	3,582	3,410	n.a.	4.9[2]
	% change	n.a.	-11.4	4.5	9.3	20.4	14.1	9.2	1.4	-4.8	n.a.	
	Receipts ($ m)	1,137	1,463	2,097	2,893	3,143	3,578	3,435	3,588	3,557	3,680	13.9
	% change	n.a.	28.7	43.3	38.0	8.6	13.8	-4.0	4.5	-0.9	3.5	
Spain	Arrivals (000s)	25,459	25,956	27,704	29,775	32,477	34,085	34,181	36,492	37,268	39,341	4.8
	% change	3.7	2.0	6.7	7.5	9.1	5.0	0.3	6.8	2.1	5.6	
	Receipts ($ m)	8,151	12,058	14,760	16,686	16,174	18,593	19,126	22,180	19,741	21,465	10.8
	% change	5.6	47.9	22.4	13.0	-3.1	15.0	2.9	16.0	-11.0	8.7	
UK	Arrivals (000s)[1]	14,449	13,897	15,566	15,799	17,338	18,013	17,125	18,535	19,398	21,034	4.4
	% change	5.9	-3.8	12.0	1.5	9.7	3.9	-4.9	8.2	4.7	8.4	
	Receipts ($ m)	7,120	8,163	10,225	11,017	11,389	14,940	13,070	13,932	14,031	15,190	9.5
	% change	16.4	14.6	25.3	7.7	3.4	31.2	-12.5	6.6	0.7	8.3	
USA	Arrivals (000s)	25,399	26,008	29,500	34,095	36,564	39,539	42,986	47,261	45,779	44,982	6.5
	% change	-5.7	2.4	13.4	15.6	7.2	8.1	8.7	10.0	-3.1	-1.7	
	Receipts ($ m)	17,762	20,385	23,563	29,439	36,250	43,007	48,384	54,284	57,621	60,001	14.5
	% change	3.4	14.8	15.6	24.9	23.2	18.6	12.5	12.2	6.1	4.1	
World	Arrivals (millions)	326	339	365	400	429	456	463	503	512	532	5.5
	% change	3.3	3.4	7.8	9.6	7.4	6.2	1.6	8.6	1.9	3.8	
	Receipts ($ bn)	116	141	172	200	216	261	268	305	307	337	12.6
	% change	4.6	21.1	22.6	15.7	8.1	21.1	2.5	13.8	0.6	10.0	

n.a. Not available
1 Includes all international inbound visitors, not only tourists (which involve overnight stays) – see definitions in note 2 to Chapter 1 (p. 329)
2 1985–93
Source: World Tourism Organisation

Table 2.2 International tourist arrivals and receipts from Third World countries

		1985	1986	1987	1988	1989	1990	1991	1992	1993	1994	1985–94% mean annual change
Grenada	Arrivals (000s)	52	57	57	59	66	76	85	88	94	109	8.5
	% change	30.0	9.6	0.0	3.5	11.9	15.2	11.8	3.5	6.8	16.0	
	Receipts ($ m)	26	29	30	29	31	38	42	38	45	55	8.7
	% change	30.0	11.5	3.4	-3.3	6.9	22.6	10.5	-9.5	18.4	22.2	
Guatemala	Arrivals (000s)	252	287	353	405	437	509	513	541	562	537	8.8
	% change	31.3	13.9	23.0	14.7	7.9	16.5	0.8	5.5	3.9	-4.5	
	Receipts ($ m)	67	77	103	124	152	185	211	243	265	258	16.2
	% change	17.5	14.9	33.9	20.4	22.6	21.7	14.1	15.2	9.1	-2.6	
Thailand	Arrivals (000s)	2,438	2,818	3,483	4,231	4,810	5,299	5,087	5,136	5,761	6,017	10.6
	% change	n.a.	15.6	23.6	21.5	13.7	10.2	-4.0	1.0	12.2	4.4	
	Receipts ($ m)	1,171	1,421	1,947	3,120	3,753	4,326	3,923	4,829	5,014	6,592	21.2
	% change	n.a.	21.3	37.0	60.2	20.3	15.3	-9.3	23.1	3.8	31.5	
Tunisia	Arrivals (000s)	2,003	1,502	1,875	3,468	3,222	3,204	3,224	3,540	3,656	3,856	7.6
	% change	26.8	-25.0	24.8	85.0	-7.1	-0.6	0.6	9.8	3.3	5.5	
	Receipts ($ m)	551	488	672	1,234	933	953	685	1,074	1,114	1,302	10.0
	% change	19.0	-11.4	37.7	83.6	-24.4	2.1	-28.1	56.8	3.7	16.9	
Zimbabwe	Arrivals (000s)	331	357	372	449	474	606	664	738	955	1,010	13.2
	% change	18.2	7.9	4.2	20.7	5.6	27.9	9.6	11.1	29.4	5.8	
	Receipts ($ m)	26	29	43	54	55	64	75	105	103	114	17.9
	% change	0.0	11.5	48.3	25.6	1.9	16.4	17.2	40.0	-1.9	10.7	
World	Arrivals (millions)	326	339	365	400	429	456	463	503	512	532	5.5
	% change	3.3	3.4	7.8	9.6	7.4	6.2	1.6	8.6	1.9	3.8	
	Receipts ($ bn)	116	141	172	200	216	261	268	305	307	337	12.6
	% change	4.6	21.1	22.6	15.7	8.1	21.1	2.5	13.8	0.6	10.0	

n.a.   Not available
Source: World Tourism Organisation

Table 2.3 The world's top twenty tourism spenders
(excluding transport), 1995

Rank	Country	% of total	
1	Germany	13.6	
2	United States	13.0	
3	Japan	10.4	48.6
4	United Kingdom	7.0	
5	France	4.6	
6	Italy	3.5	
7	Austria	3.3	
8	Russian Federation	3.3	16.2
9	Netherlands	3.2	
10	Canada	2.9	
11	Belgium	2.6	
12	Taiwan	2.4	
13	Switzerland	2.2	10.5
14	South Korea	1.7	
15	Poland	1.6	
16	Sweden	1.5	
17	Singapore	1.5	
18	Australia	1.3	6.8
19	Spain	1.3	
20	Denmark	1.2	

Source: World Tourism Organisation, 1997

> The triumph of the West, of the Western *idea*, is evident first in the total
> exhaustion of viable systematic alternatives to Western liberalism . . . and
> can be seen also in the ineluctable spread of consumerist Western culture.
> . . . What we may be witnessing is not just the end of the Cold War . . .
> but the end of history as such: that is, the endpoint of mankind's
> ideological evolution.
>
> (1989: 3–4; quoted in MacCannell, 1992: 62)

In summary, globalisation may be an emotive and powerful idea, allowing
each of us to invoke Big Macs, Coke and holidays in Thailand as our individual
testimonies in support of that globalisation. But it is also an attempt to place
ethnocentric conformity over the way we conceive change and is a reflection
of unequal development in its own right. It represents an interesting story but
a poor basis for analysis. Rather more ominously, it is a term that has allowed
First World politicians, scholars, cultures, interest groups, business people and
so on to impose an inevitability which is firmly premised upon the West – a
global economy, global culture, global politics, a global environment.

The reality of globalisation is considerably more complex and is characterised
by uneven and unequal development. A careful examination of the economic,

21

political and cultural processes is also central to a more in-depth analysis of the 'globalisation' of contemporary tourism. The next section advances and elaborates upon this discussion by considering the content and meaning of sustainability within the context of global change. Globalisation, it is suggested, is also useful in stretching our comprehension of sustainability.

SUSTAINABILITY AND GLOBAL CHANGE

The second key word in our analysis of tourism is *sustainability*, a notion that at its most basic encapsulates the growing concern for the environment and natural resources. Just as processes of globalisation are implicated in the drawing of Third World destinations into the sphere of tourism, so too notions of sustainability are closely related and disproportionately reflected in the Third World as concern for the health of the planet has resulted in the emergence of a globalised environmental politics.

The end of the Cold War – the end of the East–West conflict between communism and capitalism – also meant that the so-called 'international community' (which is heavily imbued with the influence and power of the First World) looked elsewhere for *causes célèbres* within the last decade. In particular, the powerful nations of the First World have redirected their attention on the Third World, and a concern for the global environment and 'development' now promises to become one of the principal focuses in political action and in the rhetoric and thinking of development studies (Adams, 1990: 1). Just as globalisation is characterised by unequal relations, however, so too has the global environment debate already resulted in conflict between the First and Third Worlds: conflict that came to the fore at the Rio Summit (the United Nations Conference on Environment and Development, also known as the Earth Summit, UNCED or UNCED '92) at Rio de Janeiro in June 1992 (see Box 2.3). As David Lascelles comments:

> The old East/West confrontation has shifted to one between North and South and environment and development have become fixtures on this new agenda. Some of the most acrimonious debates during the UNCED process focused on the North/South debate and the marked divisions of wealth and poverty between the two.
>
> (1992: 42)

It is necessary initially to trace how such debates over environment, development and sustainability are reflected in the issues of contemporary tourism. The first important point to establish about sustainability is that it is a word that is defined, interpreted and imagined differently between individuals, organisations and social groups. For the 'admen' of transnational corporations an interpretation of sustainability (how to entice customers to buy their product on the basis of their concern for the environment) is significantly different from that of those communities and activists claiming to resist the destruction of the countryside

Box 2.3 *The Rio Summit*

In 1989, the United Nations expressed deep concern at the 'serious degradation of the global life-support systems' (Resolution 44/228) and convened the United Nations Conference on Environment and Development in Rio de Janeiro in June 1992. It was attended by 178 governments including 120 heads of state.

The purpose and content of the conference were to 'elaborate strategies and measures to halt and reverse the effects of environmental degradation in the context of strengthened national and international efforts to promote sustainable and environmentally sound development in all countries'.

The results of the conference were foreseen in six parts:

- an 'Earth Charter' or Declaration of basic principles;
- agreements on specific legal measures;
- an agenda for action – 'Agenda 21' (see Box 4.10) and the means to implement this agenda through:
- new and additional financial resources;
- transfer of technology;
- strengthening of institutional capacities and processes.

The immediate results – the Rio Declaration, non-binding treaties on climate change and biodiversity, forest principles, Agenda 21, and meagre financial commitments – fell far short of the envisaged aims of the conference. Most of the treaties were non-binding, the declarations were vague enough to please everyone, and the commitment of resources was paltry ($2.5 billion compared with an estimated cost of programmes of $600 billion a year). Despite its size, the travel and tourism industry was not included as a separate item on the conference agenda.

A year before the conference, Maurice Strong*, the UNCED Chairman, appointed Swiss billionaire Stephan Schmidheiny to promote the international business community's standpoints on environmental issues. Schmidheiny formed the Business Council on Sustainable Development. Carothers describes the Council as a coalition of some fifty multinationals, including some of the worst polluters on the planet, whose 'goals were predictable: "voluntary" rather than legislated reduction in toxic emissions, the right to corporate privacy and wholesale support for "free trade"' (1993: 14–15).

References to the over-consumption of the rich countries were removed from treaties, mention of corporate conduct was watered down, the poorest countries barely had a say, and, despite objections from all the environmental groups in attendance, the conference was used to endorse the General Agreement on Tariffs and Trade (GATT). As a Friends of the Earth briefing paper (which also gave equal mention to the positive achievements of the conference) described it:

An overwhelming majority of the world's leading politicians have backed short-term economic expediency – 'business as usual' – instead of an integration of environment and economy. They have succumbed to

23

Box 2.3
continued

lobbying by excessively powerful business groupings intent on safe-guarding their narrow vested interests. The North has done little to address the issue of its over-consumption and its unfair share of the limited 'ecological space' on this planet. Instead, much of the burden of the environment and development crisis has been left on the shoulders of the world's poorest countries in the South.

(Friends of the Earth, 1992)

* According to Nelson (1993) and Sklar and Everdell (1980), cited in Fernandes (1994), Strong made his personal wealth from 'oil and gas, minerals, pulp and paper, and other developments'. He has served as Chairman of Petro-Canada, Executive Director of UNEP, Treasurer of the Rockefeller Fund and Alternate Governor of the Asian Development Bank, as well as sitting on major environmental committees and organisations.

(how best to 'save' nature and Mother Earth). In this book, rather than taking sustainability as given or as relatively easily defined (which is mostly done in terms of the highly ambiguous Brundtland definition, given in Box 2.4), sustainability is considered a contested concept, a concept (as Chapters 3 and 4 argue) that is 'socially constructed' and reflects the interests of those involved.

Box 2.4   The Brundtland definition of sustainable development

In 1983 the World Commission on Environment and Development was set up, with Gro Harlem Brundtland as its Chair, in response to a United Nations General Assembly resolution. The Commission's report, *Our Common Future* (the 'Brundtland Report'), was submitted to the United Nations in 1987.
Its often quoted definition of sustainable development is:

development which meets the needs of the present without compromising the ability of future generations to meet their own needs.

Protagonists of the report point out that it incorporates the essential principles of intra-generational and inter-generational equity and that it persuaded many governments to endorse the notion of sustainable development.
Its critics would argue that it contains inbuilt assumptions about the need for continued expansion of the world economy and that it failed to stress the radical changes in lifestyles and society that would be required to overcome the problems inherent in the western model of development.

The second important point is that, in addition to acknowledging and assessing the different interpretations of sustainability, we are also interpreting the debate over sustainability in a broader context. This provides for a more free-ranging discussion, allowing us to consider the ways in which different ideas of sustainability are used, for example, to sustain profits in the tourism industry (Chapter 7), or are used by social classes to retain distinctive holidays (Chapter 5), or are used by 'host' communities to exclude outsiders (Chapter 8).

Sustainability, then, is a concept charged with power. We will be turning to the notions of ideology and hegemony a little later in order to help in the exploration of tourism and sustainability. The critical questions must remain: Who defines what sustainability is?, How is it to be achieved?, and Who has ownership of its representation and meaning? It will be argued that, for the greater part, the answers to these questions are found in the First World: in businesses, governments, transnational institutions, scholars, environmentalists and new socio-environmental organisations.

Sustaining profits

Arguably, global economic restructuring and development are the most pertinent factors in the study of globalisation. Indeed the motor behind global economic change is the need for the growth of capitalism – new opportunities, new markets, and for tourism, new destinations – in other words the imperative for sustained growth and profitability. It is necessary to outline the most important features of global economic change before looking in more detail at why these changes have occurred and how they are reflected in contemporary tourism development.

The first feature has been a rapid growth in the world market, a process of internationalisation that has resulted in the emergence of a global economic system. This is most clearly reflected in the expansion and reorganisation of the global financial market with the emergence, for example, of a global stock market. Economic globalisation is also reflected in 'footloose' capital and the growth of less nationally regulated industrial, banking and commercial sectors, a process that is also clearly represented in the global tourism industry's principal economic sectors with mergers and buy-outs between international airlines and hotels.

A second feature is the relatively rapid First World de-industrialisation, with an equally rapid growth in the service sector. De-industrialisation has been necessitated by the long-term fall in manufacturing profit margins in the advanced capitalist economies (Lash and Urry, 1987), which have been forced to compete with the more cheaply manufactured goods in the so-called 'newly industrialising countries' such as the *little tigers* of Asia (Hong Kong, Singapore, South Korea and Taiwan). On the other hand, there has been a dramatic increase in service sector industries and varying degrees of reorientation in the First World to so-called service sector-based economies. This can best be understood as part of an ongoing process involving the international division of labour at a global scale. These shifts also provide important clues to the nature of attendant changes in the consumption of services (such as holidays), a grasp of which is essential for understanding the growth and development in the First World consumption of tourism.

Third, and very much interrelated with globalisation and the international division of labour, capitalism has increasingly penetrated the Third World and 'integrated' or drawn these countries into a global capitalist system. Consequently there has been a considerable increase in the number of countries which are implicated in capitalist production (Lash and Urry, 1987: 6). We shall return to the consequences of economic interdependency at various points in this book, and in the next chapter will consider some of the parallels that have been drawn between cash crops (such as bananas, coffee or minerals) and tourism.

Post-Fordism

Post-Fordism provides one way of capturing the processes of global restructuring and the qualitative changes in the organisation of both production and consumption (Allen, 1992), changes which have been alluded to above and will be expanded upon below. The regime known as Fordism (taking its name from Henry Ford's assembly lines making mass-produced cars) characterised the major capitalist economies for the best part of this century. It expresses how economies of scale are ensured by goods which are mass-produced and mass-consumed. Under conditions of post-Fordism (or neo-Fordism), however, which many commentators suggest represent the current economic regime, there is a qualitative shift from mass production and consumption to more flexible systems of production (often referred to as economies of scope or batch production) and organisation (such as flexible work patterns). Post-Fordism also makes tentative links to changes in the way that goods and services are consumed, with rapidly changing consumer tastes and the emergence of niche and segmented markets.

The applicability of these ideas to the changes in tourism have now been acknowledged. It would appear that, just as the emergence of Butlins holiday camps or packaged holidays are indicative of services mass-produced and consumed under a regime of Fordism, so too are small group tours to Bolivia or truck journeys across sub-Saharan Africa indicative of post-Fordism. Further examples of tourism types associated with post-Fordist consumption are given in Table 2.4. Lash and Urry (1994) in particular point to the demand for 'independent holidays' and the increasing environmental planning and control of tourism in places such as Belize or Bermuda as examples of new alternatives to mass tourism. They also recognise the importance of Third World countries in these changes, arguing that 'the development of "alternative tourism" in some developing countries' (1994: 274) is a clear example of post-Fordist tourism. We will return to this argument in Chapter 4 in discussing the shape of new tourisms that have emerged.

Time–space compression

So far some of the most salient features in contemporary global change have been described. But what forces drive these global economic processes? The most

Table 2.4 Post-Fordism and tourism

Post-Fordist consumption	Tourist examples
Consumers increasingly dominant and producers have to be much more consumer-oriented	Rejection of certain forms of mass tourism (holiday camp and cheaper packaged holidays) and increased diversity of preferences
Greater volatility of consumer preference	Fewer repeat visits and the proliferation of alternative sites and attractions
Increased market segmentation	Multiplication of types of holiday and visitor attractions based on lifestyle research
Growth of a consumers' movement	Much more information provided about alternative holidays and attractions through the media
Development of many new products, each of which has a shorter life	Rapid turnover of tourist sites and experiences because of fashion changes
Increased preferences expressed for non-mass forms of production/consumption	Growth of 'green tourism' and of forms of refreshment and accommodation which are individually tailored to the consumer (such as country house hotels)
Consumption less and less 'functional' and increasingly aestheticised	'De-differentiation' of tourism from leisure, culture, retailing, education, sport, hobbies

Source: Lash and Urry, 1994: 274

penetrating analysis has been offered by the Marxist geographer David Harvey, who presents the concept of *time–space compression*, which is of considerable interest in unravelling the growth and development of Third World tourism. It is worth spending a little time considering his ideas, for they move an understanding of global change beyond the descriptive analysis and prescription of *how* global capitalism works and seek instead to demonstrate *why* capitalism changes in the way it manifestly does. Furthermore, time–space compression provides an opportunity to overcome the problems recognised by Britton: 'The geography texts on tourism offer little more than a cursory and superficial analysis of how the tourism industry is structured and regulated by the classic imperatives and laws governing capitalist accumulation' (1991: 456).

Harvey presents the time–space compression thesis in his now widely cited *The Condition of Postmodernity* (1989b). The present phase of globalisation, Harvey argues, involves a marked increase in the pace of economic (and everyday) life and a phenomenal acceleration in the movement of capital and information. For example, we can sit in a travel agency in Buenos Aires and buy an airline ticket for a Paris to Lusaka flight, with a transfer to Windhoek.

Boxes 2.5 and 2.6 also demonstrate the way in which information is moved more rapidly within global networks and the way that travel experiences are advanced to such a degree that it is no longer necessary to leave your home or office. (See Appendix 1, which lists some of the travel-related bulletin boards and World Wide Web sites that can be accessed over the internet.) Box 2.5 may appear an exaggerated example but it underlines the way in which places are increasingly drawn into (or excluded from) a global system.

Box 2.5 Have mouse, will travel

Dawn is breaking in the Costa Rica rainforest. Above me, a white hawk flits silently through the treetops. A pale-billed woodpecker alights softly on the side of a soaring trunk. The branches are swathed in mosses, ferns and orchids, and hung with a triangle of vines and lianas. The Braulio Carrillo National Park is less than an hour from the capital, San José, but it could be in another world.

Gazing up through the forest, I can make out in the distance the rows of small houses, running up and down the steeply sloping hills of Tegucigalpa.

Looking down, the wide blue waters of Montego Bay stretch out invitingly from the Jamaican shoreline, with the luxurious 'cottages' of the Sandals Hotel perched on the edge of their private beach. It is an inviting prospect, but one which, for the moment, I decide to pass on . . .

Yes, sad to say, it's not for real, it's the Internet – that sprawling worldwide computer network. Armchair travel has a whole new meaning when, with a lap-top atop your lap and a modem by your side, you can skip across the globe on the World Wide Web from Abidjan to Zanzibar, alighting on everything from the Eurostar timetable to the latest malaria alerts for East Africa . . .

In some ways, it's the ultimate in eco-tourism, dropping in on distant lands without ripping holes in the ozone layer . . .

(Wright, 1996)

Time–space compression seeks to encapsulate this intensification as capitalists seek to overcome the barriers of distance and stretch their economic relationships to all parts of the globe. It is, in other words, part of an ongoing expansion of capitalist relations of production where the primary objective is to reduce the turnover time of capital and to quicken the circulation time of capital and to sustain profits. Both new markets and new products are sought in order to achieve this and the process is clearly reflected in the way in which an increasing number of holiday destinations are drawn into the global tourism industry.

The need to accelerate the circulation of capital has necessitated an economic transition, or qualitative shift, and Harvey seeks to explain this in terms of the regime of capital accumulation, a central tenet of Marxist inquiry. The key, and perennial, question is how best to deal with the over-accumulation of capital

Box 2.6 *Tourism on the internet*

Box 2.5 gave a few examples of the tourism-related items we can find on a 'surf' through the internet. The telecommunications revolution allows teleworkers and computer users to take advantage of the global information 'super-highway', via which they may receive information and messages from others with access to the network. This takes various forms:

- *email* (electronic mail);
- *bulletin boards* and virtual conferences (see Appendix 1);
- *the World Wide Web* ('the Web' or 'WWW'). The WWW has spread well beyond the academic and scientific communities and now includes many organisations in the private, public and voluntary sectors – some relevant web sites are given in Appendix 1.

A number of smaller networks, such as GreenNet, are linked to the internet. GreenNet is part of a global computer network assigned specifically for environment, peace, human rights and development groups. As Box 2.5 pointed out, the internet can be quite a travel experience in its own right. It should come as no surprise, then, that there is now a *Rough Guide to the Internet.*

Control of the internet is still largely in the hands of the users, individuals and organisations. It is also still relatively cheap to gain access, which is one reason it has begun to replace the fax as a means of, especially, international communication. Moreover, it is the stated aim of the telecommunications industry to provide access to the technology for all the population – a laudable aim. In practice, however, the promotion of its widespread use may serve to widen the unequal and uneven nature of access to information. In this respect it mirrors the effects of access to travel.

It is also feasible that increasing regulation of the networks will serve to concentrate power in the hands of those who can afford it. Information is power and this could well lead to a struggle for control over the internet and its networks in the coming years.

Even in its unregulated infancy there is already much concern about the use of the internet. Witness this view of the Lonely Planet WWW site, expressed by Marcus Grant in *In Focus*:

Visiting the Lonely Planet site is very similar to browsing their publications in a book shop. Relatively impersonal accounts of local culture, history and travel information spiced up with lots of up-to-date personal tips and anecdotes. The hosts get little look in. I decided to see what they had on offer for Burma.* It made my blood boil to find nothing about the recent human rights abuses in order to improve the tourism infrastructure. Just the usual potted history. . . . There were six high quality postcard images, mainly of temples and landscapes. Perhaps some of us should send them the images of forced labour they seem to have overlooked.

(1996: 16)

Leaving aside the merits of the *Lonely Planet* approach to Burma compared with the Marcus Grant approach, the above quotation is illustrative of the different political uses to which the internet can be put.

*The example of tourism to Burma is examined in greater depth in Chapter 9.

and how it can be 'expressed, absorbed or managed' (Harvey, 1989b: 131). Reflecting the post-Fordist debates discussed above, Harvey contends that in terms of production the problems encountered in achieving satisfactory productivity increase and the intense competition faced from, for example, 'newly industrialising countries' has forced the transition (in the First World) to a more flexible mode of accumulation in the post-1970 period, a 'new dynamic phase of capitalism' as he terms it (1989b: vi). At the heart of this change lies flexibility. As Figure 2.3 demonstrates, a capitalist system based upon a Fordist mode of production has given way to more flexible modes of capital accumulation, a regime labelled 'flexible accumulation'. We will return to this diagram a little later. As Harvey argues, this comes into 'direct confrontation' with the rigidities characteristic of Fordism. Flexibility is introduced in a range of respects including labour markets and processes, products (new and different forms of tourism) and patterns of consumption (such as new tourism).

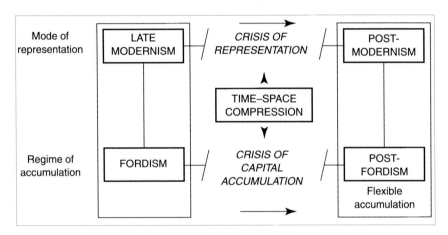

Figure 2.3 Transition in late twentieth-century capitalism
Source: Adapted from Gregory, 1994

Harvey's discussion of capital circulation and the acceleration in the pace of our everyday lives has clear reflections in the rapid expansion of tourism in the Third World. It is not only capital that is circulated at an accelerated rate, but places too, as destinations come in and out of fashion and tourism moves on elsewhere. Indeed, the growth and development of Third World tourism may be another manifestation of time–space compression with the logic of capital accumulation driving tourism to the four corners of the earth. Equally, however, it is factors such as natural disasters, political instability, the environment and so on (and how these are represented and perceived in the First World) which play a significant part in governing the ebb and flow of Third World destinations on and off the tourism map. And it is to one of these other factors, sustaining culture and lifestyles, that we turn next.

Sustaining culture and lifestyles

The dramatic economic restructuring that has occurred under the latest and extremely intense phase of globalisation is, some argue, closely related to far-reaching cultural changes (Jameson, 1984; Harvey, 1989b). Figure 2.3 suggests that the fierce round of time–space compression reflected in an accumulation crisis and the emergence of 'flexible accumulation' are equally reflected in a crisis of cultural representation (our ability to make sense of a rapidly changing world). So, while Harvey's ideas are underpinned by economics, he also stresses the importance of cultural change within global restructuring:

> While simultaneity in the shifting dimensions of time and space is no proof of necessary or causal connection, strong *a priori* grounds can be adduced for the proposition that there is some kind of necessary relation between the rise of postmodernist cultural forms, the emergence of more flexible modes of capital accumulation, and a new round of 'time–space compression' in the organisation of capitalism.
>
> (1989b: vii)

Potentially, this argument provides some important insights into how and why Third World tourism has become so popular. In other words, it is with changes in the First World, and with the factors that produce new forms of holidays and tourists, that analysis must focus first.

Postmodernism

In the above quotation David Harvey refers to 'postmodernist cultural forms'; returning to Figure 2.3 for a moment, it will be noticed that the predicted outcome of the crisis of representation is *postmodernism*. Very broadly, postmodernism refers to the emergence of new cultural styles (in art, architecture, music and the objects and experiences we buy and consume), and postmodernity to the idea that we now live in a new social epoch that has superseded modernity.

Postmodernism is a widely used and debated idea, but it is not our intention to provide a lengthy rendition on what postmodernism is and is not, or to provide an in-depth guide to protagonists' arguments. Rather, it is to accept that, like post-Fordism, this idea helps encapsulate the profound cultural changes that have been slowly emerging, particularly over the last thirty years (Huyssen, 1984; Featherstone, 1991). Practically, the changes we are experiencing are rooted in our everyday lives and 'can no longer be ignored' (Harvey, 1989b). Moreover, postmodernism also helps to draw our attention to the important relationships between First World consumption habits and the capitalist imperative of increasing the turnover time of capital.

Of the many developments in the arena of consumption, two stand out as being of particular importance. The mobilisation of fashion in mass

31

markets provided a means to accelerate the pace of consumption not only in clothing, ornament, and decoration but also across a wide swathe of life-styles and recreational activities (leisure and sporting habits, pop music styles, video and children's games . . .). A second trend was a shift away from the consumption of goods and into the consumption of services. . . . The 'lifetime' of such services (a visit to a museum, going to a rock concert or movie, attending lectures or health clubs), though hard to estimate, is far shorter than that of an automobile or washing machine. If there are limits to the accumulation and turnover of physical goods . . . then it makes sense for capitalists to turn to the provision of very ephemeral services in consumption.

(Harvey, 1989b: 285)

It is clear that, for most people, contemporary lifestyles involve a dramatic increase in the number of services consumed. (You may wish to consider briefly how the shift to services is related to the growth of tourism (a service) and how the circulation of fashion perhaps relates to popularity of holiday destinations over time.)

But part of the postmodernism argument tends to lead to the conclusion that, just as an increasingly globalised world has resulted in a global economy, so too have the same processes resulted in the emergence of a global culture characterised by Big Macs, Coke and Mnet. Such predictions (or forebodings) have led to a succession of tourism commentators bemoaning the inability of Third World communities to sustain their traditional lifestyles in the face of an imposition of western values and beliefs and the consequent erosion of cultural difference and authenticity. Morrow, for example, refers to tourism as a 'radioactive cloud of banalizing sameness' which, he dramatically argues, 'threatens the earth; the sacred and beautiful places, all the uniquenesses, have been invaded, desacralized, franchised for the masses, dissolved into the United Colors of Benetton' (Morrow, 1995).

While the ubiquity of baseball caps or holidays as evidence of global culture is anecdotally appealing, it was briefly suggested earlier that it is also a highly simplistic view of the world. It is a seriously deficient account in two important respects. First, it discourages acknowledgement of the power and inequality reflected through cultural globalisation (Hall, 1992b). Not only is globalisation extremely unevenly developed, affecting (or not as the case may be) different social groups in very different ways, it also represents an increasing interdependency between First and Third Worlds which is highly unequal: 'the West and the Rest', as Hall terms it (1992b). Crucially, in the context of tourism, the Third World consists ostensibly of tourist-receiving and not tourist-sending countries.

Second, postmodernism offers a position that refuses to acknowledge the way in which commodities are used and understood by different social groups and communities. Much the same criticism could be levelled at the way tourists are received and perceived, and how their practices are adopted, rejected or adapted. In other words, because tourists are disliked (for a variety of reasons) in one locality, it does not automatically mean they are considered similarly elsewhere, and vice versa. Similarly, the charge that tourism bastardises cultures, ripping them from their traditional roots and subsuming them into a global culture, fails to offer a more sophisticated understanding of how this process of cultural change may be negotiated by communities in the Third World. We return to this process of 'transculturation' in Chapter 8.

This has resulted in a highly polarised and simplified debate in the First World concerning the most appropriate way of holidaying, which equates to 'tourists = mass tourism = bad' and 'travellers = appropriate travelling = good'. Again this simplistic view emphasises the role of power, with the prospects and problems of tourism calculated largely from a First World perspective. It is we who decide what is right (untouched primitive cultures) and what is wrong (Thai hilltribes listening to a radio), what is appropriate and sustainable and what is not. Reflective of this, the American academic Tensie Whelan proclaims of ecotourism, 'If we are to save any of our precious environment, we must provide people with alternatives to destruction' (1991: 3). In this sense, First World inhabitants seek to sustain their own ideas and work towards sustaining their own lifestyles, within which only certain types of 'holiday' are acceptable.

Finally, far from global conformity, postmodernism helps capture the high degree of difference and 'fragmentation' that lies at the heart of contemporary cultural change (Harvey, 1989b). And it is a process of change that has had some important impacts. In terms of politics, for example, commentators have high-lighted the relative downfall of party politics and the emergence of new social movements focused upon issues such as women, minority rights, nuclear power or the environment. The emergence of societies that are considerably more socially pluralistic in terms of class, culture, lifestyles, ethnicity and so on has also been noted. It is worth considering this in a little more detail, for this too offers some important clues in the investigation of Third World tourism.

New middle classes

Postmodernism has been closely linked to the emergence and growth of the so-called new middle classes.[1] This has involved the relatively rapid expansion of social classes (and hence the plural, new middle classes, indicating that there is not one distinguishable middle class), especially in the First World.

An important feature of the new middle classes has been their identification as agents of the cultural change inherent in postmodernity, and they have been referred to as 'new cultural intermediaries' (Bourdieu, 1984). Their ranks have swollen dramatically with the growth of the media (advertising, lifestyle research, and so on), caring personal services (from physiotherapy to reflexology) and the emergence of service sector-oriented economies more generally. They are, therefore, key social groups in initiating, transmitting and translating these new cultural processes and consumption patterns, of which holidaying is demonstrably a significant part. Later chapters of this book suggest that these social classes are not only important consumers of Third World holidays but are also key groups in promoting and implementing notions of sustainability. This is not to claim that all Third World tourism is the product of cultural consumption preferences and political leanings of members of the new middle classes; clearly this would be ridiculous. Rather, it is to suggest that they are major initiators and consumers of new tourism in Third World destinations. As Morrow observes, the 'great, global middle class is in motion' (1995).

Crompton (1993) argues that much recent work on the growth and development of the middle classes is crucially linked to the growth of consumer capitalism and an emphasis on 'lifestyles'. Lifestyles range from the places in which people choose to live, to the things they eat, and, of course, to the types of holiday they take. With the rapidity of cultural and economic change, social groups are constantly attempting to identify and indicate to others their position in 'the new social and cultural order' (Shurmer-Smith and Hannam, 1994); or in other words, attempting to sustain a distinctive lifestyle.

The new middle classes are not only significant as cultural intermediaries, but are also heavily represented in the new political alignments (or new socio-environmental organisations) that have emerged focused on issues such as the environment, and that are concerned primarily with sustainability and sustainable lifestyles. Habermas (1981) interprets this as the emergence of a 'new politics' which includes and is centred on the anti-nuclear, peace, women's, environmental, minority liberation, religious and alternative lifestyle movements. (Chapter 6 will develop this argument by assessing the relevance of environmental politics, and what might be termed the sustainable lifestyle movement, to the development of new forms of Third World tourism.) Drawing upon this recent social theory and acknowledging the significance of contemporary cultural processes will help to move the analysis beyond the somewhat restrictive surveys of tourist motivation and satisfaction and the rather static typologies of tourists that have characterised much work in the field of tourism in the Third World.

Sustainable politics

If globalisation is marked by profound economic and cultural changes, equally it also alerts us to important political processes and the emergence of what is commonly referred to as global politics. A more globalised political environment has become part of our everyday experience – the European Union, the International Monetary Fund (IMF), the World Trade Organisation and international environmental organisations (such as Greenpeace) are all examples of organisations whose political relationships stretch beyond the boundaries of nation states. And the decisions of these institutions are felt thousands of miles from where they are taken, a fact that re-emphasises the power of unequal and uneven development.

Supranational institutions

Supranational institutions and agencies, which involve both a degree of political integration between states and a transcending of power of individual nation states, have grown in stature and influence in the post-1945 period. These institutions have had an important impact on global development and their influence in terms of tourism is considered in later chapters.

The most widely cited supranational actors and arguably the most significant in terms of their economic and social impacts, have been the IMF and World Bank. Together they have imposed so-called structural adjustment policies and programmes upon Third World states which have been forced to adjust their economies (mostly in terms of de-nationalisation, privatisation and a massive reduction in state services) in order to secure further loans. But these are by no means the only institutions involved. Equally important to the discussion are the policies of other supranational institutions such as regional development banks, the European Union, the World Tourism Organisation or the United Nations which convened the Rio Summit in 1992.

Beyond these bodies there are other emergent intergovernmental structures that transcend the independence of the nation state in a variety of ways. Economic regionalism, for example, has resulted in a dramatic impact in this respect and it is necessary to consider the effects of trading blocs such as the North America Free Trade Agreement (NAFTA) or the Association of South East Asian Nations (ASEAN). Such arrangements can underscore and reinforce the inequalities between countries and regions and, in effect, foist certain choices upon weaker countries both within and beyond the agreement.

There are two important observations to be made here. First, not only have the political relationships and actions forged by these institutions increased in their global reach, they have also been couched increasingly in a language of sustainability, of which the Rio Summit is the most dramatic example. Clearly, for whom these 'sustainable' policies, priorities and programmes are intended, what they entail and how they impact upon the Third World are important parts of our enquiry. Second, it appears that the product of globalised politics is the emergence of a euphemistically entitled international community, something of a globalised consciousness, a seemingly popularised and benevolent global collective or ombudsman that acts in the best global interests intervening to solve a myriad of problems from civil war and international crime cartels to the killing of whales and the destruction of rainforests. Once again, many of the *causes célèbres* of the 'international community' are environmental, and sustainability has emerged as an overarching international goal. It is necessary to examine the composition of this international community, and enquire for whom it speaks, who it represents, who it is controlled by, who it seeks to influence and in what ways.

Transnational institutions

Transnational institutions have also emerged on the global stage and possess varying degrees of influence and power with variable impacts upon the Third World. Unlike supranational institutions they are composed of bodies other than governments (Allen and Massey, 1995). Private global consortia, such as the World Travel and Tourism Council (WTTC) and the World Tourism Organisation (WTO), provide excellent examples of such institutions and their relationship to tourism.

Established in 1990, the WTTC is a global coalition of 70 chief executive officers from all sectors of the travel and tourism industry. The WTTC's 'millennium vision' is to encourage governments, in cooperation with the private sector, to harness the industry's economic dynamism and increase overall growth and job creation. The WTTC aims to do this through the promotion of open markets, the elimination of barriers to growth, deregulation and liberalisation, while at the same time pursuing sustainable development through industry environmental initiatives such as the Green Globe scheme (see Box 7.5). Again, centred in the First World and with mostly First World members, such institutions help reflect global inequalities. Some of the WTTC's campaigns and work are examined further in Chapters 7 and 9.

International non-governmental organisations

There has also been a marked growth in non-governmental organisations (NGOs) both within nation states and transnationally. International non-governmental organisations (INGOs), including such household names as Amnesty International, Greenpeace and Friends of the Earth (some of which have an indirect relationship with tourism) have increased in number from 832 in 1951 to 4,649 in 1986 (McGrew, 1995). Another notable fact is the way in which these organisations, especially those focusing on ecological and environmental problems, have invoked the global nature of the problems. Indeed, names such as Friends of the Earth, the World Wide Fund for Nature or Earth First explicitly emphasise the global nature of environmental concerns in the very names of the organisation (Yearley, 1995). In other words, the last two decades have witnessed the emergence of transnational environmental politics, a large part of which is devoted to saving Planet Earth.

It is necessary to consider how these INGOs have reacted to the growth of global tourism and equally how the activities of such INGOs have been received and are perceived by Third World countries and communities. Just as economic growth and change or cultural transformation are differentially experienced in different places, so too are global environmental politics. For example, conservation measures designed to maintain ecological biodiversity undertaken by organisations such as Conservation International have frequently been 'contradicted' by the priorities and aspirations of local communities attempting to secure their livelihoods. And yet it is the power of benevolence reflected by environmental INGOs that is the most striking feature of such cases. Through membership of such organisations or through a general empathy with their aims, the global concerns and consciousness of First World citizens are played out at a local scale; their 'will' is imposed upon communities thousands of miles away. Global environmental politics, therefore, also appear to confirm, at least in part, the contours of power that have already been mapped in the discussion of global economic and cultural processes.

In the preceding sections, it has been suggested that sustainability has a much broader currency than is acknowledged in current academic literature. Sustainability is as much to do with ensuring continued profits through more flexible patterns of capital accumulation, or maintaining middle-class lifestyles in the First World and the ability of these social groups to experience (sustained) indigenous cultures while holidaying in the Third World, as it is to do with ecology and environment. Figure 2.4 provides a reminder of the focus of this book and the range of factors which must be brought to bear in a discussion of Third World tourism. It also re-emphasises the importance of the interrelationships of the themes of sustainability, globalisation and development. It is with a consideration of these questions of power and how concepts of power can enhance our understanding of these themes and their interrelationships that the chapter continues.

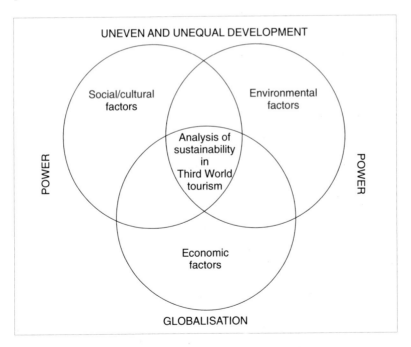

Figure 2.4 Tourism, sustainability and globalisation

THE POWER JIGSAW

The preceding discussion has sought to emphasise the importance that 'power' will play in this book. As the Indian academic, Nina Rao, argues, 'Tourism takes place in the context of great inequality of wealth and power' (quoted in Gonsalves, 1993: 8). In pursuing this argument, we are seeking to address an identifiable weakness in much work on tourism. On the one hand, much tourism analysis has played down relationships of power, which remain implicit, or are absent. Such studies have largely consisted of identifying structural and deterministic models of tourism. These are examined more appropriately in Chapter 4. On the other hand, where power is invoked in a discussion of tourism it has tended to be in passing; references to ideology, discourse, colonialism, imperialism and so on, appear on the page in a rather unstructured, even anecdotal, fashion. Although such analysis is commendable in signalling the importance of power relationships in the study of tourism, the treatment of power needs to be approached more thoughtfully. As Crick (1989) concludes from a wide-ranging review of social science literature, there is an inadequate representation of the complexities of tourism.

The discussion of power below presents three useful concepts that will be employed at various points in the book, each offering an insight into power; these are ideology, discourse and hegemony, and factors of relevance to each are summarised in Figure 2.5. In addition, the following discussion suggests how power is embodied in the idea of sustainability.

Ideology

While ideology is a complex term, one profound trait stands out: namely, its concern with the 'bases and validity of our most fundamental ideas' (McCellan, 1986: 1; quoted in Dobson, 1995). In using the term ideology, we will be referring not only to the sustaining of relationships of domination in the interest of a dominant political power (the USA, for example) or social thought (religion, for example), but also to interests that are opposed to dominant power (the anti-nuclear movement, environmentalists, feminists, and so on) that are themselves capable of forming ideologies in the pursuit of power. Although, as Eagleton (1991) notes, this may signal a degree of contradiction in the meaning of ideology, it is nevertheless fundamental to the notion that ideology is about the way relationships of power are inexorably interwoven in the production and representation of meaning which serves the interests of a particular social group. As Dobson concludes, ideologies 'map the world in different ways' (1995: 7), and it is the intention of this book to map the way in which different interests are implicated in the uneven and unequal development of tourism.

Sustainability is ideological in the sense that it is largely from the First World that the consciousness and mobilisation around global environmental issues have been generated and in the sense that sustainability serves the interests of

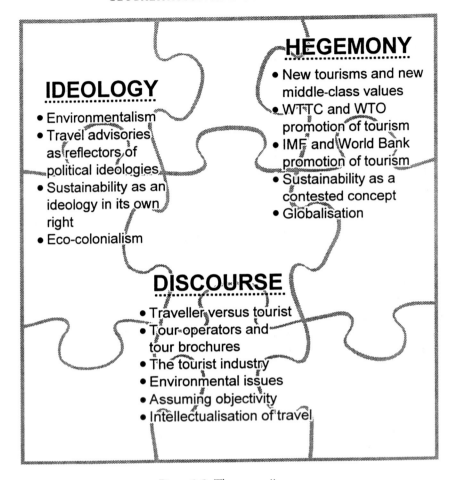

Figure 2.5 The power jigsaw

the First World. Adams (1990), for instance, refers to the ideology of sustainable development, and in the context of tourism in southern Mexico, Daltabuit and Pi-Sunyer (1990) refer to the 'ideology of environmentalism'. The power implicated through First World environmentalism has led increasingly to the 'charge' of eco-imperialism and eco-colonialism, a charge that is considered in Chapter 3.

Implicit in these criticisms is the idea that sustainability is ostensibly ethnocentric. Reconsider for a moment the quote from Robins (p. 18) where he talks about the export of western values and priorities. Such observations can also be applied to the current debate on sustainability. For the most part, it is a discussion 'framed' in the West and imposed on the 'Rest', and hence the acrimonious debates between First and Third World countries at the Rio Summit.

Discourse

Discourse is a concept closely related to ideology. Ideology is perhaps best thought of as a discriminator between power struggles which are central to a 'whole form of social life' (socialism, feminism, ecologism, perhaps) and those which are, for whatever reason, relatively less important. Prioritising the most important forms of struggle may be an exercise of power itself, but it is important to signal which struggles are ideological and which are not.

Discourse can be considered as complementary to ideology. Indeed ideology is a matter of 'discourse', a 'question of who is saying what to whom for what purposes' (Eagleton, 1991: 9). But discourse can also be non-ideological; in other words, it is not reducible to ideology.

The celebrated French philosopher, Michel Foucault (1980) suggests that discourse expresses how 'facts' can be conveyed in different ways and how the language used to convey these facts can interfere with our ability to decide what is true and what is false. For example, as Chapter 4 discusses, the term 'carrying capacity' (an important tool in the study of sustainability) can be subdivided into different types: ecological, social, economic, physical, real, effective, aesthetic; and all of these can be interpreted and measured in different ways by different people at different times and in different circumstances. In contrast, carrying capacity is often treated as if it were a 'neutral' ecological term. Zaba and Scoones challenge this neutrality: 'most of us have no problems with the notion of the carrying capacity of Botswana (pop. 1.3 m, area 567,000 sq km), but would be incredulous at the idea of calculating the carrying capacity of Birmingham (pop. 1.1 m, area 300 sq km)' (1994: 197). Not only does the notion of ecological sustainability carry some kind of scientific validity with it, but it also suggests that some places (in this case Third World environments) are more suited to its application than others. In this way, carrying capacity as discourse transmits and translates power.

There is no agreement over the exact nature, content and meaning of sustainability. It is a contested concept in all senses of the word. Different interests – supranational and transnational organisations, INGOs, socio-environmental organisations, social classes and so on – have adopted and defend their own language (discourse) of sustainability. The new socio-environmental organisations mobilised around issues of environment, for example, are not in power, and yet their ability to influence the meaning of sustainability for our everyday lives has been marked.

Similarly, consider the power to interpret and represent the Third World through travel books and brochures. On the one hand, we have the highbrow, intellectual accounts of best-selling travel writers such as Paul Theroux and Eric Newby, authors noted for the 'authoritativeness of their vision' (Pratt, 1992: 217), and the serious travel pages of broadsheet newspapers. On the other hand, we have glossy high street tourist brochures selling destinations from the Caribbean to Thailand, and which are the subject of much highbrow, intellectual criticism. These are simply different ways of outsiders representing and interpreting the Third World to their audience, each claiming authenticity and truth, albeit in very different ways.

Foucault's ideas may lead to the conclusion that knowledge in tourism is produced by competing discourses. Discourse, therefore, is a useful concept in emphasising how a certain subject or topic is talked and thought about and how it is represented to others. Most importantly, discourses are 'part of the way power circulates and is contested' (Hall, 1992b: 295).

Hegemony

Discourse is also an essential property of hegemony, the last piece in the power jigsaw. Hegemony was a concept developed by the Italian Marxist, Antonio Gramsci, to emphasise the ability of dominant classes to convince the majority of subordinate classes to adopt certain political, cultural or moral values; a more efficient strategy than coercing subordinate social groups into conformity (Jackson, 1992). Hegemony, therefore, is essentially about the power of persuasion and is immediately differentiated from ideology, which by contrast may be imposed forcibly (Eagleton, 1991), as in the former apartheid system in South Africa or through the imposition of IMF structural adjustment policies. The best way to conceive of hegemony is as a 'broader category than ideology' which '*includes* ideology, but is not reducible to it' (Eagleton, 1991: 112).

The real innovativeness of Gramsci's thinking is the conclusion that hegemony is never fully realised in capitalist societies – that it is continually contested (Jackson, 1992). As Williams concludes, hegemony must be 'renewed, recreated, defended, and modified' (quoted in Eagleton, 1991: 115) and is 'inseparable from overtones of struggle' (Eagleton, 1991: 115); a relationship that does not necessarily hold true for ideology.

The three key words that frame this book, Third World, sustainability and tourism are examples of hegemony in practice. Tourism, as we shall see in later chapters, is replete with examples of hegemonic strategies ranging from tourism codes of conduct to the advocacy of more responsible, appropriate or sustainable forms of tourism. It is also evident in the way in which tourism is contested between different social groups (traveller versus tourist, for example) and between different places (Thailand versus Chile, for example). Hegemony is especially useful for its dynamism and practical usage encompassing and focusing attention on a wide range of practical strategies that are adopted by a variety of interests. Such characteristics place notions of struggle and contestation at the centre of the inquiry.

The advocacy by environmentalists of the need to act globally, for example, is an interesting aspect of the persuasiveness of sustainability and how it ties in both the global and local dimensions and stresses the interdependency of places. Residents of distant places are asked to 'consider' other places; in the dictum of Friends of the Earth, 'think globally, act locally'. Conservation measures in southern Africa and rainforest preservation in Central America can be lobbied for and financed from the First World; and a degree of control and influence over Third World affairs is exercised through First World conscience-prodding. Sachs refers to this as the 'hegemony of globalism':

> Until the 1980s, environmentalists were usually concerned with the local or national space. . . . But in subsequent years, they began to look at things from a much more elevated vantage point: they adopted the astronaut's view, taking in the entire globe at one glance. Today's ecology is in the business of saving nothing less than the planet.
>
> (1993: 17)

Testimony to the hegemonic properties of sustainability, perhaps, is the rapidity with which the word has entered public usage on a seemingly global level since its use by Brundtland in 1987, along with the large number of texts that are devoted to dissecting, interpreting, defending or reclaiming the idea of sustainability. For some it is a means of sustaining much more than just environment. It is about 'sustainable development' and incorporates indicators such as income, employment, health, housing, human welfare indicators that are concerned with a 'more rounded policy goal than "economic growth"' (Jacobs and Stott, 1992: 262). For others sustainability is to be reclaimed within a far more radical agenda of political ecology where ecological issues and questions of social justice are paramount (for example see Hayward, 1994; Shiva, 1988; Lipietz, 1995). Characteristic of hegemonic positions, sustainability is contested within a continuum of viewpoints (Adams, 1990) ranging from 'reformism' (often referred to as light green, conservationist or environmentalist) to 'radical-ism' (referred to variously as dark green, deep ecology or, in Dobson's (1995) phraseology, ecologism).

Similarly, sustainability and its application to tourism should not be considered a once-and-for-all position – a neutral, scientific term to which techniques can be applied and upon which policies and programmes can be implemented and evaluated and blueprints, ideal types and models catalogued and advocated. Rather, it constantly changes as the broader influences and interests change reflecting a dynamic situation and concept.

CONCLUSION

This chapter has introduced a number of ideas and concepts which will be elaborated and expanded in the rest of the book. It is best to think of these concepts as a bag of tools that are useful for building up an understanding of the complexity of tourism in the Third World and most especially for dealing with the new forms of tourism that find particular expression there.

There are two notions that are fundamental to the analysis of new tourism: globalisation and sustainability. The starting-point is the accelerated pace of global change. We examined three aspects of globalisation – economic, cultural and political – in order to explore the key changes in late twentieth-century capitalism. In terms of economic globalisation, it was argued that an intense phase of time–space compression has necessitated a move to post-Fordist modes of production and a regime of flexible accumulation. The emergence of a global economy as a result of time–space compression was linked with the expansion of capitalist relations of production, especially to the Third World, where these changes are seemingly reflected in the rapid increase in tourism and the rise of different and more flexible forms of new tourism.

Time–space compression was also invoked in the discussion of cultural globalisation. Here the focus was on the emergence of so-called postmodernist cultural forms and of new patterns of consumption and the relationship of these forms to the growth of the new middle classes. It was suggested that these classes are major consumers of Third World tourism and that political globalisation serves to highlight the emergence of a range of global institutions and organisations, especially concerned with issues of sustainability, that are likely to have a major impact upon the development of Third World tourism. Finally, despite the implicit suggestion within the idea of globalisation that everywhere is becoming the same, the discussion of globalisation alerted us to the possibility of increasing differences and to the uneven and unequal nature of development.

An attempt was also made to fix the notion of sustainability in a broader analytical framework. Again, this was traced through economic, cultural and political processes, ranging from the need to sustain profitability and economic growth through 'sustainable development', to the sustaining of cultural life-styles and the politics and programmes of global institutions and organisations. Overall it has been argued that while a discussion of environmental or ecological notions of sustainability is central to the arguments about Third World tourism, it provides too narrow a focus, and the importance of cultural and economic applications of sustainability must be acknowledged.

Finally, the uneven and unequal nature of global capitalist development and the way in which this is inherent in the development of Third World tourism were emphasised. In order to underline these critical features the ways in which concepts of power – ideology, discourse and hegemony – can assist our comprehension of tourism development were discussed and it was suggested how these relationships of power are reflected through sustainability. In short, we require what Massey (1995) terms a 'geography of power' to make sense of Third World tourism developments.

3

POWER AND TOURISM

Throughout Chapter 2 the uneven and unequal nature of development was emphasised and it was argued that an effective analysis of tourism must acknowledge the importance of relationships of power. In this chapter we begin to consider the way in which power is reflected through tourism in more detail. First, the most systematic attempt to explain the unequal nature of tourism development – the political economy of Third World tourism that seeks to emphasise the dominance and control of tourism from the First World – is reviewed. The discussion moves on to trace other ways in which power has been implicated in the analysis of Third World tourism, particularly through the use of imperialism and colonialism. It is argued that these relationships of dominance have also emerged in new forms of tourism with the citation of 'neo-colonialism' and 'eco-colonialism'. This chapter also provides a review of the importance of 'authenticity' to the study of tourism. It is argued that a consideration of authenticity is a further way in which relationships of power can be traced.

While the political economy of Third World tourism and subsequent work is of considerable interest and applicability (and indeed remains the principal explanation of unequal development, especially of mass tourism) it is argued that it does not provide such a penetrating critique of new forms of tourism in the Third World. Indeed, political economy approaches suggest that the dominance of the First World over the Third World can be overcome, in part, by the creation of new 'alternative' forms of tourism. We challenge this suggestion.

The final section of the chapter suggests an alternative critique through four key characteristics of much new tourism. The first emphasises that all forms of tourism are tied into the growth and expansion of capitalist relations of production. We call this characteristic 'intervention and commodification'. It builds upon the economic aspects of globalisation from the previous chapter and stresses the way in which holiday destinations are drawn into a global system of interdependency, or are by-passed by it. Given the context of global inequality and unevenness of development, the second characteristic stresses the 'subservience' that lies at the heart of tourism in the Third World, regardless of the form it takes. The final characteristics seek to provide a more nuanced critique of new forms of tourism, referred to as 'fetishism' and 'aesthetisation', which seek to demonstrate the way in which the reality of the Third World is either hidden or is used to create a special aura of travelling in Third World regions.

THE POLITICAL ECONOMY OF THIRD WORLD TOURISM

By the early to mid-1970s it was already acknowledged that tourism did not necessarily offer a panacea to Third World countries struggling for economic growth (Turner and Ash, 1975; Turner, 1976). A number of highly critical studies focusing, in particular, on the fate of the small island economies in the Caribbean (Bryden, 1973; Hills and Lundgren, 1977; Perez, 1974, 1975) began to highlight the unequal economic and social impacts associated with tourism. Of special importance was the observation that Third World economies drawn to tourism as a way of earning foreign exchange witnessed the leaking of much of the money made, straight back out of their national economies. This leakage, as it is now commonly known, was seen to arise primarily as a result of the First World ownership and control of the tourism industry in the Third World: from hotels to tour operators and airlines (see the section in Chapter 7 on size and structure of the tourism industry).

These early studies also began to hint at the relationship between tourism and 'underdevelopment'. It was not until Stephen Britton's analysis of Fiji, however, that a more thorough attempt was made in applying dependency theory to the study of tourism (1981a, 1981b, 1981c, 1982). The importance of Britton's analysis is that he stresses the need 'to place tourism firmly within the dialogue on development' (1982: 332) and investigate why tourism so often perpetuates uneven and unequal relationships between the First and Third Worlds.

The theory of dependency is best understood as a riposte to the *laissez-faire* (free market economics) approach to economic development and international trade. As Chapter 2 has suggested, the global expansion of capitalism has drawn the Third World into increasingly tight economic relationships with the First World, and tourism, now the largest global industry, has been a significant component in this process. Dependency theory has sought to demonstrate how and why these tightening relationships are highly unequal.

Dependency theory argues that western capitalist countries have grown as a result of the expropriation of surpluses from the Third World, especially because of the reliance of Third World countries on export-oriented industries (coffee, bananas, bauxite, and so on) which are notoriously precarious in terms of world market prices. The theory uses the notion of centre–periphery (or core–periphery) relationships to highlight this unequal relationship, where the core is the locus of economic power within a global economy.

The most celebrated of the dependency theorists, André Gunder Frank, takes matters one step further in his notion of the 'development of underdevelopment', which stresses that it is the underdevelopment of the structures in Third

World countries created by First World capitalist development that creates dependency (Kay, 1989). Above all else, theories of dependency are in general agreement that the interdependence resulting from global economic expansion and the inability for autonomous growth results in unequal and uneven development.

Britton applies this body of theory to tourism. Centrally, he argues, dependency involves the 'subordination of national economic autonomy' (1982: 334) as a direct result of the unequal relationships inherent in the world economy and that within the present structure of international tourism, Third World countries can assume only a passive role (1981a). Britton summarises his approach as follows:

> Underdeveloped countries promote tourism as a means of generating foreign exchange, increasing employment opportunities, attracting development capital, and enhancing economic independence. The structural characteristics of Third World economies, however, can detract from achieving several of these goals. But equally problematic is the organisation of the international tourist industry itself.
>
> (1982: 336)

But Britton's research and narrative are very much part of the analysis of the mainstream – mass – tourism industry. As such, he argues that tourism in Third World economies is best conceptualised as an enclave industry (referred to by Turner (1974, 1976) as the 'golden ghettoes' and by Krippendorf (1987) as 'holidays in the ghetto') where tourists only occasionally venture beyond the bounds of their hotel compounds (what is referred to as an 'environmental bubble' in Box 3.6 – see later in this chapter). While Britton's critique is widely cited and provides valuable insights into the unequal structure of Third World tourism, we must ask how useful his analysis is for a critical understanding of new forms of Third World tourism that seek to escape the 'ghettoes'. We return to this consideration a little later.

TOURISM AS DOMINATION

Given the arguments advanced by an increasing number of tourism commentators that the Third World is structurally dependent on the First World, there is little surprise in finding a wide range of references to the principal forms of global domination: colonialism and imperialism. While these terms are often used loosely and interchangeably, colonialism is best conceived as a special form of imperialism (that is, the imposition of power by one state over another) involving the occupation of territories. The following sections begin to build up a picture of how these relationships of power are reflected in the analysis of tourism.

Colonialism and imperialism

The inference of colonialism and imperialism that underlies the theories of underdevelopment and dependency has a special appeal to writers on tourism. Both the characteristic First World ownership of much Third World tourism infrastructure and the origin of tourists from the First World have for many become an irresistible analogy of colonial and imperial domination. Indeed, the distinction drawn in dependency theory between a First World core and Third World periphery is part of a more general theory of imperialism. Nash argues that tourism only exists in so much as the metropolitan core generates the demand for tourism and the tourists themselves. He concludes 'it is this power over touristic and related developments abroad that makes a metropolitan center imperialistic and tourism a form of imperialism' (1989: 35). Similarly, van den Abbeele (1980) laments tourism as doubly imperialistic both in turning Third World cultures into a commodity and providing hedonistic practices for wealthy First World tourists. Clearly, this is more than just an academic concern or critique. Take, for example, Box 3.1, which provides the background to Survival International's campaign on tourism and tribal peoples.

For Third World critics in particular, as Gonsalves observes, it is the very presence of tourists that leads to the 'view that modern tourism is an extension of colonialism (with all the attributes of a master–servant relationship)' (1993: 11). It is an increasingly widely shared opinion in the Third World. Chung Hyung Kyung's observations are illustrative of the passion and conviction with which these are expressed: 'Colonialism has many faces. Third World tourism, an advanced form of "post-colonialism", is a disease which destroys people's bodies and souls. . . . Third World tourism carries a major symptom of colonialism: "Domination and Subjugation"' (1994: 21).

It is this notion that tourism is implicated in the maintenance of colonialism that is so important here. Perez (1974), for example, argues that 'Travel from metropolitan centres to the West Indies has served historically to underwrite colonialism in the Caribbean' (1974: 473). And Bruner insists that, however much we attempt to deny or evade the relationship, 'colonialism . . . and tourism . . . were born together and are relatives' (1989: 439). They are, Bruner contends, driven by the same social processes involving the occupying of space (by tourist infrastructure and ultimately by tourists) opened through the expansion of power.

It is not just academics that are drawing parallels between tourism and colonialism. Srisang, a former Executive of the Ecumenical Coalition on Third World Tourism (ECTWT), the world's largest tourism NGO, suggests that:

> tourism, especially Third World tourism, as it is practised today, does not benefit the majority of people. Instead it exploits them, pollutes the environment, destroys the ecosystem, bastardises the culture, robs people of their traditional values and ways of life and subjugates women and children in the abject slavery of prostitution. In other words, tourism

Box 3.1 The 'new imperialism'

The majority of the world's tourists are from the industrialised countries: 57 per cent from Europe, 16 per cent from North America. 80 per cent of all international travellers are nationals of just 20 countries. Thus it is largely the tourist industry in the affluent tourist-generating countries that determines the nature and scale of tourism. These tour operators are interested primarily in short-term benefits and realising a return on capital and investments . . .

Much of the money generated by tourism is sent abroad. 60 per cent of Thailand's $4 billion a year tourism revenues leave the country. Some critics of the tourist industry have called it the 'new imperialism' . . .

In response to the more obvious negative effects of tourism, many tour operators have now proclaimed themselves to be 'green' and jumped on the ecotourism bandwagon. . . . Ecotourism hopes to change the unequal relationships of conventional tourism. Thus it encourages the use of indigenous guides and local products. Ethical tours claim to combine environmental education with minimal travel comfort, help protect local flora and fauna, and provide local people with economic incentives to safeguard their environment. For example, the ecotourists can now join a rainforest research project, visit African mountain gorillas or take a water divining tour to the Sahel . . .

Yet even small groups of people, or for that matter the lone traveller, no matter how sensitive, may have a disruptive effect on local culture . . .

While tourism usually promises to provide employment to the local community, the jobs are most often unskilled, menial and poorly paid. . . . More often than not, the needs and rights of indigenous peoples are ignored. For example in west Nepal, the Chhetri people were moved from their lands to make way for Lake Rara National Park, . . . In Kenya's Shaba reserve, scarce water is diverted from the spring once used by local Samburu herdsmen to water their cattle, in order to fill the swimming pool of the Sarova Shaba Hotel . . .

Survival does not claim to be able to resolve the debate surrounding ecotourism. However, when tribal communities are the destination for tourists, it is right and appropriate that the wishes of these communities are respected. The key word is control. Not only do tribal peoples have a right to their lands, they also have the right to decide what happens on their lands, to determine their future and way of life . . .

Ecotour operators are busy selling 'rainforest tourism' to the environmentally interested traveller by promoting the image of tribal people as 'noble savages'. Rather than patronising tribal peoples . . . we need to see them on their own terms as dynamic and complex societies . . .

The need to bring in foreign currency is used to justify this abuse of tribal peoples' rights and denial of their dignity. Clearly, tour operators and governments are often willing collaborators and perpetrators of this form of exploitation. This can be stopped if tribal peoples are given control over the . . . development of tourism in their communities.

> (extracts from Survival International (1995) *Tourism and tribal peoples*, information sheet). Survival is a worldwide organisation supporting tribal peoples, and more examples of its research are given in Chapters 6 and 8.)

epitomises the present unjust world economic order where the few who control wealth and power dictate the terms. As such, tourism is little different from colonialism.

(1992: 3)

The ECTWT itself is equally outspoken, referring to the majority of Third World tourism as 'an expression of neo-colonialism contributing to racism, erosion of moral values, economic impoverishment and cultural degradation' (ECTWT leaflet, undated).

For other writers, however, the relationship between colonialism and tourism to which they allude amounts to little more than a casual or anecdotal observation, often on the tourists themselves. Hence, the analogies between the affluent middle classes and 'scavengers' (MacCannell, 1976), the 'easy-going tourist' and the 'conqueror and colonialist' (Cohen, 1972), and the suggestion that 'for many tourists, aggressive – almost colonialist – behaviour becomes a norm while on holiday' (Shaw and Williams, 1994: 80). As Krippendorf concludes, in the absence of changes, tourism will remain for the host 'a special form of subservience' (1987: 56).

While such observations are understandable, even justified in the way in which tourism seems to reawaken memories of a colonial past (Crick, 1989), they represent a reaction to tourism based more upon an emotional response. In this vein, as Allen and Hamnett conclude, it 'can be argued . . . just as some "Third World" countries have thrown off the yoke of colonialism, they have taken up the yoke of tourism' (1995: 252).

Two observations arise from this review. First, is the rather ambiguous fashion in which the charge of imperialism and colonialism is often made. Because tourism is a conduit for relationships of power, it has been easy for authors to use terms for these forms of domination to describe a vast array of relationships involved; from multinational hotel chains to a waiter/diner exchange. It has, therefore, become an attractive comparison to make, with the words imperialism and colonialism immediately invoking certain images and responses in our minds. Second, as we observed with Britton, a good deal of the equation *tourism = imperialism + colonialism* still arises as a critique of the mainstream mass tourism industry. It is somewhat blunt or crude in dealing with new forms of tourism whose claim is to escape these very relationships of domination. In the section 'Alternative critiques for alternative tourism?' (pp. 63–83), therefore, we attempt to analyse (or disaggregate) these forms of power. This provides a clearer picture of how relationships of domination are manifest and suggests how these observations might be applied to new forms of tourism.

Neo-colonialism

One of the ways in which the discussion can be reframed, is through the reference to 'neo-colonialism'. As Thomas argues, although colonialism as a pervasive

49

moment in history has all but gone, 'the persistence of neo-colonial domination in international and inter-ethnic relations is undeniable' (1994: 1).

In the context of tourism, the charge of neo-colonialism has already emerged as a principal way of describing the retention of former colonies in a state of perpetual subordination to the First World, in spite of formal political independence. Hence, Britton (1981c) refers to Fiji as a neo-colonial economy and seeks to demonstrate why tourism reinforces the pattern of spatial organisation which evolved during colonialism. The tourist industry, he argues, is a 'neo-colonial extension of economic forms present in pre-independent Fiji' (1981c: 149). Similarly, Shivji argues of Tanzania, 'Since the success of tourism depends primarily on our being accepted in the metropolitan countries, it is one of those appendage industries which give rise to a neo-colonialist relationship and cause underdevelopment' (Shivji, 1973).

However, while such analysis is clearly vital in constructing a broader critique of Third World tourism, the discussion has too often been restricted to a consideration of economic impacts. So, for example, the complexities of class and race have been largely neglected and the spectre of neo-colonialism engendering a subtle, but pervasive racism, has remained largely unexplored. Again, we will return to such considerations below. Most notably then, a discussion of neo-colonialism allows us to think in terms of the existence of a number of colonialisms (Thomas, 1994), as opposed to one all-embracing form of 'colonialism', and explore the many different ways in which power is expressed in a so-called post-colonial world.

Eco-colonialism as neo-colonialism?

One way in which we can immediately see the relevance of thinking in terms of different forms of neo-colonialism is its application to sustainability and environmentalism, and ultimately to development. This challenges the moral high ground so effectively adopted by 'environmentalists' over the last twenty years.

On the one hand, there are emerging critiques of environmental organisations themselves, not far removed from the critical attacks launched on other supranational agencies such as the World Bank and IMF. Phillipson, an internal auditor of World Wide Fund for Nature, for example, accuses them of 'egocentricity and neo-colonialism' (quoted in Fernandes, 1994). Central to these criticisms is the way in which organisations seek to impose policies and programmes on Third World countries.

On the other hand, a more thorough critique of environmentalism and ecologism as a movement has begun to emerge. Such criticisms have tended to focus on the morality vested in the environment – it is our duty to save the earth! – and the crusade-like fashion with which environmental issues are pursued. To pursue the analogy, there is a sense in which an army of eco-missionaries, or as some would argue, eco-fundamentalists, have fanned out across the Third World to green the Earth's poor.

As noted earlier, the moral basis for environmentalists' claims has emanated from the symbol of interdependence and 'oneness' of the Earth which is founded upon the notion of a global ecosystem. Wolfgang Sachs neatly draws out the relationships of power and domination from this 'systems language'. It is worth quoting Sachs at length, for his ideas are both powerful and thought-provoking. Box 3.2 presents a number of his views from several sources.

Box 3.2 *Environment and domination*

The terms 'ecosystem' or 'global system' cannot shake off the legacy of engineering; the language is committed to an interest in regulation and control. (1992a: 22)

. . . it is this concept ecosystem that gave to the ecology movement a quasi-spiritual dimension and scientific credibility at the same time . . .

. . . For many environmentalists now, ecology seems to reveal the moral order of being . . . it suggests not only the truth, but also a moral imperative and . . . aesthetic perfection. (1992b: 31–2)

Satellite pictures of the earth . . . evoke a spurious universalism. In these pictures, human beings and what nature means to them and their lives are missing. Global resource management tends to disregard the local context. Such disregard used to go under the name of colonialism. (1992a: 22)

In the face of the overriding imperative to 'secure the survival of the planet', autonomy easily becomes an anti-social value, and diversity turns into an obstacle to collective action. Can one imagine a more powerful motive for forcing the world into line than that of saving the planet? Eco-colonialism constitutes a new danger for the tapestry of cultures on the globe. (1992b: 108)

(Sachs 1992a; 1992b)

From Sachs' (and others') writings, we are quickly led to question the intention and outcome of much environmentalism and the way it is advanced through notions of sustainability. As the environmental critic, Vandana Shiva, asks rhetorically: 'Global environment or green imperialism?' (1993: 151). The importance, once again, is the way in which such discussion reflects back on the global changes considered in Chapter 2. Globalisation drives us towards the logical conclusion that there is only one world: a global economy, a global culture, a global environment. It is the violent imposition of this idea from the First World, an imposition clearly reflected in the Rio Summit (see Boxes 2.2 and 4.10) which creates the 'moral base for green imperialism' (Shiva, 1993: 152).

These critiques of environment and ecology have clear and wide-ranging ramifications for the study of tourism too. It is the power invested in the concepts such as sustainable tourism and environmental tourism that have been central to critical responses. Reflecting the discussion of discourse in Chapter 2,

Herman argues 'sustainable tourism is rooted in much "double-speak". This double-speak needs to be recognised for what it is: "The misuse of words by implicit re-definition, selective application of . . . words, and other forms of verbal manipulation"' (Fernandes, 1994: 28, quoting Herman). Similarly, writing of tourism development in Quintana Roo, southern Mexico, Daltabuit and Pi-Sunyer (1990) refer to environmentalism as a 'powerful rhetoric' (p. 10).

Beyond dependency?

The discussion so far indicates that relationships of power have been at the heart of critiques of Third World tourism. Consequently, responses and suggested solutions to these problems have been couched in the need to move beyond dependency and to break off the subordination and domination that is said to characterise mass tourism. Britton's approach is indicative of these arguments. As we have seen, Britton's analysis is situated firmly within a typical mass pack-aged tourism environment. Predictably his prescription for a reform of such tourism rests on the need to move beyond mass tourism – or what he refers to as 'conventional' tourism – and develop alternative forms (Britton and Clarke, 1987). Among other things, this shift supposedly responds to the problems he identifies, promotes the local ownership of tourism resources, creates local employment and helps stem the haemorrhaging of foreign exchange from the economy.

A second, and very different, tack is offered by Auliana Poon (a World Tourism Organisation and United Nations Development Programme consultant attached to the Caribbean Tourism Organisation). (We return to the policies of these organisations in Chapter 9.) Like Britton, Poon sees dependency as the key question. But for Poon 'dependency' is defined in a very different context and she challenges the general agreement that small island states are compelled to participate in a world economy beyond their control. Of the Caribbean, Poon argues, 'The innovativeness of many indigenous tourism enterprises . . . coupled with new developments in the world tourist industry, warrants a re-conceptualisation of the world question of dependence' (1989a: 74). The conclusion drawn by Poon is that it is not the metropolitan core singled out by Britton (and others) nor the lambasted activities of multinational and trans-national companies upon which Third World countries are dependent. In a linguistic (and conceptual) twist reminiscent of free market economics, Poon argues that tourism in the Third World is dependent on 'innovation', the 'foster-ing of indigenous skills, creativity and innovativeness, rather than perpetual reliance on MNCs [multinational corporations], which hold the key to the future survival' (Poon, 1989a: 74). It is a conclusion that lays both onus and blame squarely on the shoulders of tourism providers in the Third World and exonerates the mass tourism operators from culpability in the highly uneven and unequal development of global tourism. Moreover, it is an interpretation of dependency that anticipates the policies and programmes of donor bodies. These agencies impose their own vision of Third World tourism on the countries and projects to which they lend.

Others have followed suit in advocating a range of tourisms prefixed with descriptions that indicate their qualitative difference from current forms of mass tourism. The traditional mass packaged holidays, typically described with the Ss – sun, sea, sand and sex – are challenged with the three Ts of new forms of tourism: travelling, trekking and trucking. These are examined in more detail in Chapter 4. Box 3.3 summarises these shifts in contemporary tourism.

Box 3.3   *Shifts in contemporary tourism*

OLD	NEW
FORDIST	POST-FORDIST
Mass	Individual
Packaged	Unpackaged/flexible
Ss (sun, sea, sand, sex)	Ts (travelling, trekking, trucking)
Unreal	Real
Irresponsible (socially, culturally, environmentally)	Responsible
MODERN	POSTMODERN

Similarly some authors have chosen to advocate the role of independent travellers in promoting socially just forms of tourism. Kutay rather eulogises travellers referring to them as 'Peace Corps-type travellers looking for a meaningful vacation' (1989: 35), and Jeremy Seabrook sings the praises of those dissatisfied with the packaged form of travel: 'the success of the Lonely Planet Guides testifies the hunger of young people for a deep exploration of the countries they visit and that includes a growing curiosity about social and living conditions' (1995: 22). Seabrook moves towards what could be argued is at best wishful thinking (or a flight of fantasy) and at worst an apologist's approach to the acknowledged problems of all forms of tourism. His approach is problematical not only in assuming to speak for Third World communities, but also to speak on behalf of tourists themselves. His new-found activity is social tourism:

> Social tourism can never be a mass movement of people. . . . It is a response to a growing number of people who wish to deepen their understanding of north-south relations. . . . What they wish for is contact with

rootedness, true diversity, and an extension of understanding, rather than more escapism, with which their own culture amply supplies them.

(1995: 23)

While, therefore, mass tourism has attracted trenchant criticism as a shallow and degrading experience for Third World 'host' nations and peoples, new tourism practices have been viewed benevolently and few critiques have emerged. But do these alternatives offer viable responses and solutions to the existing problems as so many authors imply? Or are these new tourism practices further evidence of the way in which the ebb and flow of tourism is conditioned and controlled from the First World?

Returning to Poon's analysis, we find it not only wanting in terms of explanatory power, but also in that it assumes an over-simplistic binary transition, or transformation and metamorphosis as Poon terms it, from an old tourism which was 'not only mass, but . . . standardised and rigidly-packaged' (1989a: 74) to new tourism based 'upon a new "commonsense" or "best practice" of Flexibility, Segmentation and Diagonal Integration' (1989a: 75) – a point that will be expanded upon in Chapter 4. Box 3.4 summarises Poon's position.

Box 3.4   Tourism in metamorphosis?

Old tourists	New tourists
Search for the sun	Experience something new
Follow the masses	Want to be in charge
Here today, gone tomorrow	See and enjoy but not destroy
Show that you have been	Just for the fun of it
Having	Being
Superiority	Understanding
Like attractions	Like sport and nature
Reactions	Adventurous
Eat in hotel dining room	Try out local fare
Homogeneous	Hybrid

(Poon, 1993: 10)

Poon does not envisage that mass tourism will disappear altogether; rather that it will fade into relative unimportance. Drawing upon an analogy of typewriters and computers she concludes:

While there will continue to be a market for typewriters (mass tourism), the growth of new computers (new tourism) will be far greater. Having used typewriters and then been exposed to the power of computers, users will be unwilling to go back to the old way. This exact logic holds for old and new tourism.

(Poon, 1993: 23)

What Poon describes as new tourism is what Lash and Urry refer to as 'post-Fordist consumption' (Lash and Urry, 1994: 274). Although Poon's analysis is less conceptually elaborate, both approaches strike the same chord. Lash and Urry build their ideas upon a transition from one stage of tourism to another, although their ideas are set within a deeper theoretical seam that runs through two principal works, the *End of Organized Capitalism* (1987) and *Economies of Signs and Space* (1994), which set out a political economy.

For Lash and Urry the development of capitalism takes place in three main phases: liberal capitalism, organised capitalism and disorganised capitalism. In terms of this last stage, they do not contemplate a shift to some form of 'random disorder' but rather a 'fairly systematic process of disaggregation and restructuration' (1987: 8) of capitalism and capitalist countries.

So how do these stages relate to the emergence and development of tourism? Liberal capitalism, they argue, is characterised by individual travel by the wealthy, usually stereotyped by the Grand Tour of the eighteenth and nineteenth centuries. The second phase is best understood as one of organised mass tourism that stretches from the development of Blackpool to the emergence of the Costas (in Spain) and package destinations elsewhere. Their last phase of disorganisation is a little more difficult to understand (and justify) and they claim it is characterised by the 'end of tourism'. Here they are signalling sociologically that tourism has lost its specificity as an activity and that people 'are tourists most of the time' (1994: 259), an argument that stems from the 'analysis of the social relations actually involved in tourism' (1994: 270). While this is a rather messy and convoluted approach to the debate over contemporary tourism, their observations are useful in setting the emergence of alternative (and by inference sustainable) tourism in a broader conceptual context. Box 3.5 summarises the changing features of tourism with the development of capitalism.

Authenticity – looking for the real thing

Authenticity is a critical concept within the sociology of tourism and is the subject of one of the most enduring debates (for example, see Cohen, 1989; MacCannell, 1973, 1976, 1992; Pearce and Moscardo, 1986; Turner and Manning, 1988). It is now commonplace to hear references to authenticity in everyday usage. How often do we hear people refer to their holiday experiences as real or unreal, authentic or fake, for example, or are we presented with a range of programmes and publications – the Real Holiday Show, the *Independent*'s Guide to Real Holidays, *Rough Guides*, and so on – that are motivated by the desire for real, authentic experiences? As Poon contends, tourists are moving away from 'tinsel and junk' in the search for 'more real, natural and authentic experiences' (1989a: 75). Authenticity is central to our discussion, therefore, because it is fundamental to much of the debate about the content (real, ethnic, off-the-beaten-track, and so on) and appropriateness (eco-, alternative, sustainable) of new forms of tourism in the Third World. As Daltabuit and Pi-Sunyer (1990)

Box 3.5 Tourism and the development of capitalism

Mass consumption	*Post-Fordist consumption*
Purchase of commodities produced under mass production	Consumption rather than production dominant
High or growing rate of expenditure on consumer products	New forms of credit and indebtedness
Individual producers dominate particular industrial markets	Almost all aspects of social life become commodified
Producer dominant	Consumer dominant
Little differentiation between commodities	Greater differentiation of purchasing patterns
Relatively little choice/producers' interests reflected	Consumer movements and politicisation of consumption
	Consumers react against the 'mass' and producers more consumer-driven
	Many more products and shorter lives
	New kinds of specialised commodity emerge

(Urry, 1990a: 14)

---

argue, for example, '"The environment", as commodity or experience, is no less a fantasy than any other image elaborated by the leisure industry. That it has so broad an attraction probably owes much more to the postmodern quest for authenticity' (1990: 11). And, as we have suggested, authenticity also has strong resonance in the debates over sustainability – environmental and cultural.

It can be argued that 'real' among the new middle classes indicates a desire for authenticity, for 'honesty', however circumscribed that may be in reality. Kestours, a private travel agent formed in 1956, specialising in Caribbean holidays, exemplifies this and is indicative of the transformation in contemporary travel. The company's two-brochure package symbolises the transition from a derided mass tourism to a new, possibly postmodern, and more truthful tourism that characterises certain segments of the industry. The first brochure, a traditional glossy with large-format coloured photographs of sun-drenched beaches, shady palms, luxury hotels and loving couples, contains minimal text. These images say it all, this is paradise. The second brochure, printed on manilla, seemingly recycled, paper, contains only line drawings (and two passport-size black-and-white photographs on the back cover). This is reflective of new tourism and it is here that the Caribbean paradise is introduced:

> The Caribbean is a tropical destination. . . . Ants, mosquitoes and cockroaches thrive in hot climates and while usually harmless are sometimes a nuisance. In the long hours of sunshine, lack of rain can mean erratic water and electricity supplies – really hot water is rare. The Caribbean is a long way away. Your flight takes at least 8 hours and even in modern wide-bodied jets, it can be tedious as can protracted entry formalities when you arrive. . . . The Caribbean is very different to Britain. To enjoy your holiday to the full, you must accept its shortcomings as a challenge and as an enriching experience rather than as a reason for complaint.
>
> (Kestours: *The Caribbean Holiday Guide, 1991/92*: 1)

The evolution of this debate is mapped out in Box 3.6.

As we indicated in Chapter 2, however, discussions have tended not to explore authenticity within the context of wider cultural consumption practices. This is especially so among the new middle classes who are major consumers of new forms of Third World holidays. (We attempt this exploration in Chapter 5.) On the one hand, there are rhetorical exchanges over the invidious tourist– traveller continuum, coupled with the more rigorous attempts to typologise different types of 'tourists' and tourist experience. On the other hand, there are those accounts that rely ostensibly on motivation, attitude or satisfaction surveys (see Dearden and Harron, 1991, 1993, for example). Consequently, broader issues such as how social class and tourism embody class differentiation tend not to be addressed.

Arguably the most significant point to note about authenticity in tourism is the manner in which it reflects the wider global processes already discussed in Chapter 2. So, for example, it can be seen to reflect the suggestion that we are moving towards an undifferentiated global culture and the accelerating desire to experience real cultures through travel. MacCannell captures the importance of authenticity within this broader framework:

> By now, almost every individual on Earth has been informed that he or she is related to every other individual in 'The New World Order'. The message is usually accompanied by a sinking feeling that all our actual relationships, even with former intimates, are falling apart. A secondary effect of the alleged globalisation of relations is the production of an enormous desire for, and corresponding commodification of, *authenticity*.
>
> (1992: 169–70)

But while this debate is of interest, it has tended to mask other, perhaps more invidious, processes that are at work. Authenticity, it will be argued, must be understood within a broad frame – it is not just about 'real' tribes in Thailand, Kenya or Bolivia, it is about the consumption of 'real' lives too and this includes poverty, civil struggle, and so on.

Box 3.6 Travels in authenticity: the evolution of the concept

PSEUDO-EVENT

Daniel Boorstin (1961) attempts to chart the shift from individual traveller to the emergence of mass tourism and tourists. In *The Image* he argues that contemporary Americans are unable to experience 'reality' directly and instead thrive on what he terms 'pseudo-events': *inauthentic* contrived attractions. After a while this pleasure becomes a self-perpetuating system of illusions – *an environmental bubble*. This concept of tourists being cocooned from reality in a bubble is a widely held idea today. In their famous *The Golden Hordes* (1975), Turner and Ash follow Boorstin's lead and place tourists at the centre of a strictly circumscribed world, with travel agents, carriers, hotel managers and so on taking the role of surrogate parents. Analyses of this kind are underlined by issues of class (see Chapter 5).

STAGED AUTHENTICITY

MacCannell detects the class bias in Boorstin's analysis. For MacCannell the essence of all tourism is the quest for authenticity that he considers absent from everyday modern life. In this quest for authenticity local communities – *hosts* – become cheesed off with being gawped at, but also see the chance to make a quick buck! Applying the work of the Canadian sociologist, Erving Goffman, MacCannell argues that front and back regions (lavish dining room/grimy kitchen, for example) can be identified in tourism spaces. This results in contrived tourist spaces where front regions are made up to look like the 'real' thing; and hence *staged authenticity.*

COMMUNICATIVE STAGING

Erik Cohen is in general agreement with MacCannell but disagrees that all tourists seek authenticity: a highly idealised global view of tourists. He identifies four types of 'touristic situations': authentic, staged authenticity, denial of authenticity, contrived. Cohen suggests we need to consider the role of *communicative staging* where tourist sites have not been transformed but the sites are presented and interpreted as authentic to tourists by their guides.

THE GAZE

Urry concedes authenticity may be an important component of tourism but only because there is some sense of a contrast with everyday experiences. He agrees with Cohen that some tourists delight in inauthenticity (what he terms post-tourists). The gaze, Urry argues, 'is as socially organised and systematised' and in any historical period is constructed in relationship to its opposite (that is, non-tourist forms of social experience and consciousness). A particular type of tourist gaze will therefore depend on what it is contrasted to. Tourist sites can be classified in terms of three dichotomies: historical/modern; authentic/inauthentic; romantic/collective.

Useful sources: Boorstin, 1961; MacCannell, 1973, 1976; Cohen, 1974, 1989; New Internationalist, 1984: 24–5; Urry 1990a

---

Othering and otherness

Through a brilliant analysis of western literary representations of the Middle East, *Orientalism*, Edward Said (1991) has demonstrated how western intellectuals lovingly created the countries to which they were devoted in an insidiously romanticised fashion. Through their writings, they portrayed how these 'other' cultures were exotic, sensuous, erotic (to which we could add today: simple and sustainable), at one with nature. This process of creating and interpreting other peoples and places is referred to sociologically as a process of othering.

There has been a marked growth of interest in mass and minority (non-western) cultures, religious traditions, ethnicity, and environment and ecology: aspects of otherness that find special representation in the Third World and are reflected in the First World in alternative markets, charity shops, and so on. Short (1991) refers to some homes in the First World resembling 'ethnographic museums' and Third World literature, food, cinema and music are increasingly demanded, especially by the new middle classes. This phenomenon is clearly steeped in relationships of power (what we referred to in Chapter 2 as an example of hegemony). And this has clear reflections in the idea of the creation and representation of authenticity. As the satirical quote with which Said opens his book encapsulates: 'They cannot represent themselves; they must be represented.'

In addition, it is a process, as Hall argues, that involves representing an 'absolutely essentially different, *other*; the Other' (1992b). In Hall's analysis of the West (or First World) and the Rest (essentially the Third World), othering produces a series of sharp binary opposites: for example, the West as democratic, free, developed, peaceful, and the Rest as despotic, undeveloped, violent, barbaric, fundamentalist, and so on.

There are two observations to be made here. The first is the way in which this process of othering is reflected through the debate on environmental sustainability: it is the Third World, the First World has implicated, that is over-populated and is an increasing drain on natural resources. Second, othering clearly involves a process of reflection. Other cultures and environments are everything that our cultures and environments are not. Thus western lifestyles can be denigrated as empty, culturally unfulfilling, materialistic, meaningless, while, on the contrary, Third World cultures can be bestowed with meaning, richness, simplicity and, of course, authenticity. Indeed, the search for authenticity implied here has been seen very much as a response to the dissatisfaction with 'modern' living – a postmodern reaction to the decline of difference (different cultures, places and so on) and diversity. Both MacCannell (1976) and Cohen (1979a; 1985) refer to 'postmodern travellers', or what Cohen calls *experimental* and *experiential* tourists.

It is not difficult to see how this vein of analysis can be applied to contemporary Third World travel. In tourism othering is a key process of socially constructing and representing other places and peoples. As Explore, a small company formed in 1982 offering 'exploratory holidays' to those 'keen on discovering real qualities and real places', promises, the intention is to know the 'other' side so rarely seen. And as Shurmer-Smith and Hannam argue, such an approach 'informs much of the genre of travel writing and people's choice of holidays' (1994: 19). Equally, it also demonstrates the power to bestow and expropriate meaning. But as Figure 3.1 suggests, seen from a Third World perspective this process of reflection can be read very differently. Box 3.7, which produces extracts from a tour brochure and broadsheet newspaper, provides examples of our discussion.

Box 3.7 Representing the Other

Welcome to South America

'A paradise on earth, a different world.'
CHRISTOPHER COLUMBUS upon reaching South America in 1498.

The adventurous traveller longs to visit faraway places and discover the unknown. South America is, for many, the last frontier – a wild, exotic corner of the world mostly unexplored and largely unspoilt. From the Caribbean coast in the north, through the rich rain forests of Brazil, the high plains of Bolivia and the boundless prairies of Argentina, to desolate Tierra del Fuego, South America stretches nearly five thousand miles with the bristling Andean Mountain range straddling the entire continent from north to south. Sheer canyons, crystal clear lakes, majestic volcanoes, ice-blue glaciers and cascading waterfalls combine to make it a land of unparalleled beauty and splendour.

('Passage to South America' brochure)

Too good for tourists

Belize – the former British colony of British Honduras – is now a much desired ecological destination favoured by ecotourists. Martha Gellhorn simply deems it 'too good for tourists'. Her exposition is reminiscent of bygone travelogues: 'untainted by tourism, Belizeans are lovely.' It is an unsophisticated and ulti-mately racist ethnography that Gellhorn offers – ranging from her observation of Mayan 'Spanish speaking bird-like women, all married at 14, all giggly and happy with hordes of children', to the Garífuna 'matt black, with sharp strong Indian features, and a reserved ungiggly manner'. But she also celebrates the 'achievements' of colonialism. 'These self-confident people are loyal subjects of the Queen, since Belize is a member of the Commonwealth with a lady Governor General. They don't know how lucky they are past and present.'

(quotations: Gellhorn, 1990)

Writing in the *Guardian*, Ros Coward alludes to the relationship between the middle classes, authenticity and otherness:

the middle classes smugly believe that . . . problems are not created by their sort of holidays. They travel independently, [and] visit ever more

Figure 3.1 Reflections: the West and the Rest
Source: Artist Norna Beadle

remote places. The moral superiority of this tourism comes from the idea
that it provides an experience of the authentic culture of the host country
rather than its destruction. The . . . problems are blamed on the kind of
holiday taken by less affluent members of the affluent west . . . who do not
understand that the true purpose of travel is to experience otherness. . . .
As a result, the discerning have to travel further afield.

(1996: 11)

Otherness and authenticity are united in a desire to ensure that culture
and ethnicity are preserved. It is the promotion of primitiveness within which
authenticity becomes the principal commodity (Cohen 1979a, 1979b, 1989;
Errington and Gewertz, 1989). This is perhaps best conceived, albeit in a rather
different context, as part of wider postmodern nostalgic yearnings (Jameson,
1984, 1991). In the context of Third World tourism, this is translated through
a range of new tourisms, from colonial tourism in the Caribbean to the excite-
ment of *discovering* Zimbabwe, 'like the great missionary and explorer Dr David
Livingston [sic] 150 years earlier' (Africa Exclusive). Not only is there nostalgia
for ancient traditions and environments, most visibly evoked in nature
documentaries, but also a nostalgia for the travel styles of yesteryear.

In Chapter 2 it was argued that sustainability can be regarded as a transmitter
of power relationships concerned, in part, with the sustaining of Third World
environments and cultures: the preservation of the other. Leading alternative
tour companies are indicative of this strand of sustainability. (In the following
pages the tour company brochures quoted are from the years 1992–96.)
Dragoman, for example, extols Survival International as the leading authority on
tribal peoples and donates £2.00 for every direct booking. Encounter Overland
actively supports World Wide Fund for Nature, Save the Gorillas, Project Tiger
and rainforest preservation movements in South America. And Guerba
Expeditions cites over £12,000 to be contributed to World Wide Fund for
Nature and its commitment with WWF to lessen the impacts of tourism. High
Places highlights its support of the Río Mazán Project; and Himalayan Kingdoms' pitch is: 'Being "green" is a popular bandwagon, but we do whatever we
can to minimise the impact of tourism. We have been instrumental in working
out a code of conduct with Tourism Concern, and are currently spearheading a
new concept of "mountain friendly" climbing expeditions' (1994 brochure).

To dismiss these expressed environmental concerns as merely marketing
ploys is to understate not only a valuable research resource but also the
importance of these concerns expressed both within certain social groups and
by the companies themselves; many small companies have an undeniably
genuine commitment to environmental issues. But for some commentators
the intimate relationship between tourism and environment has amounted to
a hegemony or what Daltabuit and Pi-Sunyer (1990) refer to as an ideology
of environmentalism. At the height of ecological reductionism, tourism itself
is considered an 'ecological phenomenon' (Attenborough, 1986).

In addition, as Chapter 5 will suggest, the 'environmental other' both reflects the need of the new middle class to sustain travel experiences that are capable of maintaining cultural capital (and the conferring of social status) and represents the manner in which social groups (as well as political institutions and businesses) establish boundaries of power. The travel company Explore exemplifies this in its 1996 brochure:

> Our approach from the start has been to travel in small groups, causing as little environmental damage and cultural disturbance as possible. We achieve this mainly by utilising local resources and services. And also by making our itineraries individual enough to be a sustainable alternative to the juggernaut of mass tourism.
>
> (3)

A further example of the way in which sustainability can be seen as hegemony in tourism is the way in which cultural preservation is promoted as an end in itself. Such concerns, as will be argued later, appear to reflect the furious debates within tourism about cultural authenticity, of indigenous or real Third World cultures under siege from global capitalism and western values. This signals a second interpretation of sustainability focused upon the desire to protect Third World indigenous cultures from dissolution.

ALTERNATIVE CRITIQUES FOR ALTERNATIVE TOURISM?

The preceding discussion has highlighted what we consider to be the main constraints in the analysis of Third World tourism. It is evident that much criticism has centred upon the traditional mass forms of tourism. While a critical analysis of the activities of supranational institutions, such as the IMF and World Bank, and multinational companies is evidently necessary, we have suggested this fails to tell the whole story. It is not just the actions of the 'big players' that we must place under the microscope (they are in a sense the obvious 'culprits'). It is equally necessary to provide a critique of the actions of environmental organisations or the armies of backpackers whose actions are largely seen as benign or benevolent. Thomas touches on this need for a more sophisticated discussion because it is

> easy to denounce government policies and bodies such as the IMF, but perhaps more difficult to explore constructions of the exotic and the primitive that are superficially sympathetic or progressive but in many ways resonant of traditional evocations of others.
>
> (1994: 170)

In particular, it is essential that we challenge the tacit assumption that the emergence of new forms of tourism is both designed for, and will result in, surmounting the problems that have been identified. Returning to our starting-point, exactly what is sustainable tourism seeking to sustain and for whom? We must then attempt a response to Paul Gonsalves' call – 'Wanted: a Third World political economy of tourism' (1993: 11).

The following sections provide one way of beginning to assess and criticise the other – alternative – parts of tourism, to which reference will be made later in the book. They set out what we consider to be the principal processes through which power is conveyed and reflected in tourism. This framework assists in a more critical understanding of new forms of tourism, although it is not meant as a definitive, critical account of Third World tourism. The framework is split into four sections (intervention and commodification; subservience; fetishism; and aestheticisation) with each section anticipating the main arguments that will be developed in later chapters. Box 3.8 provides a short definition of each of these factors and it may be useful to refer back to this box as you read through the sections. Each factor is not independent of the others. On the contrary, they are overlapping and interrelated.

Intervention and commodification – controlling the goods

In Chapter 2 we suggested that the spread of capitalist relations of production throughout the Third World is one of the most notable global economic processes in the period from the 1960s. It was suggested that time–space compression provides a conceptual understanding of why this expansion has taken place and why services, such as tourism, have become increasingly attractive as capitalists attempt to speed up the turnover time of capital. This is of considerable significance, to repeat Britton's observation, because 'geography texts on tourism offer little more than a cursory and superficial analysis of how the tourism industry is structured and regulated by the classic imperatives of and laws governing capitalist accumulation' (1991: 456).

Intervention, however, has not only involved the activities of capitalists. Equally significant, as Truong notes in her study of South-East Asia, 'there has been a high degree of external influence on the formation of tourism and leisure policies' (1991: 99), a pressure exerted by foreign governments, global financial institutions such as the World Bank, and by 'development' agencies.

These processes have been most obvious in the development of international mass tourism. With the spread and intervention of capitalism into Third World societies, tourism has also had the effect of turning Third World places, landscapes and people into commodities. In other words, we consume these elements of a holiday in the same way as we consume other objects or commodities. As Drew Foster, Chairman of one of the leading UK tour operators to the Caribbean (Caribbean Connection) puts it: 'The Caribbean is a great product' (author's transcripts, 1995).

This process of intervention and commodification continues as mass tourism ventures explore new Third World opportunities, but is now joined by

Box 3.8 *Elements of a new tourism critique*

Intervention and commodification

These ideas attempt to capture the rapid expansion of capitalist relations of production in the Third World (a critical factor of economic globalisation as discussed in Chapter 2) and the way in which the spread of tourism has led to destinations, local cultures and environments (such as national parks, wildlife, flora and fauna, and so on) being transformed into commodities to be consumed by tourists. Examples of commodification are the way in which an Amboseli lion is calculated to be worth $27,000 a year in tourism revenue, or the way in which cultural traditions and ceremonies are packaged and sold to tourists, and the timing of rituals is altered to fit tourist schedules (see MacCannell, 1992, for example).

Subservience (domination and control)

As First World tourism expands and commodifies Third World destinations there is a tendency for Third World communities and individuals to assume unequal or subordinate relationships to both First World tourism interests and the interests of 'local élites' (which is discussed further in Chapter 8). It is a reflection of unequal and uneven relationships of power and development.

Fetishism

The fetishism of commodities (or commodity fetishism and the associated concept of reification) is a concept that embodies the way in which commodities hide the social relations of those that have contributed to the production of that commodity (be it a good or a bad experience) from the consumer (such as the tourist). In a nutshell, tourists are generally unaware of the conditions of life experienced by the waiters, cooks, tour guides and so on, the people who service their holidays and the other people who form part of their tourist gaze.

Aestheticisation

This represents the process whereby objects, feelings and experiences are transformed into aesthetic objects and experiences (of beauty and desire). Aestheticisation is a notable characteristic of the way in which the new middle classes construct their lifestyles and is well represented in the ascendency of new forms of tourism as important cultural goods (discussed further in Chapter 5). But aestheticisation must be interpreted broadly. Not only is there a desire to experience primitive cultures, environments and wildlife, but also a desire to experience 'real' poverty and really dicey situations that new tourism sometimes presents.

a considerably more subtle and benevolent form of external influence. As we have suggested, support for mass tourism developments has tended to give way to new, 'softer' forms of tourism that appeal to notions of sustainability. The smaller and more diversified tour operators, together with new social movements mobilised around issues of environmental and cultural difference, that are now commonplace in the First World are seeking out rather different experiences from their excursions into the Third World. In part they are seeking to ensure the maintenance of these experiences and attractions (or commodities), by translating their desire for environmental conservation and cultural preservation to Third World communities. Tourism is being upheld as both cause and effect of environmental, and to a lesser extent, cultural preservation.

Through a range of initiatives, from the 'debt-for-nature' swaps pioneered by the US Government (Enterprise for the Americas Initiative Act, Department of Treasury press release, 27 June 1990) to the phenomenal growth in 'green tourism' operators and to lending agencies placing environmental conditions and caveats on their loans and grants, a *greening* of social relations is being promoted. It is a kind of eco-structural adjustment where Third World places and peoples must fall into line with First World thinking. For example, the Domestic Technology Institute (a US-based non-profit organisation), which is planning a 'multinational organisation to develop and co-ordinate low impact tourism in Third World countries', has warned, 'These [Third World] countries can be forced into establishing natural and cultural resource policy before they can get World Bank loans' (Pleumarom, 1990: 14).

As we have already suggested and will see in later chapters, there is a range of ecological concepts that are capable of spearheading and justifying interventionist policies. *Bio-diversity*, for example, a concept adopted by organisations such as the World Wildlife Fund for Nature (WWF), the International Union for Conservation of Nature (IUCN) and Conservation International (see Chapter 6), seeks to impose conservationist regimes on Third World countries. Their efforts also seek to link up Third World communities with US corporate buyers – clear evidence of the commodification process.

At their most powerful, donor agencies and environmental organisations join forces to pursue conservationist policies in which tourism is a major economic factor. As Gordon argues of Namibia, 'Adjusting the sails of tourism Nature Conservation began to explore the viability of élite expeditions to "natural and wild" parts' (1990: 7), a context Gordon refers to as little more than 'welfare colonialism'.

Other interventionist methods have also emerged among environmental organisations and are being successfully employed, such as appeals to adopt animals and parts of nature (see for example Conservation International). In what would appear to be the antithesis of local control, the multinational Programme for Belize, which has established a private reserve in this Central American country, sells certificated acres of rainforest.

Buy an acre for a loved one, or for yourself . . . the best way of spreading the word is to buy those near and dear to you an acre for themselves. Nearly half the donations we receive are made as gifts on behalf of partners, friends and family and we particularly enjoy sending off packages as christening gifts, wedding presents, anniversary celebrations etc. We know from the response that, whatever the occasion, it is the perfect present.

(UK *Newsletter*, January 1990)

Independent tour operators have also sought to support or become actively involved in conservationist measures. Many operators cite their membership of environmental organisations (such as the WWF) and contribute financially to their work. Other operators have become more proactive in intervening in conservationist measures; Worldwide Journeys and Expeditions (1992), for example, claims management involvement in the only 'privately' managed national park in Africa, at Kasanka (Zambia). In either case the ultimate objective is the same – the need to protect their principal commodity: 'pristine environments' and 'wildlife'.

In Chapter 5 it will be argued that individual representatives of the new middle classes (travellers, backpackers, and so on), are no less interventionist. The scale and scope of their activities may be qualitatively different from the power wielded by tour operators or large environmental organisations, but their culpability in the process of commodification is no less. For such tourists it is the supposed desire to experience 'indigenous cultures' – the Third World otherness – that is a major driving force of their travels and results in the search for 'off-the-beaten-track' or 'less visited areas'.

Speaking of trekking in northern Thailand, Trailfinder's 1992 travel brochure warns that as 'the popularity of this type of holiday experience increases it is more difficult to find . . . "untouched" or "traditional villages"'. Of course, solutions to such problems are always on hand and Trailfinders promises treks to villages 'which are only visited a couple of times a month. These groups are welcomed as a refreshing diversion to normal village life.' Such activities underscore a nostalgic desire for an imagined, 'real' and 'authentic' primitiveness and support the drive for cultural preservation; a method of imposing our desires on other people.

The whole language of new forms of tourism is also premised upon intervention and commodification. Returning to Box 3.7, 'Welcome to South America', the opening paragraph from the Passage to South America brochure, expresses a neo-colonial process of discovery, penetration and expropriation – of last and ultimate frontiers; it should come as little surprise that the early colonists are celebrated by many tour operators. Figure 3.2 illustrates some of the ways in which this colonial past is reproduced by tour operators offering holidays to the adventurous and relatively affluent new middle classes. For tours to Africa, Dr Livingstone is frequently invoked to assist in these hedonistic discoveries. For example, of Malawi, J. & C. Voyageurs promise: 'Following in the footsteps of Dr Livingstone, gliding quietly up the Shire River, unchanged since the first

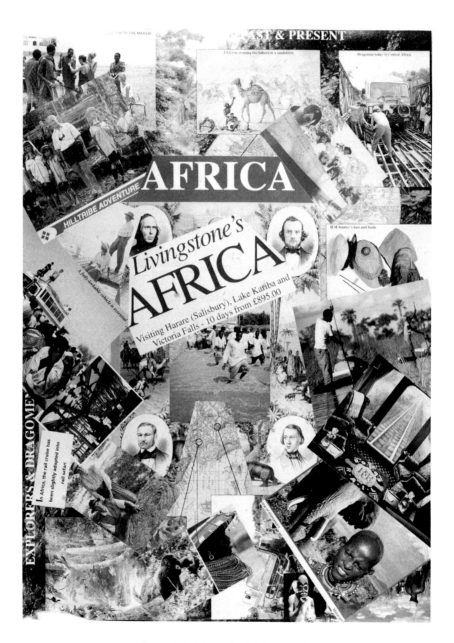

Figure 3.2 Neo-colonial discovery?

appearance of the white man' (1992). Or Africa Exclusive in expressing their thoughts on Zimbabwe: 'when we first discovered this beautiful country – like the great missionary and explorer Dr David Livingstone 150 years earlier – we were very excited' (1992). And Jacobs (1995), writing of Mali, testifies: 'Mungo Park, the late 18th century explorer, would have recognised the view', a land he tells us is 'bypassed by history'.

Even against the backdrop of the five hundredth year of exploitation since the arrival of Christopher Columbus in South and Central America, those presenting a supposedly more enlightened form of travel still find room wryly to celebrate his conquest and assure us that the age of discovery has not passed: 'Given the right advice and [a] . . . Travel Round the World ticket, Christopher Columbus might have saved himself a lot of trouble. He might, for example, have tried both eastern and western routes around the world. But the age of exploration is not dead. There's the whole world waiting to be discovered afresh' (STA Travel, 1995).

It is not just the language that is used, however, but the manner in which some new forms of tourism have attempted to capture the aura and mystique of colonial forms of travel and holidaying. There has, for example, been a resurgence in the popularity of colonial rail travel, train cruises as Eames describes them, 'elegantly packaged tours with a taste of tradition' (1994: 88), and luxury safaris: 'Classic Kenya' – 'an escorted private safari in the old style tradition' in a 'series of luxuriously appointed mobile tented camps' (Worldwide Journeys and Expeditions). Again, Figure 3.2 illustrates some of the images found in tour brochures and promotional material.

It is a romanticism for travel modes of the colonial periods which, unwittingly perhaps, recreates the subordination of Third World peoples in an invidious aura. And it has invoked a nostalgic longing for untouched, *primitive* and native peoples who are there to meet the demands of tourists: both in terms of service and as an object to be enjoyed and photographed. It would seem that racism has become commodified and is now part of the travel experience.

Subservience – enslaving the goods

In the 1970s Turner wrote that a 'significant part of tourism involves shipping rich white pleasure seekers into some of the world's poorest black societies' (Turner, 1976). Of course, nothing much has really changed in this respect, as Tables 2.1–2.3 suggest. But as has been argued, there has been a widely held belief that changes in the forms of tourism will produce qualitatively different experiences. But different experiences for whom, we must ask?

As already suggested, traditionally the analysis of the political economy of Third World tourism has tended to focus on the three principal components of the industry: international airlines, tour operators and hotel chains. Critical commentary has confined itself largely to observing that all three components are owned and controlled from the First World. As the political economy of tourism demonstrates, Third World countries are subordinated to the flows of tourists, tourism capital and resources from and to the advanced capitalist economies. Logically, such analysis has reached the conclusion that 'alternatives' that avoid such ownership characteristics represent either a considerable step towards the creation of a much fairer basis for international tourism or, for some, the answer to unequal and uneven tourism development. In part this may be true. However, it also represents an over-simplistic conclusion in two important respects.

First, it is a conclusion that is based upon too narrow an analysis of tourism within a global context. In this book we will argue that, despite changes to the ownership of tourism resources, tourism in the Third World will remain a special form of domination and control. However, this is not to argue that there will be no individual success stories (of places and even countries). Rather, it is the global inequality between First World and Third World in terms of power that is of such significance as to warrant an analysis of much tourism as 'a special form of subservience' (Krippendorf, 1987: 56); or that the widely cited Ss of tourism – sun, sea, sand and sex – are matched by the Ss of the content and out-come of tourism – subjugation, servility and subservience. So it is important that we walk a tightrope here, mindful that too global an analysis ignores local lessons and too local an analysis ignores global questions.

Second, existing studies have tended to blind commentators to the coalface of tourism: the relationship between tourists and those they are visiting. It is this basic encounter that remains the most significant activity in tourism. Rao captures this and reflects on the way in which many First World commentators refer to this as a relationship between host and guest:

> Can one really describe the encounter between the tourist . . . and the Other in the so-called voluntary relation of guest and host? Such a relation is again dictated by the tourism discourse which seeks to sweep away the basic commercial nature of the encounter.
>
> (Rao, 1991; quoted in Gonsalves, 1993)

Thinking critically about this 'relationship' opens new doors in our analysis. For example, it questions the holiday consumption practices and preferences of the new middle classes and should help us understand processes external to many Third World countries where tourism has produced a helpful, smiling and servile tourism class, serving the interests and economic preferences of business and local élites.

Even after training courses and 'tourism weeks', Jean Holder (the Secretary General of the Caribbean Tourism Organisation) complains of Caribbean tourism, 'aggressive attitudes . . . often emerge' which 'have their basis, to some extent, in the fact that a significant number of employees are not proud of what they do, and harbour resentments rooted in the inability to distinguish between service and servitude' (Holder, 1990: 76). Arguably it is easy for those in positions of influence and power and those professional local élites assisting in the running of the industry to be able to make this distinction, but perhaps far harder for the army of labour that is required to service the industry and whose structural disadvantage (in terms of class, gender, race, income, and so on) is compounded.

Whereas under industrial capitalism the workforce confronted its subordination in the industrial structure, within tourism workers in many cases must daily confront the symbolic representation of their servility – the tourist. Holder continues: 'there appears to be a deep-seated resentment of the industry at every level of society – a resentment which probably stems from the historic socio-cultural associations of race, colonialism and slavery' (1990: 76). This is an important point, for it strikes at the heart of the perception (arguably psychology) of tourism that reflects back on deep-seated historical inequalities, and a number of Caribbean writers have drawn the parallels to the dawn of a 'new slavery' (see Kincaid, 1988; Fanon, 1967; Naipaul, 1962; for an up-to-date discussion of the Caribbean, see Pattullo, 1996).

For new tourism to claim that these forms of servility and subordination are overcome is somewhat disingenuous. The number of tourists involved and whether they are conventional mass packaged tourists or an off-the-beaten-track type of traveller, is secondary. A subservient relationship persists and the comparisons with colonialism do become difficult to ignore – witness the scepticism exhibited in the headlines shown in Figure 3.3, referring to both mass tourism and new forms of tourism. We have already noted the neo-colonial aura that is re-created through eco-safaris, for example. Box 3.9 shows a single black-and-white photograph offered in the 1992 J. & C. Voyageurs brochure (all other photographs are large colour photographs of wildlife) of a trail of porters (thirty-five they tell us) shown tramping through an 'eco-colonial landscape', carrying the supplies for the group of six: 'most of the comforts of home – iced drinks, spacious sleeping tents, loos and showers'. It is an image and aura that is re-created by many so-called new small independent tour operators, whether it be luxury safaris or treks and expeditions for the young and adventurous; an army of global porters trot behind or ahead (via different routes) to ensure that these new, ethical, tourists are regularly refreshed.

Kutay, former director of the Ecotourism Society, poses the critical question: 'Those who view tourism as the final, most humiliating stage of human domination condemn it. . . . But is modern tourism any more dominating or humiliating than religious imperialism or European mercantilism?' (1989). Indeed, it is not, and this suggests we must consider whether tourism in its new forms is still a highly interventionist and subordinating activity.[1]

Fetishism – hiding poverty

The concept of fetishism is a useful way of getting us to think about the relationships that lie behind the things we buy and consume. A central precept of Karl Marx's thinking ('historical materialism') expresses the way in which commodities hide or veil the social relations embodied in their production. Take

Figure 3.3 To be or not to be a tourist?

Box 3.9 *Global porters*

You have to live off the land when on safari in the Selous Game Reserve. This is Africa as it used to be. . . . The walking safari normally lasts for 6 days, and a group of 6 guests is supported by 35 porters. . . . The crew undertake the domestic chores, such as cooking and laundry, etc., and pitch and break camp each day. . . . A typical day would be: up at six in the clear cool dawn, walk until the midday heat builds up, then lunch and a siesta in the shade; finally, after a cup of tea, an evening walk, with the countryside serene and the animals out in force.

(J. & C. Voyageurs brochure)

the last piece of fruit you ate that may have come from the Third World. There will have been a multitude of people and relationships responsible for getting this fruit to your grocer or supermarket shelves. And you are able to consume your fruit without considering, for example, the conditions faced by the workers who picked it and what aspirations and dreams these hidden workers have. It is necessary to get behind these surface appearances and unveil the fetishism of commodities. This experience will enrich our comprehension of the tourism process.

When we talk of the fetishism embodied in the production of commodities, we are also implying that the same processes are identifiable in the production of services, such as tourism – a task of applying Marx's ideas first undertaken in the context of leisure by Veblen (1925). There are two points at which notions of fetishism are particularly interesting in the study of tourism.

The first is the way in which tourism is an issue so hotly contested between different social classes (a point discussed at greater length in Chapter 5). Travel acts like any other commodity in expressing who we are, what we consider our status to be, what we believe in, and so on. It is a commodity through which we are able to express ourselves culturally. Take this example from Borzello who attempts to tell us what the difference is between tourists, travellers and those who choose to take a 'truck' expedition (passengers Truck Africa prefer to call 'would-be backpacker[s]'): 'Tourists just graze through a country. Travellers live and integrate into it. Truck travellers are not travellers but a very peculiar sort of tourist' (1991). The implication is that 'travellers' seek to roll back the fetishism that Borzello accuses 'tourists' and 'truck travellers' of experiencing (see Box 3.10). And yet at the same time the search (or competition) for the most virtuous way to take a holiday becomes a fetishistic cultural game in its own right; it is culture as the 'supreme fetish' (Bourdieu, 1984).

Second, travel in many cases brings the tourist in direct contact with labour (hotel staff, tour guides, and so on). Superficially then, the fetishistic 'mist' that enshrouds, masks and objectifies labour dramatically clears within much tourism, as the 'guest' is faced with the 'servant'. And, indeed, as suggested above, it is the claimed ability to transcend the fetishism characteristic of mass tourism and to meet real people in real places producing real things, that lies at the heart of new forms of tourism. It would not be unreasonable to expect new practices to achieve this and smash through such fetishism. However, we have argued that travel is a commodity, and one that is consumed principally as an image. This has led to new, more ingenious and intellectualised ways of creating an aura of travel (a point Figure 3.3 seeks to convey) and a re-ordering of the fetishism involved.

So it is that a range of less savoury realities of some parts of the Third World today – inequality, poverty and political instability – are also there to be enjoyed as part of the tourism experience. They are called upon to both titillate and legitimate travel, to help distinguish these experiences from mere mass tourism and packaged tourists. It is perhaps an extreme method of seeking authenticity through travel with many backpackers, for example, simultaneously bemoaning and celebrating the existence of such characteristics.

Exodus exemplifies the glib juxtapositions. Of Latin America, for example, its 1994 brochure states: 'some of the most sophisticated and wealthy cities in the world . . . but along with the wealth there is appalling urban poverty.' Predictably, no more is to be said on urban poverty and instead Buenos Aires is sold as

Box 3.10 Focus on truck travels

Truckers

There are a number of companies that now offer expeditions to most regions of the Third World travelling in 'trucks'. They are operationalised versions of overland trips that have become increasingly popular for 'would-be backpackers', as Truck Africa (1995 brochure) puts it, and are aimed at the 18–45 age bracket. Exodus Stipulate: 'This style of travelling is more suited to younger people. . . . We therefore apply an age limit of 17–45 on all trips of 4 weeks' duration and over. We will occasionally waive this for people who are older, and have demonstrated to us that they have the right attitude and have completed similar journeys before.' (1995: 4)

Risk and adventure

An essential ingredient. As Dragoman contend, this is for 'those who want the thrill of "real travel". . . . We will be crossing areas of the world that do not adhere to western safety standards and may have inherent political and economic instability.' (Dragoman)

But civil wars, revolutionary insurrections and suppression become part and parcel of the travel experience. Obstacles to be got round and overcome. Exodus simply states, 'we have seen and coped with numerous coups, wars and revolutions in all three continents, and our leaders have proved themselves to be pretty good at finding their way round the problems that politics can place in their way.' (4) A serious interrogation of the meaning of the civil struggles experienced in many Third World countries is replaced by macabre celebrations that destinations are back in business: 'For many people Peru is the most interesting of all Andean countries. With the leaders of the Shining Path guerilla movement now firmly behind bars, Peru is back to business as usual, and once again on our list of top destinations for the discerning traveller.' (Exodus newsletter, April 1994)

Exploration

Invoking the aura of exploration (see Figure 3.2), it is the ease of penetration and a truck's panoptic qualities that become pre-eminent selling points. Exodus refer to its trucks as 'particularly suitable . . . where contact with the outside world is of paramount importance.' It must provide for optimum surveillance so the prey of this post-colonial gaze is captured and greeted. Encounter Overland claim of their vehicles, 'With the sides up, the all-round views are superb and there is a great sense of being in close with the world around.' (8) Exodus claim 'The open sides . . . give the best views and the best contact with the people and countries that we visit. . . . They allow all-round vision, ease of entry and exit, good off-road capability.'

Paradoxically, others point out, it is the ridiculed, enclave, mass tourist infrastructure and resorts, from which truckers attempt to distance themselves, that is mimicked by the truck. The charges of environmental bubbles from which

Box 3.10
continued

tourists need not leave if they desire are replicated by the truck. Simultaneously it allows both a 'closeness with nature' but guarantees 'everyone can remain well within the framework of the vehicle body'. (Encounter Overland, 1995: 8)

Travel brochures present the neo-colonial consumption of black landscapes by white truckers. Where natives are able to assimilate whiteness, it is in ridiculing, fleeting and patronising representations: natives momentarily peering through a camera, or situated directly within their colonial legacy pulling an ailing truck from muddied waters. Natives drawn into the 'white man's' world. 'Lunch with the Maasai', for example, captions a photograph illustrating the ease with which this other world can be consumed. Lunch is prepared beside the truck and the Maasai have their trinkets examined by truckers in the foreground.

a 'huge, sophisticated capital' with 'plenty to see' and where 'carnivores can wash down one of the giant steaks with excellent local wine'. Similarly, *Backpacker's Africa* (Bradt, 1989: 10), sweeps aside a politically repressive environment; Kenya, it is argued, is 'blessed with outstanding geographical and cultural variety and her relatively stable government and capitalist economy is an added attraction for visitors' (quoted in *In Focus* 2, 1991). Or as Tucan, a company specialising in adventure tours to South America 'to meet the needs of younger travellers', suggests of the 'real' Latin America experience: 'Enormously wealthy people in their impressive mansions and estates live virtually cheek by jowl with the very poor in their squalid adobe huts and yet the populace as a whole show a special friendliness coupled with a real sense of hospitality' (1992 brochure). Widescale repression of human rights, deeply rooted racism and intense class political struggle are null and void in the brave new world of adventure travel, and provide for a rather different itinerary of attractions in regions such as Central America (see Figure 3.4).

New forms of tourism, as we have argued, seek to penetrate the less visited parts of the Third World and commodify what is there. It is a form of commodity racism (McClintock, 1994), which has its roots in the diaries (or travelogues) of nineteenth-century Victorian 'explorers'. The desire to consume these strange, other, worlds that they had 'discovered' became a fetishistic ritual which tourism has maintained. And it is this discovery of these other people and places that is so striking, especially, of new forms of travel today.

It is with the stream of touristic images, the trophies of these discoveries, that fetishism is most visibly maintained. Out, or at least marginalised, are the images of long, sandy, palm-fringed beaches. In are the images of unlimited wildlife, adventuristic landscapes and painted indigenous cultures. In fact the treatment of animals and peoples has become dramatically inverted with Cara Spencer Safaris proclaiming: 'A bonus for us was the warmth, concern and courtesy of the people. . . . And their approach to wildlife conservational land management is highly progressive' (1992). At best they are, it would seem, of

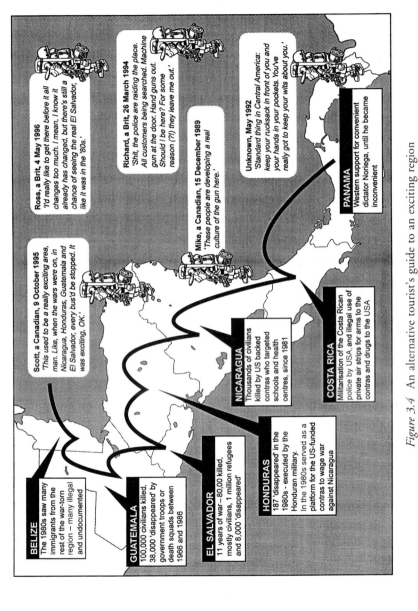

Figure 3.4 An alternative tourist's guide to an exciting region

BELIZE
The 1980s saw many immigrants from the rest of the war-torn region – many illegal and undocumented

GUATEMALA
100,000 civilians killed, 38,000 'disappeared' by government troops or death squads between 1966 and 1986

EL SALVADOR
11 years of war – 80,00 killed, mostly civilians, 1 million refugees and 8,000 'disappeared'

HONDURAS
187 'disappeared' in the 1980s - executed by the Honduran military.
In the 1980s served as a platform for the US-funded contras to wage war against Nicaragua

NICARAGUA
Thousands of civilians killed by US backed contras who targeted schools and health centres, since 1981

COSTA RICA
Militarisation of the Costa Rican police by USA, and illegal use of private air strips for arms to the contras and drugs to the USA

PANAMA
Western support for convenient dictator Noriega, until he became inconvenient

Scott, a Canadian, 9 October 1995
'This used to be a really exciting area, man. Like, when the wars were on, in Nicaragua, Honduras, Guatemala and El Salvador, every bus'd be stopped. It was exciting, OK.'

Ross, a Brit, 4 May 1996
'I'd really like to get there before it all changes too much. I mean, I know it already has changed, but there's still a chance of seeing the real El Salvador, like it was in the '80s.'

Richard, a Brit, 26 March 1994
'Shit, the police are raiding the place. All customers being searched. Machine gun at the door. Hand guns out. Should I be here? For some reason (?!) they leave me out.'

Mike, a Canadian, 15 December 1989
'These people are developing a real culture of the gun here.'

Unknown, May 1992
'Standard thing in Central America: keep your rucksack in front of you and your hands in your pockets. You've really got to keep your wits about you.'

equal value and are there to be photographed and embraced. The search and expropriation of authenticity, it appears, has become a mission (Wheat, 1994) of a neo-colonial exploration:

> Both explorers and dragomen journeyed through far-flung lands in search of long-forgotten civilisations and empires. They explored hostile deserts to find nomadic tribes and ancient cities. They went in search of legendary mountains deep in the heart of jungles and brought back stories of fabulous wildlife.
>
> (Dragoman, 1995: 1)

It is this observation that Said so forcefully expresses in his study of *Orientalism* (1991) which we introduced earlier. Exploration, travel writing, tourism, and so on, are means of representation. But they are ways of representing the world that also amount to, and maintain, a 'formidable structure of cultural domination': a system of truths, as Said argues, that has 'rarely offered the individual anything but imperialism, racism and ethnocentrism for dealing with "other" cultures' (1991: 18).

Aestheticisation – enjoying poverty

Not only is the consumption of tourism fetishistic, it is also intensely *aestheticised*. By this is meant the way in which travel and tourism are used to express 'good taste', a cultural accoutrement indicating to others what it is we find of beauty or note, and what we are seeking to get from a holiday. This recalls Wolfgang Sach's reference to the 'environment' as aesthetic. For many ecotourists it is just that, a thing of beauty that also expresses a belief in green issues. Similarly, it has been suggested above that many Third World cultures have been aestheticised, as a thing worth experiencing and preserving.

It is necessary, above all, to provide evidence and verify what has been 'experienced' and it is here that dominance is further expressed. The use of photography and its relationship to tourism has been the subject of much critical commentary (Albers and James, 1988; Barthes, 1981; Sontag, 1979; Urry, 1990a); it is as if things have come full circle, for as Sontag argues, 'From the beginning, professional photography typically meant the broader kind of class tourism' (1979: 57). While holiday 'snaps' are supposedly symptomatic of shallow tourist experiences, photography (especially monochrome) supposedly captures both the historical ambience and the closeness of tourists to the other.

It is this need to accumulate authentic images through the stream of portraiture of 'natives' or the embracing of the tribal child that has become an enduring image. Figure 3.2 portrays the innocence, authenticity, naturalness and nativeness of the Third World at the same time as aestheticising the inequality of development. The momentary encounters captured in some of these photographs symbolise the inherent power that travel embraces. The passivity with which natives and wildlife are portrayed in photographs has become indistinguishable. Passive, they are to be discovered, sighted, viewed and, ultimately, 'shot' (Sontag, 1979).

But it is not just places and peoples that are the subject of aesthetic representation. Our discussion has pointed towards a process where it is suffering and poverty that become aestheticised. Consider the 'alternative' tourism guide to Central America in Figure 3.4. Travel to 'dangerous' Third World countries, to regions suffering civil war and insurrection has become attractive to the bearers of new tourism who have become increasingly preoccupied with the need to distance themselves from tourists.

> Michael, a personable 28 year old on a three month leave from New York finance business is out of sorts over breakfast. Things are not going according to plan. He has come to Calcutta to find poverty – and, like so many travellers to India, to find himself. Both are proving elusive. Sure, there are beggars, he says. Sure, there are people sleeping rough. But, quite frankly, it's not real poverty, is it? It's not – he toys thoughtfully with his poached egg – it's not swollen-bellied poverty. . . . In the dining room, Michael from New York is a changed man. After breakfast, he told a taxi-driver to find him real poverty. After half-an-hour they found it. Real squalor. Real swollen bellies. Michael beams over the mulligatawny. He is well on the way to finding himself.
>
> (McClarence, 1995: 55)

Where thieves prosper

Miles Warde, writing in the *Guardian Weekend* (Box 3.11), captures the way in which romanticism has seeped into contemporary Third World tourism. A romantic – aesthetic – aura feeds into the retelling of these travel accomplishments.

This piece also suggests that the term aestheticisation incorporates the brutal reality, rather than stepping aside from it; that reality is then glorified and becomes intrinsic to the uniqueness and quality of adventure travel, a bitter twist for those young adventure travellers visiting unstable regions such as Central America, who can now include army checkpoints, state repression and civil strife on their list of things to tick off.[2]

> As Vietnam undergoes a new invasion from the West, you can be sure that returning travellers will have to endure the incessant cries of 'You should have been there a few years ago,' from the pioneers, each claiming to have been there before it was spoilt by tourism. 'Vietnam? Oh, it's not the same without the smell of napalm.'
>
> (Morrow, 1995)

Box 3.11   Where thieves prosper

Someone produced a bottle of whisky, and the stories began . . .

. . . It is a custom among travellers in South America to put at least two hours a day for telling stories. They are rarely pleasant. Some are cliched. . . . A Canadian begins.

'I heard this in Belem, about an English guy, first day in Brazil, clean off the plane from London, and he was on this bus looking for somewhere to stay when four guys jumped him by the turnstile. Pinned him down on the ground – and get this – they rammed a fork in his arse and cut his pack off his back and threw it out the window to another guy. Jeez, can you believe it? A fork – they robbed him with a fork.'

More whistles of admiration. Even I liked that story, and was very impressed with the way it had remained so accurate as it travelled 6,000 kilometres up the coast from Rio. . . . You see, it was my story, and I still have the fork marks to prove it . . .

Finally, a note of encouragement. Just before I flew home from Colombia I was in a bus queue. The bus arrived and the queue surged forward. I put both hands in my pockets – it becomes second nature – and waited for someone to try something. Then I noticed a loss of weight from the shoulder around which my camera had been hanging, hiding on the inside of my jumper. Wheeling round I caught the first throat which came to hand. It belonged to a young lad; he was holding my camera in one hand and a pair of wire snips in the other. I pulled my knee up in his crotch and he gasped, dropping the camera into my outstretched hand. The whole queue broke into applause. It was a sweet moment.

(Warde, 1992)

Weekending in El Salvador

The article in Box 3.12 was accompanied in the *Independent on Sunday* by a Foreign Office travel advisory, 'Areas to avoid in Central America', of which it is advised that El Salvador is 'one of the most dangerous countries to visit in the region': advice, as Calder argues, some 'backpackers interpret . . . as a challenge rather than a warning' (Calder, 1994a). It becomes a barometer of success. Reading like the script to Oliver Stone's *Salvador*, it begins to express the sickly aestheticisation that is supported by new forms of tourism.

How then can we make sense of this ultimate aestheticisation of reality, from which racism and class struggle actually seem to be enjoyed? Or as Kincaid has commented, people 'visiting heaps of death and ruin and feeling alive and inspired at the sight of it' (1988: 15). As Krippendorf (1987) argues, it is not that people do not, or cannot, 'see through the clichés', but their complicity in being 'seduced by them, again and again'. Perhaps the answer should be sought in the nature of tourism itself as a commodity, especially in its new wafer-thin disguise as a more ethical and moral activity.

Box 3.12   Weekending in El Salvador

The tension-filled journey to El Salvador was not one that we wished to prolong. . . . Apprehensiveness hung in the air like smoke. . . . Five hours later we arrived at the Salvadoran border. . . . We had been careful not to bring in anything remotely *subversivo* – Graham Greene, Joan Didion and Patrick Marnham left reluctantly behind. . . . Finally satisfied that we weren't revolutionaries or agents of the left Farabundo Martí Front for National Liberation, they let us through . . . we started to count all the men in round helmets and pale green uniforms. . . . Everywhere we looked there they were: American M16 machine guns slung casually round shoulders . . .

A station-wagon pulled out of a side road, seven plywood coffins strapped on top. At a cafe in the bus terminal we sat next to a young man with two shining claws instead of hands. He was the first of many amputees we were to see over the next couple of days. . . . We took a yellow taxi. . . . 'How is San Salvador these days?' we asked the driver innocently, aware that most taxi drivers in El Salvador moonlight as paid informers. 'Muy pacífico,' came the confident reply. But every time he stopped at a red light, his head would turn slowly to the left and then the right, and his steady gaze was invariably returned . . .

. . . Salvadoran television news showed reports of 'guerilla activity' in Chalatenango, an area I had hoped to see. Later I learnt that the Salvadoran army . . . had fired mortars into the nearby village . . . and machine-gunned three women and two small babies. Unable to get to Chalatenango we went north to Tonacatepeque. . . . Emerging into the bright sunshine, we found a young man face down in the gutter. We thought he was dead, in El Salvador a fair assumption. But he was merely dead drunk. We then drove southwards, to the beach at La Libertad, to see El Salvador at play. . . . Just half a mile away was the high black lava field known as El Playón, favoured for years by the death squads as a convenient dumping ground for mutilated bodies.

. . . We went back to San Salvador. . . . Top on our list of essential visits was the tomb of Archbishop Oscar Romero, gunned down by the army while saying mass in March 1980.

(Wolff, 1991)

CONCLUSION

This chapter has sought to elaborate the relationship between tourism and uneven and unequal development. First, it reviewed the various ways in which critics of Third World tourism have placed relationships of power at the heart of their analysis. It argued that the political economy approach still provides the most systematic critique of Third World tourism. An increasing number of subsequent commentaries have drawn analogies between tourism and colonialist and imperialist forms of domination as a way of expressing the unequal power relationships between the First World and Third World.

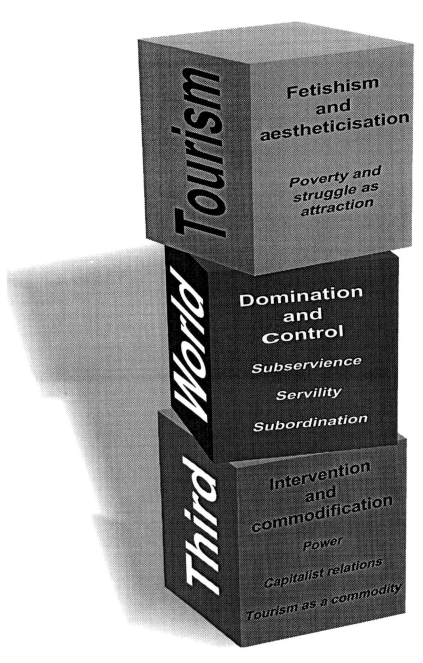

Figure 3.5 Developing a critique of new forms of tourism

But it has also enquired how effective such critiques are in handling the new forms of tourism that have begun to emerge in the Third World. Forms of tourism with prefixes such as 'alternative', 'appropriate', responsible', 'acceptable' and, of course, 'sustainable' attempt to challenge the notion that all forms of tourism necessarily draw the Third World into a highly unequal relationship with the First World. Indeed, it was argued that the political economy of Third World tourism implies that such forms of tourism can overcome this unequal relationship.

The final part of the chapter sought to demonstrate the way in which new forms of tourism can actually maintain unequal power relationships, albeit in less obvious ways. In doing so, it has suggested an alternative critique of the new forms of tourism, a critique which seeks to take the analysis beyond the characterisation of tourism as a form of colonialism. It is not suggested that such a characterisation of the relationships involved in tourism is wrong or inappropriate or irrelevant; rather, that these relationships cannot be fully explained simply by the notions of dependency and domination.

This critique is summarised diagrammatically in Figure 3.5. It builds on the former notions of power and capitalist relations. These are not unimportant – indeed, take this block away and the whole critique falls down. Likewise, the ideas of neo-colonial domination are just as important; again, the critique does not stand up without this level of understanding. But, critically, we add a final set of factors, fetishism and aestheticisation, which are necessary to explain many of the recent features, interpretations and characteristics of the new forms of tourism.

The next chapter examines the relationships we have set out by focusing more specifically on the association between tourism and sustainability.

4

TOURISM AND SUSTAINABILITY

The previous two chapters have argued that the term 'sustainability' is now an essential item in the vocabulary of modern political discourse. It is an ideological term. It can be used, and is used, by all the mainstream political parties, and those outside the mainstream, to illustrate and describe policies, the implications of which can be shown to be anything but sustainable in a number of ways. And it is widely used in a meaningless and anodyne way.

Sustainability is perceived and described as an essential part of the ideology of the New World Order and all the trends and tendencies that are associated with it. These tendencies, almost movements, include a 'new' consumerism, whose semantic ally is sustainability. The two notions have developed hand in hand to give mass consumption a more acceptable justification to the new middle classes who can afford to consider sustainability. But as Selwyn remarks, 'supported by a chorus singing of the joys of economic plenty in a world scarred by scarcity, the tourist often appears as a shining hero of the most pervasive myth of our time: that which tells of the omnipotence and untrammelled sovereignty of the individual consumer' (Selwyn, 1994: 5).

In the field of tourism, the term 'sustainability' can be and has been hijacked by many to give moral rectitude and 'green' credentials to tourist activities. And it is by no means just the tour operator and other profit-making companies standing to gain from the activity who have used the term for their own ends. Conservationists, government officials, politicians, local community organisations and tourists themselves have all misused and/or misunderstood the term. Some illustrations of these instances are given in Chapters 5–9.

But whereas the paradigms of political discourse have changed little as a result of the new word 'sustainability', the study of tourism has had to adapt itself to the creation of a whole new branch of the discipline, 'sustainable tourism'. (This term is, incidentally, used in conjunction with many other descriptive labels, and these are discussed in the section of this chapter entitled 'Resulting problems and the rise of new forms of tourism', on pp. 95–8.) The practice of tourism is also modifying itself to take account of the new types of tourism; and opportunities to indulge in sustainable tourism are springing up all over the world. These too are examined in Chapters 5–9.

This chapter examines the notion of sustainability as it is applied to tourism. It begins with a consideration of the development of the mass consumption of tourism and its lead into a new form of consumerism in the industry. A look at the terminology of the new forms of tourism is followed by an analysis of a range of definitions of these new forms. This leads on to examinations of, first, a number of principles often applied to sustainability in tourism and, second, the tools and techniques commonly used to measure and describe sustainability. Finally, we take a look into the intertwined futures of sustainability and tourism.

THE GROWTH IN MASS TOURISM

The rise and rise of the importance to our lives of holidays taken collectively at a distance from our home is well-documented in many texts – see, for example, Lavery (1971), Murphy (1985) and Krippendorf (1987). The early association of this feature of our lives with industrial capitalism is also well noted.

The stimulus given to the holiday industry by technological developments in the field of transport is clear from the histories of the railways (Great Britain in the nineteenth century), the motor car (widespread ownership in the First World after the 1950s) and the wide-bodied jet (post-1960s).

A potted history of some of the major historical influences on the tourism industry is given, in rather glib style, in Figure 4.1. We do not wish to make light of these developments and influences, but, rather than chronologically documenting the history of the growth of tourism, these three chapters are used to offer a critique of the current political and developmental contexts of the tourism industry.

The models

Such a critique cannot be made without at least a brief examination of the explanatory models of tourism development. We believe that these are generally deficient as they fail to account for the distribution of power; and the need to set an understanding of tourism in the context of its structures of power has already been presented in Chapters 2 and 3.

Very broadly, explanatory models can be grouped into those which explain the tourist's motivation, those which explain the role of the tourist industry, and those which explain the development of the destination community. But such a categorisation is simplistic. Some models, for instance, such as Butler's Product Life Cycle Model (Butler, 1980), attempt to explain the behaviour of both the industry and of the destination community. Moreover, these categories and the models which follow them fail to explain the relationship between the different elements of the industry (tourist, service provider, and local populace at its simplest) and the wider context of development processes. Chapter 2 has already made reference to these processes and Chapter 3 examined them in greater depth.

Figure 4.1 History of the tourism industry
Source: Biff Products, the *Guardian* 25 July 1994

Probably one of the simplest models, which serves also as a definition of tourism, is the equation provided by Smith (1989):

$$T = L + I + M$$

where:

$T =$ *tourism*;

$L =$ *leisure time*, which has increased for the majority of workers in the industrialised, technocratic, western, capitalist nations since the Second World War;

86

$I =$ *discretionary income* (often referred to as surplus income), which is now commonly used up in the pursuit of instant happiness rather than in the savings for future security associated with the work ethic which was prevalent in industrialised nations in the first half of the twentieth century;

$M =$ *positive local sanctions* (or *motivation*), which are those factors that prompt the tourist to tour – these are many and varied, but often spring from an escapist motivation. (Thus the local sanctions or conditions may be positive in the sense of prompting travel, but negative in the sense that they reflect unchosen or unhappy aspects of the life of the traveller.)

The model should more correctly be expressed as a functional relationship which varies with a range of factors rather than as a mathematical equation, but it is obviously not intended as a precise representation. Despite its intention, however, its use, even as an aid to understanding, is restricted by its vagueness, and it offers no explicit recognition of the issue of power over the activity.

Krippendorf (1987) also referred to the surge in the importance of leisure time in modern western life: 'Most people in the industrialised countries have been seized by a feverish desire to move. Every opportunity is used to get away from the workday routine as often as possible' (1987: xiii). His simplistic Model of Life in industrial society (Work – Home – Free Time – Travel) reflects the

Table 4.1 Murphy's growth factors in the evolution of tourism

Era	Motivation	Ability	Mobility
Pre-industrial	Exploration and business Pilgrimage/religion Education Health	Few travellers; those involved were wealthy, influential or received permission	Slow and treacherous
Industrial	Positive impact of education, print and radio Escape from city Colonial empires	Higher incomes More leisure time Organised tours	Lower transport costs Reliable public transport
Consumer society	Positive impact of visual communication Consumer society Escape from work routine	Shorter work week More discretionary income Mass marketing Package tours	Growth of personal transport Faster and more efficient transport
Future	Vacations a right and necessity Combined with business and learning	Self-catering Smaller families Two wage earners per household Demographic trends favour travel groups	Alternative fuels More efficient transport Greater use of public transport and package deals

Source: Murphy 1985: 22

historical change in the balance of work and leisure in the lives of industrialised and urbanised populations. Again, an historical inevitability underlies the model rather than a political perspective. As has already been argued, the latter is crucial to an understanding of the nature of tourism. And it is no less crucial to an understanding of the new forms of tourism.

Murphy (1985) cites the three crucial growth factors of motivation, ability and mobility as explanations of the evolution of tourism. His characterisations of each of these factors in each of four chronological eras of development are given in Table 4.1. While these factors are clearly germane, neither the historical developments themselves nor tourism can be understood without an analysis of their relationship with the prevailing power structures. And all of these models fail to offer such analysis.

An attempt to summarise the major features of studies in tourism is offered in Box 4.1. This simple summary emphasises two crucial points. First, there are many textbooks that describe in detail the existing approaches to tourism analysis. Second, it is striking how few concepts there actually are in the tourism debate and the hold that these ideas have retained in directing subsequent research. Many of them are endlessly repeated or contested in case study material.

Box 4.1   Studies in tourism

Structure of the tourism industry

Attempts to identify the main actors and structures in the tourism industry usually take the form of a flow diagram. The most widely cited is Mathieson and Wall's (1982) conceptual framework of tourism. Other texts worth consulting are Shaw and Williams (1994) and Burns and Holden (1995).

Impacts of tourism development

Seemingly a favourite among academics, who list as many impacts as they can under three headings: environmental, economic, social-cultural. Mathieson and Wall (1982) started the trend which many others have followed (see for example, Pearce, 1989, 1995; Shaw and Williams, 1994; and Burns and Holden, 1995). The best and most accessible work of this nature dealing with the Third World is John Lea's (1988) *Tourism and Development in the Third World.*

Models of tourism development

Another 'method' with a vice-like grip on academics is to look at the way in which holiday destinations move from boom to bust. The most famous examples are Doxey's (1976) 'index of irritation' (with destination communities moving from 'euphoria' to 'antagonism') and Butler's (1980) *resort life cycle model* (with destinations moving from initial 'exploration' to 'stagnation'). An army of others have followed in testing these models out.

Box 4.1
continued

There is also an increasing number of texts devoted to providing blueprints – or models – for appropriate tourism development. These offer both methods and case studies. For example, see Whelan (1991) and Gunn (1994).

Tourist typologies and motivational characteristics

Interesting attempts, usually from anthropologists and sociologists, to place tourists into boxes. The leaders have been Erik Cohen (1972, 1979a) whose tourists range from 'recreational' to 'existential' and Valene Smith (1989) whose tourists range from 'charter' to 'explorer'. Many others have set off 'tourist spotting' too.

The 'ethics'

Models can explain tourist behaviour patterns through the contexts of ideology and/or political developments. One example of this is the work ethic. The notion of the work ethic has been a contentious point in political discourse for many decades (see, for instance, Marx (1965) and Harvey (1973)), but as an explanation of the work patterns and practices of industrialising and industrialised societies it has been and is commonly used and referred to, even if not accepted by all. In essence, it relates the pursuits of moral rectitude and economic survival as the principal motives for people's actions.

The waning of the strength of the work ethic as a variable explaining the development of western capitalist trends in behaviour has been shadowed by the appearance on the scene of the notion of the leisure ethic. The two ethics are naturally closely related, for an important justification for the leisure ethic in many minds is that it is *not* work. The leisure ethic grows in importance as the work ethic wanes when the pursuit of hedonism replaces the pursuits of moral rectitude and economic survival associated with the work ethic. But it is important to note that the two ethics are not mutually exclusive.

It is also important, however, to point out that the leisure ethic appears to have two distinctly different faces at present – the face of the urban salaried westerner and the face of the local people who have to cater for the leisure ethic of the former. In 1991 Butler wrote of the 'leisure ethic which is at least parallel to, and in some cases more powerful than, the work ethic' (Butler, 1991: 201). The work ethic still holds strong in many societies. The leisure ethic has not yet overtaken it, save in a few wealthy societies. But it is gaining ground, not so much as a result of the spread of wealth as something personally experienced by populations around the world, but rather more as a result of the spread of service to the wealthy around the world.

Since the early 1980s, the leisure ethic as an explanation of First World tourist behaviour has been joined by the notion of a conservation ethic which has begun to have a bearing on patterns of travel and tourism, especially in the Third World. The two recent ethics, leisure and conservation, are closely associated. The former reflects the economic power of the individual tourist, and the latter reflects their ability or desire to impose that power on the areas, communities and populations that they wish to visit.

In a Third World which is rapidly being stripped of many of its natural assets and resources, it is imperative for the new middle classes who travel that those assets and resources which they wish to travel to are preserved for that purpose. Hence, the conservation ethic: introduced by environmental and conservation organisations (the World Wildlife Fund, the World Conservation Union (IUCN), Friends of the Earth International, the Ecotourism Society, the Audubon Society, and private organisations such as the Caribbean Conservation Corporation and the National Association for the Conservation of Nature (ANCON in Panama)), which suggest debt-for-nature swaps, and entice northern populations to donate towards or buy an acre of tropical rainforest (see Figure 6.4) in the belief that they will be contributing towards the conservation of the planet's biosphere. According to Ryel and Grasse (1991), the conservation ethic

> provides the framework within which all marketing and travelling should take place and includes several basic components: increasing public awareness of the environment, maximising economic benefits for local communities, fostering cultural sensitivity, and minimising the negative impacts of travel on the environment.
>
> (164)

Figure 4.2 represents the three ethics and associates them with types of tourism and prevailing economic and cultural power relationships. The models outlined earlier say little of the relationships of power between different elements of society with respect to the tourism industry, while it is these relationships of power which underlie the different 'ethics'. The structures of power in the industry are a crucial explanatory variable of the growth, development, patterns and types of tourism practised, and these are alluded to in an understanding of the ethics.

Cultural trend	Economic trend		Tourism	Power
Modernist	Fordist	THE WORK ETHIC	Mass and package	Merchants and new service providers
		THE LEISURE ETHIC	Package, exploration adventure	Transnational corporations + lending organisations
Postmodernist	Post-Fordist			
		THE CONSERVATION ETHIC	Nature and sustainable	Socio-environmental organisations + lending organisations

Figure 4.2 Ethics and the industry

The growth

There is of course no doubt that the facility to take a holiday (that is, leisure) has spread, especially in the last four decades. The linking of the package holiday with increasing opportunities for large numbers of people to travel overseas has had profound effects on many areas of the world which now serve as receivers of tourists. The growth and nature of the package holiday business are conveniently and humorously summarised in the *Guardian's* Pass Notes in Box 4.2.

This global spread of tourism is explained by Prosser as the 'tidal wave of the pleasure periphery' (Prosser, 1994: 24–5). Assuming an origin of most tourists in western Europe and eastern USA, Prosser identifies five peripheral regions of the world which have been successively commodified for the tourist industry over the last hundred years. Beyond the origin region, which represents the first periphery, these are successively:

- the western Mediterranean and Florida;
- the eastern Mediterranean, North Africa, California and the Caribbean;
- Africa, Asia, Latin America, the Pacific Basin, and Australasia;
- and finally Antarctica and remote areas of all other continents and oceans.

This is a purely descriptive model of the spread of tourism. But Prosser offers an analytical model to explain the changing consumption patterns and trends through the concept of successive class interventions:

> over time, a particular mode of consumption, fashion or lifestyle will spread downwards through the socio-economic class structure of a society. An admired élite inspires or propagates a fashion which is then aspired to by progressively broader sections of society, who as they become able, attempt to emulate the behaviour and style of the perceived élites. . . . As this process continues, the discoverer and élite groups, driven by the desire for novelty, uniqueness and exclusivity of experience, seek out fresh destinations and move on . . .
>
> (1994: 24)

Projections for the year 2000 predict a continuation of this spread and growth for the industry into the foreseeable future. Figure 4.3 shows a sixteenfold increase in tourist arrivals from 1950 to 1990, and a further projected increase of 50 per cent from 1990 to 2000.

It should be stressed that these figures include all international arrivals using tourist visas. This includes a proportion, which probably differs with space and time, of people who travel for reasons other than tourism. Nevertheless, the figures are clearly illustrative of the trend.

Green Flag International has stated that 'tourism [is] likely to be the world's leading industry by the end of the decade' (1990). The World Travel and Tourism Council (WTTC) claims that travel and tourism is already the world's

Box 4.2 Package holidays

Pass Notes

No 223: Package holidays

Born: In modern form, sometime in the 1950s, although the original Thomas Cook hatched a rudimentary version in the mid-19th century.

Father: Fascist dictator.

What? General Franco hit on tourism as a short-cut to economic prosperity; fishing villages filled with concrete hotels and BEA supplied the Brits.

Distinctive features of a package holiday? Watney's, chips with everything, pot-bellied men with tattoos . . .

You know what I mean: All right, the distinctive features of a package holiday are, first, that flight, hotel and everything else are bought in one deal off the shelf and . . .

What about the reps? . . . second, that a holiday company representative is on hand to provide assistance at the resort itself.

Don't these reps drag people out of bed and make them go cycling in the Atlas mountains and take part in hokey-cokey competitions? The more pro-active rep is, indeed, a legendary figure.

So the package holiday boomed? Yes. By the seventies it was an integral part of the British way of life, inspiring its own pop songs, like . . .

No, please! . . . Soleil Soleil, Y Viva Espana, The Birdie Song . . .

Enough! Why did the Brits go such a bundle on Spain? Same reason posher Brits went a bundle on France and Italy.

Which is? The uniquely civilised European way of life.

Pardon? Sex and low-duty liquor.

But now package holidays are finished? Far from it. True, Thomson, Thomas Cook and others are discounting 1994 holidays, but that's recession. This year saw 12.5 million packages sold, matching the 1988-90 peak years.

Where are they going? The old favourites — Spain, Greece, France.

What about Thailand, Egypt, South America? They're not package holiday destinations. They're experiences.

Sorry? A middle-class package holiday, as in "The Nile Experience".

What's the difference? None, except (a) the price, (b) the respectability and (c) you come back on the Orient Express.

How is the package holiday business run? Like the rest of British industry. Manic expansion one minute, psychotic cutbacks the next.

That's why so many have gone bust? Intasun et al? Yes, 'fraid so.

Their epitaph?: Y Vi-va! Esp . . .

Stop! What's a surviving package-holiday tycoon most likely to say? It's a great place for the kids, and a great place for you.

Where is? Anywhere.

Least likely to say? You'd have a better time walking in Yorkshire.

(*Guardian* October 1993, 3)

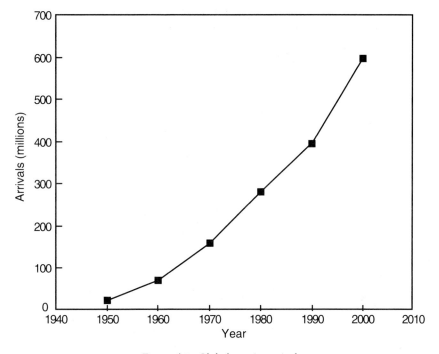

Figure 4.3 Global tourist arrivals
Source: World Tourism Organisation, *Compendium of Statistics* 1990 and 1995,
Madrid: WTO

largest industry, contributes 5.9 per cent of world GNP, provides 7 per cent
of global employment, and grew by 260 per cent between 1970 and 1990
(WTTC, 1991). The WTTC includes more than just the tourism industry in
its figures, and it is in effect a lobby group for the industry; but even though
their claims may be exaggerated, they are nevertheless indicative of a large-scale
and fast-growing industry.

This growth in the importance of tourism has not passed by the Third World.
Box 4.3 illustrates this using the examples of Zimbabwe and the region of
Eastern Africa. This point is further emphasised by Tables 2.1–2.3, which
include data on the growth of tourism from and to a number of Third World
countries. Regionally, the decade from 1985 to 1994 saw rates of increase in
tourist arrivals to Africa of 89 per cent, South America 86 per cent, Central
America 91 per cent, the Caribbean 71 per cent, East Asia and the Pacific 142
per cent, and South Asia 48 per cent. All bar South Asia experienced growth at
rates above the world average of 62 per cent over the same decade (all figures
from WTO, 1995a).

Mass tourism, then, has increased remarkably in recent years and, despite
recession and recent downturns, most projections show a continuation of this

93

Box 4.3 The growth of tourism in Zimbabwe and Eastern Africa

Figure 1 shows a steep and steady growth in tourist entries into Zimbabwe between 1985 and 1994 – a threefold increase over the ten-year period. Apart from South Africa, Zimbabwe was the only sub-Saharan country to have received over a million arrivals in 1994. Its arrivals account for nearly 30 per cent of all international tourist arrivals to the region of Eastern Africa. Over the same period, Zimbabwe's monetary receipts from tourism have massively increased.

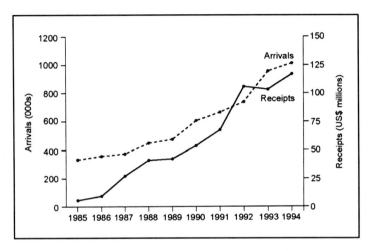

Figure 1 International tourist arrivals and receipts in Zimbabwe, 1985–94

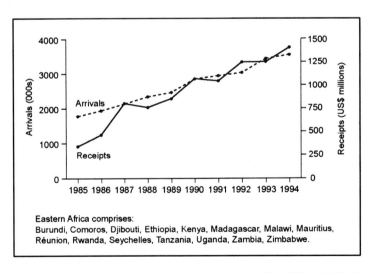

Eastern Africa comprises:
Burundi, Comoros, Djibouti, Ethiopia, Kenya, Madagascar, Malawi, Mauritius, Réunion, Rwanda, Seychelles, Tanzania, Uganda, Zambia, Zimbabwe.

Figure 2 International tourist arrivals and receipts in East Africa, 1985–94
Source: World Tourism Organisation

94

Box 4.3
continued

Although not as spectacular a growth as Zimbabwe's, Figure 2 shows a steady growth in both international tourist arrivals and monetary receipts from tourism for the region of Eastern Africa as a whole. This occurred despite the problems of politics and violence in Burundi and Rwanda and their overspill into neighbouring countries.

Between 1985 and 1994, arrivals of tourists in Eastern Africa rose at an average of 8 per cent per annum compared with a world average rise of 5.5 per cent. Receipts in Eastern Africa increased by 14 per cent per annum compared with a world average increase of 12.6 per cent.

trend. The ability to holiday anywhere in the world has become an essential part of modern professional life in the wealthy world. Not surprisingly, the growing middle class in the middle-income economies of the world are also increasingly keen to participate in this pursuit of hedonism. The potential for more growth is, therefore, great.

But can the planet sustain this growth? Is the current practice of tourism suitable for us to pass down to our future generations as a model of economic development which will guarantee them a source of income without the destruction of the environment from which they make it? The next section outlines just a few of the growing litany of social, cultural, economic and environmental problems created by the industry and its practices and conduct.

RESULTING PROBLEMS AND THE RISE OF NEW FORMS OF TOURISM

The phenomenon and growth of mass tourism has led to a range of problems, which have become increasingly evident and well publicised over recent years. They include environmental, social and cultural degradation, unequal distribution of financial benefits, the promotion of paternalistic attitudes, and even the spread of disease. These have been described in many publications (Bugnicourt (1977), Harrison (1979), Hong (1985), Krippendorf (1984), Lea (1988), *New Internationalist* (1988, 1993, 1994), *In Focus* (various), *Equations* (various), Cultural Survival Quarterly (1982, 1990a, 1990b)) and by a variety of organisations.

Some of these problems have become matters of global concern, as in the cases of, for example, the state of the Mediterranean Sea, deforestation and consequent soil erosion in various regions of the Himalayas, litter along Nepalese mountain tracks, and the disturbance of wildlife by Kenyan safari tours.

It may be most illustrative to outline the general problems associated with mass international tourism through the words of the Reverend Kaleo Patterson, a pastor on the island of Kauia, Hawaii:

I have counselled the prostitute, the desk clerk, the maid and the bartender. I have had to counsel and pray with the whole housekeeping section of a major resort development consisting of over a hundred Filipinos and Hawaiians. I have been involved in hundreds of re-burials of ancient Hawaiian grave sites because of a new resort development or existing resort renovations. I have witnessed the desecration of our sacred places and cried over the senseless pollution of our reefs and rivers. I have held picket signs in protest and given testimony at public hearings. I have organised workshops and forums on tourism. I have even chased an obstinate tourist into the sanctuary of a local pizza restaurant in an attempt to vent my anger in confrontation. I have seen the oppression and the exploitation of an 'out-of-control' global industry that has no understanding of limits or responsibility or concern for the host people of a land. . . . All is not well in paradise.

(1992: 4)

The problems here illustrated by Patterson are tangible and identifiable and can be solved in practical ways. They are often and commonly associated with mass tourism, although there is mounting evidence from impact studies to suggest that new forms of tourism can also be held responsible for such problems.

It has often been claimed that, in part, the development of alternative forms of tourism has resulted from the need to address these problems. Other factors have also been cited as responsible. For instance, the rise of a population of tourists who are becoming increasingly sophisticated and aware in their leisure pursuits; socio-economic trends in northern countries (Krippendorf, 1987); the replacement of the work ethic with a leisure ethic (Butler, 1991) (see the previous section); and the post-Fordist production trends and postmodern cultural trends which were discussed in Chapter 2. To a greater or lesser degree all these factors explain the rise of new forms of tourism.

While we do not deny the existence of these factors, we believe that the association of the growth of new forms of tourism with the problems arising from conventional mass tourism is misplaced. They may indeed have been used at times as an excuse for this growth in new tourism. But we believe that this growth has come about more as the 'natural' continuation of the process of colonialism and control, but now in a different form (Fernandez, 1994; Munt, 1995). As Fernandes argues, much of what are now seen as new forms of tourism have arisen because 'the mainstream tourism industry has in fact merely tried to invent a new legitimation for itself – the "sustainable" and "rational" use of the environment, including the preservation of nature as an amenity for the already advantaged' (1994: 4).

Whatever the reasons for their growth, there now abound forms of tourism apparently attempting to grapple with the negative impacts of tourism and claiming to be alternative, different or sustainable. Frank Barrett's article in the *Independent*, extracts of which are given in Box 4.4, describes these new features of the industry.

Box 4.4 *Slow boats to China replace lager louts*

When the *Independent* was launched in 1986, there was some debate in the Weekend department of the paper as to whether we should call the travel section 'Independent Traveller'. Hard to believe now, but as recently as the mid-Eighties, independent travellers were still considered in some quarters to be socks-and-sandals wearing, knapsack-toting, five star eccentrics. At that time most people taking a foreign holiday bought some sort of inclusive package from a tour operator.

But if 1986 was an *annus mirabilis* for the *Independent*, it was an *annus horribilis* for the package holiday business, which after more than two decades of sustained growth suddenly suffered a downturn in bookings and embarked on an internecine price war that eventually resulted in scores of bankruptcies and brought the industry to its knees: a position from which it has never properly recovered.

As the Eighties continued, people no longer thought about package holidays being fun or good value; instead they became associated with lager louts and unseemly behaviour. Resorts like Torremolinos, Benidorm and Palma Nova emerged as the modern equivalent of a music hall joke. . . . By the start of the Nineties, independent travellers were no longer considered oddballs . . .

Stand . . . in the departure hall at Gatwick or Heathrow and you see that the world really has become our oyster. Charter flights leave regularly for Queensland, Australia; Luxor, Egypt; Banjul, West Africa; several Caribbean islands; Mombasa; Goa; and, of course, Florida, which has now become as familiar to the Nineties traveller as Majorca was to holidaymakers in the Sixties and Seventies. Such is the pattern of wide-ranging travel in 1994 that the old distinction between short-haul and long-haul has almost disappeared – now there is just travel . . .

Any visitor to the Independent Traveller's World exhibition . . . can be sure of two surprises. The first is the extraordinary range of destinations being promoted by the exhibitors, everywhere from Latvia to Lesotho – for the new breed of British independent traveller, the more exotic the destination, the better.

The other surprise will be the extraordinary variety of people visiting the exhibition. . . . there will be a lot of younger people with the time and money to see the world keen to find out more about cut-price round-the-world tickets that offer stopovers in the Easter Islands or breaks of journey in Cape Town. But a quick glance . . . will offer convincing evidence that you don't have to be 23 to want to buy a round-the-world ticket: there will be just as many fifty- or sixty-year-old people snapping up Travel Survival Kits from the Lonely Planet reps or perusing the latest Rough Guides . . .

The exciting thing about . . . [the] exhibition . . . is that they offer a rare opportunity for independent travellers to have face-to-face meetings with travel suppliers like Regent Travel, Exodus, Explore Worldwide, STA Travel and Trailfinders. And even more interestingly, it provides independent travellers with a chance to meet other independent travellers to exchange information and advice.

Box 4.4
continued

> Events like the Independent Traveller's World show that there is a demand from sophisticated travellers for information about sophisticated travel. You will have a chance to question more than 40 different travel suppliers about slow boats to China, express trains to Ulan Bator, coaches across America and rambles through the Himalayas. . . .
>
> (Barrett, 1994)

Within the industry, tourism to protected areas and pristine wilderness is one of the most rapidly growing sectors. This is what Boo describes as the ecotourism sector which 'has rapidly evolved from a pastime of a select few, to a range of activities that encompasses many people pursuing a wide variety of interests in nature' (1990: 2). It also goes under several other names or descriptors which will be examined in the following section.

Published data on the increase in the importance of new forms of tourism are difficult to come by. Where they exist, they do so only for specific sites, parks or tours, and their overall significance in the tourism industry is still impossible to measure. Box 4.5 gives a number of indications of this growth from around the world. The examples given indicate a wide range of estimates which may reflect the arbitrariness of their selection by the cited sources. In other words, none of these estimates appears to be based on hard evidence. But all of them seem to suggest an increasing share of the tourism market for types of tourism which may (or, as we shall see, may not) be referred to as 'responsible', 'sustainable', 'alternative' or 'environmentally friendly'. It is noteworthy that the estimates given are generally not dissimilar to the estimated 10–12 per cent of the First World population interested in the issues which concern the socio-environmental movement, and this link is explored again in Chapter 6 (see WTO, 1995b).

It will be clear from the examples given that one of the difficulties in measuring this growth is the uncertainty of what is being measured. The terminology associated with the type of tourism and the different definitions of these types varies, as does the debate about their degree of sustainability. The terminology and definitions of the new forms of tourism are discussed in the next two sections.

TERMINOLOGY

Box 4.6 presents some of the terms most likely to be associated with the new forms of tourism. These terms have been culled from the vocabulary of relevant academic papers, journals, advertisements and tour operators' brochures. It is not an all-inclusive listing, and other terms are frequently added, or dreamt up.

Box 4.5 *Indications of the growth in new forms of tourism*

- International passenger survey figures from the Central Statistical Office in Britain show that for 1991 the number of package holidays sold fell by 7 per cent to 10.6 million, while the number of independent travellers increased to 13.9 million (Barrett, 1992). Obviously, all 'independent travellers' cannot be classified as practising sustainable tourism, but in general more of this group would claim to be travelling ethically than the group of package holidaymakers. Their claim can be questioned of course.
- From 1990 to 1994 the number of long-haul holidays from the United Kingdom rose by 39 per cent, while those to European and Mediterranean destinations rose by only 12 per cent. Again, it cannot be assumed that all, or even nearly all, of the long-haul holidaymakers are practising sustainable tourism. Many will go for the sun, sand, sea and sex, for which they previously travelled to the Mediterranean. But it is likely that a greater proportion of the long-haul travellers are travelling with some cultural or environmental purpose in mind. (This despite the fact that air travel can hardly be referred to as 'environmentally sustainable' – see Chapter 7.)
- Whelan states that the 'WTO [World Tourism Organisation] . . . found that adventure travel enjoyed almost 10 per cent of the market in 1989 and is increasing at the rate of 30 per cent a year.' (Whelan, 1991: 4–5, citing Kallen, 1990)
- 'In Thailand, tribal trekking has grown from a handful of hikers in the 1970s to more than 100,000 in 1988.' (Eber, 1992: vi)
- 'According to recent visitor surveys, about 36 per cent specifically cite ecotourism as among their main reasons for visiting Costa Rica.' (Rovinski 1991: 54, citing Boo, 1990)
- Forbes and Forbes project an increase of more than 150 per cent between 1988 and 1998 in special interest travel revenues. (Forbes and Forbes, 1993: 130)
- 'According to the World Wildlife Fund, about fifteen per cent of the world's 450 million travellers in 1991 were taking hiking shoes and rucksacks on their holidays rather than sarongs and swimsuits.' (Pattullo, 1995: 109)
- Mason reports that Greenland (where most tourism is motivated by the natural environment) is attempting to expand its tourism in the next ten years from about 5,000 visitors in 1995 to 35,000 by 2005. (Mason, 1995: 4)
- Selecting a variety of countries renowned for their exploitation of some form of nature tourism (Belize, Costa Rica, Ecuador, Kenya, Dominica, Madagascar), Cater lists the large increases in tourist arrivals and tourism receipts over the decade of the 1980s. (Cater, 1994: 70)
- Tapper reports that 'while tourism is growing on average at about 3 per cent per year, growth in nature-based tourism is between 5 per cent–10 per cent per year.' (Tapper, 1993: 146)

Box 4.6   An A to Z of new tourism terminology

The following list is not exclusive but is indicative of the types of terms used as descriptors of the new forms of tourism.

academic tourism
adventure tourism
agro-tourism
alternative tourism
anthro-tourism
appropriate tourism
archaeo-tourism
contact tourism
cottage tourism
culture tourism
eco-tourism
ecological tourism
environmentally friendly tourism
ethnic tourism
green tourism
nature tourism
risk tourism
safari tourism
scientific tourism
soft tourism
sustainable tourism
trekking tourism
truck tourism
wilderness tourism
wildlife tourism

Additionally, terms used to describe markets include:

niche
individuated
specialised
flexible
personalised
customised
designer

It would be tempting to dismiss the terminology as insignificant and of little consequence to the notion of sustainability except inasmuch as it provides us with descriptive labels. But the use of these terms represents an attempt to distance the activities associated with the new forms of tourism from what are presumed to be the unsustainable activities pursued by the mass. Frank Barrett of the *Independent* calls this 'a reaction to the naffness of package holidays' (Barrett,

1989). But, as may be seen in later chapters (especially Chapter 7), sustainability is a goal and/or claim of various sectors of the mass tourism industry as well as the sector of the industry which can be described as 'new'.

With this burgeoning list of new terms has emerged a new range of travel agents and tour operators which offer their clients individually centred, flexible, personalised holidays. Phrases used to appeal to the tourist's desire for something different and exclusive include: 'designer' tourism from Cara Spencer Safaris and 'bespoke' holiday journeys from Journey Latin America. The markets associated with this are referred to as 'individuated' or 'specialised', as distinct from 'mass'.

It is also necessary for operators to differentiate themselves from 'conventional' travel operators. Magic of the Orient, therefore, claim their holiday brochure is 'like no other', for it is a 'collection of ideas', and Roama Travel (specialising in treks and climbs in India and Nepal) establish that they are not 'a travel agent but a specialist tour operator'. Travel consultancies, a middle-class transformation of the travel agent, have also begun to appear. Marco Polo Travel Advisory Service, for example, offers a consultation service to the 'imaginative and independent traveller looking for an extra dimension' to their holiday. The small specialist operators catering for the new middle classes who form an increasingly significant market segment can translate their desire to be a late twentieth-century adventurer, explorer or 'traveller'. Urry (1990b) argues that this represents a consumer reaction against being part of a mass; and the emergence of these specialised markets is a feature of a post-Fordist mode of consumption. In the same way that the new middle classes assume control of the 'new' activities through their exclusiveness, so the operators assume exclusivity, and therefore status, for themselves on the grounds of their specialised, individualised offerings.

The messy word de-differentiation (a key feature of postmodernism) is used to convey a straightforward idea. It involves the way new tourism practices may no longer be about tourism *per se*, but embody other activities. As Box 4.6 indicates, an A to Z of 'tourisms' has evolved. On the one hand this means combining a variety of 'activities' such as adventure, trekking, climbing, sketching and mountain biking. More significantly, on the other hand, it means the marriage of different, often intellectual, spheres of activity with tourism (that is, academic, anthropological, archaeological, ecological and scientific tourisms).

These latter forms of tourism are becoming increasingly important to the small-scale tour operator and travel agent. School, college or group tours combining elements of fieldwork and vacation offer a way of catching larger numbers of clients on a package which has to be especially designed, but for the group rather than an individual. Some examples of this type of development are given in Chapter 7, which includes illustrations of the usage of some of these terms.

In this book we use the phrase 'new forms of tourism' generically to cover all the terms given in Box 4.6. Other terms, such as 'sustainable', 'alternative', 'ecotourism', and the like are used either because they appear in quotations or because they refer to a specific type of tourism that is appropriately described only by that word.

DEFINING THE 'NEW' TOURISM

Because the study of the forms of new tourism is still in its infancy, there is no clear agreement on their definitions and conceptual and practical boundaries. This lack of consensus is at its most conspicuous in the difference between those who study new forms of tourism and those who operate tours. But disagreements are also evident among others in the field, the conservationists, government officials and service providers. The new tourisms are truly contested ideas and the tourism literature is peppered with claim and counter-claim, with mainly academics and interest groups advocating and defending particular terms and definitions. A little like tourism destinations themselves, the terminology of new tourism experiences a relatively rapid circulation as terms come in and fall out of fashion (although there is often little to distinguish one term from another).

As with any activity which involves many groups, the terms mean different things to different people, according to the role they have within the activity. Protagonists perceive and portray the activity they are involved in as 'sustainable', 'no-impact', 'responsible', 'low-impact', 'green', 'environmentally friendly', or they use some other suitable term to convey the message. We do not intend to get sidetracked into a lengthy discussion of the many and varied definitions of new tourism types. Suffice it to say they share, in varying degrees, a concern for development and take account of the environmental, economic and socio-cultural impacts of tourism. They also share an expressed concern, again with varying levels of commitment, for participation and control to be assumed by 'local people'. In this way most of these terms and their meanings underlie or echo the notion of sustainability in its varying guises. This again suggests an important link between sustainability and new tourism.

The extent to which these terms can be used interchangeably is itself a debatable point. But more significant is the extent to which the claim of sustainability can be justified, a point examined under the heading 'The principles of sustainability in tourism' (p. 105) and in Chapter 5 and subsequent chapters.

One further point of relevance to the sustainability of tourism is the acknowledgement in some quarters that mass packaged tours may be just as sustainable as some of the new forms of tourism – see, for instance, Table 7.1 (p. 198), which makes a qualitative comparison of the sustainability of a package sun-sea-and-sand holiday and a trekking tour. Acknowledgement is made by organisations such as Tourism Concern, Green Globe and by some in the industry that sustainability is not the exclusive concern of new forms of tourism. But it is the language and terminology of the new forms of tourism which have been used in the attempt to subsume sustainability.

This section sketches in turn the meaning of sustainability in the field of tourism to tourists, tour operators, host communities, national governments, regional and international organisations, and academics. The differing definitions are only briefly sketched here but much of the discussion in Chapters 5–9 reflects these varying interpretations.

The tourists themselves often define their activities in terms of practical details. Green, for instance, suggests that 'Environmentalists and nature lovers should be able to survive camp without disposable paper cups, use-only-once plastic groundsheets, and crates of Coca Cola' (Green, 1991). The desire of the new middle classes for exclusivity in their holidays and tours has already been noted in their use of descriptive language.

In countries renowned for their natural beauty, however, it is not uncommon to hear these same tourists display the same 'been-there-done-that' attitude to national parks and spectacular locations as the conventional package tourists of whom they are so disdainful. Countries and locations are there to be 'collected' – for the list, the kudos and the image. And not a few object to having to pay a higher entrance fee than local nationals for access to protected areas, thereby betraying an ignorance of the social, cultural and economic ramifications of their position of wealthy foreign visitor.

The tour operators describe new tourist activities in short advertising 'bites', such as 'spectacular views of active volcano', 'surrounded by exuberant evergreen forest', 'exciting thrills of white water rafting', and many more. Specialised eco-tour operators may develop these along the lines of 'it is our pleasure to give the visitors personalised, low-impact, rustic excursions to national parks, Mayan archaeological sites, and natural wonders of Guatemala' (Ultra-Unlimited, 1991). Chapter 7 takes a longer look at the publicity and language used by tour operators, especially with respect to their claims about sustainability.

Not surprisingly, the destination communities stress the importance of their own involvement in all stages of planning and operating tours. In its lengthy definition of ecotourism, ATEC (the Talamanca Association for Ecotourism and Conservation in Costa Rica) includes the following:

> Ecotourism means more than bird books and binoculars . . . more than native art hanging on hotel walls or ethnic dishes on the restaurant menu. Ecotourism is not mass tourism behind a green mask. Ecotourism means a constant struggle to defend the earth and to protect and sustain traditional communities. Ecotourism is a cooperative relationship between the non-wealthy local community and those sincere, open-minded tourists who want to enjoy themselves in a Third World setting.
>
> (ATEC, 1991: 1)

The debate is currently not one of whether local communities should be involved in the development of tourism to their areas, but how they should be involved and whether 'involvement' means 'control'. This struggle for power and control over the tourist activities and financial benefits is at its sharpest at the destination end. The degree of control is generally perceived as being a significant element of sustainability.

In most relevant texts these communities are referred to as 'hosts', but we try to avoid this term as it generally conveys the idea that the resident populations in these communities are willing partners in the activity. In some instances, this may be the case, but, as we shall see, in others it is not. More examples of the destination communities' own descriptions of the activities of visitors are given in Chapter 8.

Governments are more concerned with the national planning strategies required to exploit the potential of their natural environments – in some cases without destroying them. But all too often, ministers who speak radically, convincingly and frequently about protection of the nation's environmental and cultural treasures are the same people who sign the agreements which allow transnational companies to build a hotel or tourist complex whose development pays no heed to the environmental, social and cultural impacts caused. Examples abound, and a number are given in some detail in Chapter 9.

In their pursuit of sustainability, international conservation bodies, such as the WWF, see the new forms of tourism as a useful tool in ensuring the conservation of a region's natural areas. They plan 'buffer zones', 'corridors of conservation', 'biosphere reserves', and similar. And they attempt to exert pressure on national governments to implement policy to protect such areas. The Paseo Pantera Project promoted by Wildlife Conservation International (WCI) and the Caribbean Conservation Corporation (CCC) envisages the achievement of conservation goals through the promotion of ecotourism: 'Properly-planned tourism can 1) provide funds for acquisition and management of protected areas, 2) enhance the economies of local communities and 3) promote environmental education for park visitors, both national and international' (CCC, 1992). The interpretations of sustainability made by different strands of the international socio-environmental organisations are examined again in Chapter 6.

The supranational institutions discussed in Chapter 9 include the multilateral lending agencies such as the World Bank, the International Monetary Fund (IMF), the European Development Fund (EDF), other regional development funds, and organisations such as the Overseas Private Investment Corporation (OPIC). Their view of tourist developments is used to justify the creation and promotion of investment opportunities, especially those for transnational corporations. Indeed, it might be argued that the industry involved, tourism in this case, is largely irrelevant so long as it fits in with the prevailing ideological planks of privatisation, deregulation, reduction of public spending, tight money control, increasing production for export, and other policies which concentrate wealth and power in the hands of the corporate captains of industry.

Finally, academics agonise over the most appropriate term to use and often reach the conclusion that 'no consensus has yet emerged as to the precise nature of alternative tourism' (Weaver, 1991) and 'it is still not possible to be exact about whether the term "ecotourism" is meant as a pure concept or as a term for wide public use' (Evans-Pritchard and Salazar, 1992). Sustainability ⁚ tourism are rich fields of discourse and debate, potentially never⁻ academic community. Wheeller (1992) points out that, while th critics of new forms of tourism in academia, there are also many embrace and promote the notion as if it is a new idea: 'Glib, ge are frequently incorporated into policy statements without ... . being given to defining in a practical sense such phrases as, for example, "the local community", "tourism education" or "good visitor management"' (143).

With such a diverse range of standpoints, it is not surprising that, as Weaver (1991) points out, there is little agreement on the definition of the term. Nor is it surprising that there is such a plethora of terms – as we saw in Box 4.6.

THE PRINCIPLES OF SUSTAINABILITY IN TOURISM

Given that none of the definitions cited in the previous section are comprehensive and all-encompassing, an approach more frequently taken is to examine and assess tourist activities according to whether they satisfy a number of criteria of sustainability. Whether, for instance, a specific tour, lodge or wildlife reserve is operating sustainably might be assessed by reference to the criteria listed in Box 4.7. It is stressed that this list is not presented here in a prescriptive manner; rather, it has been culled from observed practice, especially the practice of those organisations which attempt to publicise lists of environmentally and ethically sound companies. As such, it stands as descriptive of these practices. It is not our view that these principles represent a 'correct' or absolute version of the meaning of sustainability. Indeed, we believe there to be no absolutely true nature of sustainability and, as the last section has illustrated, it is not definable except in terms of the context, control and position of those who are defining it. And as Box 4.7 points out, the notion of sustainability has many ramifications. These are briefly examined in the following subsections.

Box 4.7 *Criteria often used for sustainability in tourism*

IS THE LODGE, RESERVE OR TOUR:
1 SUSTAINABLE?
  • environmentally
  • socially
  • culturally
  • economically
2 EDUCATIONAL?
3 LOCALLY PARTICIPATORY?

Ecological sustainability

The condition of ecological sustainability need hardly be stated as it is often the only way in which sustainability is publicly perceived. The need to avoid or minimise the environmental impact of tourist activities is clear. Maldonado *et al.* (1992) suggest that the calculation of carrying capacities is an important method of assessing environmental impact and sustainability. In an important work on carrying capacity, they calculate the capacities for seven different tourist foci in Costa Rica, and in so doing take the notion of sustainability beyond its current fatuous interpretation by so many users of the term. Box 4.8 gives an outline of their calculations of the carrying capacity for the Guayabo National Monument, a small area of archaeological significance in Costa Rica.

While the work of Maldonado *et al.* undoubtedly takes the measurement of carrying capacity some way beyond what the scientific community has so far managed, it is important to understand that the notion of carrying capacity may be used to wrap a social or economic constraint in a cloak of scientific jargon. Where exclusivity is promoted by the operators, a low carrying capacity is likely to be publicised. And conservation organisations involved in the promotion of new forms of tourism are more likely than most to foster imaginary maximum capacities in pursuit of conservation and economic gain. A number of examples of the variability of carrying capacity calculations are given in Chapter 8.

Moreover, despite the progressive nature and importance of this work, Box 4.8 illustrates that the calculations are dependent on assumptions which are in some cases arbitrarily chosen (such as the maximum number in a group and the ideal management capacity) and in others widely variable (such as the degree of slope). Other assumptions and conditions affecting the physical and management capacity of the area (such as the availability of guides, maps, rest spots and the incidence of low cloud cover) might be thought of as relevant, but are not included. Furthermore, it will not escape your attention that a change in one value allocated to one assumption or input could have a substantial effect on the final carrying capacity calculated. Calculations of carrying capacity and the limits of acceptable change, a notion that has been developed out of the former, are discussed further in Chapter 8.

Social sustainability

Social sustainability refers to the ability of a community, whether local or national, to absorb inputs, such as extra people, for short or long periods of time, and to continue functioning either without the creation of social disharmony as a result of these inputs or by adapting its functions and relationships so that the disharmony created can be alleviated or mitigated.

Some of the negative effects of tourism in the past have included the opening of previously non-existent social divisions or the exacerbation of already existing divisions. These can appear in the form of increasing differences between the

Box 4.8 Carrying capacity calculations for the Guayabo National Monument, Costa Rica

The following calculations and conclusions were made by Calderón and Madriz Díaz of the Geography Department of the University of Costa Rica, and were reported in *Análisis de Capacidad de Carga Para Visitación en las Áreas Silvestres de Costa Rica* by Maldonado *et al.* (1992).

The Guayabo National Monument covers 217 hectares and is the most significant area of archaeological interest in Costa Rica. Its structures date back to 500 AD.

The map shows its location in the centre of the country. Within its area there are three trails along which visitors are channelled. In 1990 the monument received 12,356 visitors, 92 per cent of whom were Costa Rican nationals, and 80 per cent of whom arrived in their own vehicles.

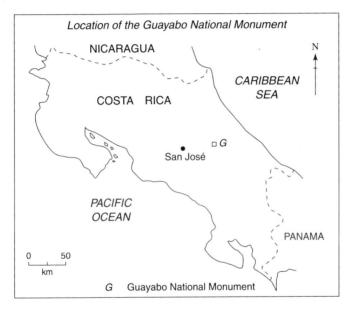

Location of the Guayabo National Monument

G Guayabo National Monument

Among the problems of the area identified by the team of geographers who conducted the study on which the carrying capacity calculations were based, were inadequate government funding for protection of the area, increasing visitor numbers, tree felling in a buffer zone around the monument, indiscriminate hunting of birds, deterioration of the archaeological remains due to visitors, and the lack of a management plan.

Carrying capacity calculations

Three types of carrying capacity were calculated for each of the three trails separately, as it wasn't possible to apply the three different concepts to the area as a whole. The three different types are:

Box 4.8
continued

- physical carrying capacity;
- real carrying capacity;
- effective or permissible carrying capacity.

They are defined by the variable used in their measurement.

Physical carrying capacity (PCC): calculated according to the space necessary for one person to move freely in a specified time and assumed to be 1 square metre per person. The average width of the trails is 1 metre, so each visitor uses 1 linear metre of the trail at any given moment. For one of the trails, Sendero Los Cantarillos, other relevant assumptions made are:

- visitors follow the trails in groups of no more than 25 (each group with a guide);
- a distance of at least 100 metres is maintained between groups;
- the trail has a length of 1,100 metres;
- an average time of 1 hour is required for a visitor to complete this trail;
- the monument and trail are open to the public for 7 hours per day and 360 days per year.

PCC = length × visitors/metre × daily duration (hrs/day)
= 1,100 × 1 × 7 = 7,700 visits per day
= 7,700 × 360 = 2,772,000 visits per year

Real carrying capacity (RCC) is the physical carrying capacity 'corrected' to allow for the following factors: precipitation (FP = 1.39%), vulnerability to erosion (FE = 38.28%), degree of slope (FS = 38.28%). Correction factors were calculated for each of these and expressed as percentages. The calculations are based on survey data, for details of which the reader should consult the original work. The real carrying capacity calculations are:

RCC = PCC × (100 − FP)/100 × (100 − FE)/100 × (100 − FS)/100
= 7,770 × 0.9861 × 0.6172 × 0.6172
= 2,892 visits per day
= 2,892 × 360 = 1,041,276 visits per year

Effective carrying capacity (ECC) is the real carrying capacity 'corrected' to allow for the difference between the actual management capacity and the ideal management capacity, and is represented as FM. The actual management capacity of the monument is given by the number of personnel (administrative staff, park guards, and guides) employed (in this case 10). The ideal management capacity is given by the number that would be required to fulfil all functions allocated to the staff of the monument (39).

FM = (39 − 10)/39 × 100 = 74.36%
ECC = RCC × FM = 2,892 × (100 − 74.36)/100
= 2,892 × 0.2564 = 741.5 visits per day
= 741.5 × 360 = 266,943 visits per year

beneficiaries of tourism and those who are marginalised by it, or of the creation of spatial ghettoes, either of the tourists themselves or of those excluded from tourism. Stonich *et al.* (1995) provide a clear example of these social divisions on the Bay Islands of Honduras (see Box 4.9).

If we accept the premise that tourism sets up an intrinsically false and fabricated social division between the server and the served in the first place, it is of course inevitable that tourist developments (resorts, enclaves, condominia) will create such divisions. It is one of the purposes of the tools of sustainability, such as carrying capacity calculations, environmental impact assessments, and sustainability indicators, to minimise the effects of these divisions to a point at which they can be excused. To this end, Clark (1990) has suggested the possibility of calculating the social carrying capacity. This is examined further in Chapter 8.

Cultural sustainability

Societies may be able to continue functioning in social harmony despite the effects of changes brought about by some new input such as tourists. But the relationships within that society, the mores of interaction, the styles of life, the customs and traditions are all subject to change through the introduction of visitors with different habits, styles, customs and means of exchange. Even if the society survives, its culture may be irreversibly altered. Culture of course is as dynamic a feature of human life as society or economy; so the processes of cultural adaptation and change are not assumed by all in all cases to be a negative effect. But cultural sustainability refers to the ability of people or a people to retain or adapt elements of their culture which distinguish them from other people.

Cultural influences from even a small influx of tourists are inevitable and may be insidious; but the control of the most harmful effects, emphasis on the responsible behaviour of the visitor, and the prevention of distortion of local culture might be assumed to be essential elements of sustainable tourism. Pratt's notion of transculturation (Pratt, 1992) encapsulates the way in which marginalised or subordinated groups select and invent from materials transmitted to them by dominant 'metropolitan' cultures. Cultural adaptation occurs in this way, and may result in change towards the wishes of the dominant culture. We return to this notion in Chapter 8.

Cultural impacts are more easily seen over the long term and are therefore more difficult to measure, although the cultural subversion of many local communities has been well documented, especially but not exclusively by anthropologists – the work of de Kadt (1979), Plog (1972) and Smith (1989), for instance, illustrates the cultural ill-effects of tourism. Organisations such as Survival International and Tourism Concern have also documented the cultural subversion of indigenous groups, which is another theme we examine further in Chapter 8.

Box 4.9 Social divisions in the Bay Islands of Honduras

An examination of the costs and benefits of tourism developments on the island of Roatán, off the Caribbean coast of Honduras, was made by Stonich, Sorenson and Hundt and reported in the *Journal of Sustainable Tourism* (1995).

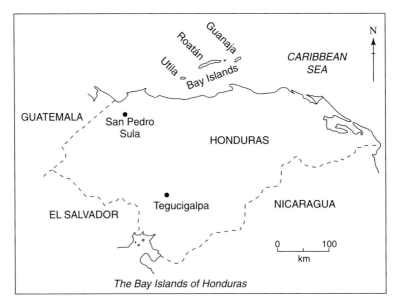

The Bay Islands of Honduras

Inhabitants on the island were categorised by:

community:
  West End
  Sandy Bay
  Flower's Bay

ethnicity:
  islanders
  ladinos

gender:
  male
  female

The degree of tourism development differs considerably in each of the three settlements.
  Variables describing income, household economic strategies and demography were analysed for each of the above categories.
  Among other findings and conclusions were:

Box 4.9
continued

- increased social differentiation as a result of tourism developments;
- the assignment of the majority of ladinos and islanders to low-status, low-paid, temporary jobs;
- reduced access for local people to the natural resources on which they depend for their livelihoods;
- escalating prices;
- land speculation;
- increased outside ownership of local resources;
- deterioration of the biophysical environment.

Economic sustainability

The condition of economic sustainability is no less important than all others in any tourist development. Sustainability in these terms refers to a level of economic gain from the activity sufficient either to cover the cost of any special measures taken to cater for the tourist and to mitigate the effects of the tourist's presence or to offer an income appropriate to the inconvenience caused to the local community visited – without violating any of the other conditions – or both. As expressed thus, it may appear as if the other aspects or conditions of sustainability are being 'bought off'. In other words, regardless of how much damage may be done culturally, socially and environmentally, it is perfectly acceptable if the economic profitability of the scheme is great enough to cover over the damage, ease the discontent or suppress the protest.

Economic sustainability, we would argue, is not a condition which competes with other aspects of sustainability. Rather, it can be seen as equally important a condition in its own right. On the other hand, it is not the only condition of sustainability, as might appear to be the case from the thoughts of numerous active agents of the industry. The condition of this as an element of sustainability in no way reduces the significance or level of acceptance or tolerance of the other conditions. Nor does it cloud the importance of the contextual issue of power over tourist activities. With this in mind, the question of who gains financially and who loses financially often sets the power and control issue in sharper and more immediate focus than all other facets of sustainability.

The educational element

It is often stated that an important difference between the new forms of tourism and conventional tourism is found in an element of educational input into the activity. This does not mean that it is necessary to reach high academic levels in order to be a sustainable tourist; but a greater understanding of how our natural and human environment works is almost always a goal, if not always stated, of the activity. It is, however, often stated as a goal without being practised. Pressure of business may render this so, but cynicism may also explain it – the flimsiest pamphlet of information for the tourist can be used as evidence of an educational input, and therefore of the 'genuine' motives of the operators and the real desire to aim for sustainability.

Again, it is important to refer this principle back to its context of power and development. Who is the beneficiary of the educational element? Does this enhance their degree of control over the activity and its distribution of benefits? This element, we would argue, has the potential to further widen the inequalities of tourism developments.

At conferences on the subject (First World Congress on Tourism and the Environment, Belize 1992; Ecotourism – A Sustainable Option, Royal Geographical Society, London 1992; Managing Tourism, Commonwealth Institute, London 1995; Sustainable Tourism, San José, Costa Rica 1995; to name just a few) education in this respect is generally taken to mean one of two things: first, the enlightenment of the new tourist in the cultural ways and norms of those they are visiting – an education for its own sake; and second, the training of the 'hosts' so that they are better able to cater for the wishes of the new middle classes who visit them.

There are very few acknowledgements of the need to educate the local populace of the destination communities about the tourists. One notable exception to this is Krippendorf (1987), who encourages the dissemination of information about the tourists to those they are visiting:

> By supplying the host population with comprehensive information about tourists and tourism, many misunderstandings could be eliminated, feelings of aggression prevented, more sympathetic attitudes developed and a better basis for hospitality and contact with tourists created. . . . Such information should aim at introducing the host population – initially, all those who come into direct contact with tourists, but also the public at large – to the tourists' background: their country, their daily life (working and housing conditions, etc.), their reasons for travelling and their behaviour patterns.
>
> (1987: 143)

Another form of educational input into sustainable tourism is the provision of 'technical information on how to do ecotourism *right*' (our emphasis) (Whelan, 1991: 4). Arrogance like that betrayed in this paternalistic attitude expresses the idea that 'we' know how to do it and 'the rest' just need to be educated in our ways. That is not to say that all forms of skills and knowledge transfer are arrogant or patronising; the programmes of Voluntary Service Overseas (VSO) and Coda International Training, for instance, are generally welcomed by and are often initiated by the local people who become their beneficiaries.

Local participation

The importance attached by many parties to the inclusion of the local populations is considerable. Indeed, there is more debate about the degree of inclusion or control to be exercised by destination communities than about the need for their involvement at all. Seven different types of participation are identified by Pretty (1995: 4–5), ranging from 'manipulative participation' ('simply a pretence, with peoples' representatives on official boards but who are unelected and have no power') to 'self-mobilization' ('people . . . taking initiatives independently of external institutions to change systems') (see Table 8.1, p. 241).

This debate is thrown into sharp contrast by the two standpoints of 'host' communities as objects of tourism or as controllers of tourism. Again, this matter is often considered to be at the heart of the difference between conventional mass tourism and supposedly sustainable new forms of tourism. But it is argued here that the issue of control is the same whether it refers to mass tourism or any of the new forms of tourism. Indeed, there may be something in the idea that the local authorities and local service providers of a mass tourism clientele have a greater degree of control and power over their activities than do those of the new forms of tourism. Chapter 8 examines this debate.

The aspects of sustainability discussed above are not presented here as prescriptive. We are not suggesting that a given lodge, tour or reserve can be assessed through these criteria for sustainability. In fact, it should be clear that no establishment would be able to meet all these criteria. If they were universally used for making judgements about whether a given practice was sustainable and if all criteria had to be satisfied, then clearly nothing would be judged as sustainable. But this raises the point that sustainability should perhaps be seen as a continuum, and should be assessed on a scale similar to that of probability, offering differing degrees of sustainability. Such an idea opens the concept up to distortion and misuse, but as we have seen and as we shall see, it is misused already.

In any case, it is our contention that sustainability is not definable by and is not reducible to a series of absolute principles. If principles can be applied to the notion, then it can only be in a relative way, relative to each other without contradiction, relative to the varying perceptions of those who use them, and relative to the values, ideological and moral, of those who apply and interpret them. 'Good' and 'bad' are relative terms, as is sustainability. With this in mind, it is worth considering the priorities for sustainable development set out by Agenda 21.

Agenda 21 and sustainable development in tourism

Agenda 21 is a global action plan endorsed by the 1992 Rio Summit (see Box 2.3) in Brazil. It sets out the priorities for sustainable development into the twenty-first century. Box 4.10 gives a brief description of the points of relevance in Agenda 21 for the tourism industry.

In its widest sense, tourism is a form of trade, not of goods perhaps, although the commodification of tourist destinations and talk of the 'tourist product' is

---

Box 4.10   Agenda 21 and tourism

Although key sections of Agenda 21 address business, industry and trade unions, it is primarily directed at governments and educators. The action taken by the former in particular have a bearing on the tourism industry, at both national and local levels. International government agreements may also affect certain tourism sectors.

Agenda 21 impinges on tourism in two ways. First, tourism is specifically mentioned as offering sustainable development potential to certain communities, particularly in fragile environments. Second, tourism will be affected by Agenda 21's programme of action because its many impacts may be altered by the legal framework, policies and management practices under which it operates.

Among other priorities given in Agenda 21, *governments* are urged to:

- improve and reorientate pricing and subsidy policies in issues related to tourism;
- diversify mountain economies by creating and strengthening tourism;
- provide mechanisms to preserve threatened areas that could protect wildlife, conserve biological diversity or serve as national parks;
- promote environmentally sound leisure and tourism activities, building on . . . the current programme of the World Tourism Organisation.

Business and *industry*, including *transnational corporations*, are urged to:

- adopt . . . codes of conduct promoting best environmental practice;
- ensure responsible and ethical management of products and processes;
- increase self-regulation.

(Stancliffe, 1995)

---

now firmly established and accepted. Shortly after the Rio Summit, Arden-Clarke argued that 'the whole of the Agenda 21 section dealing with trade amounted to an evasion of key trade and environment issues, rather than a basis for their solution' (1992: 13).

Arden-Clarke's arguments about Agenda 21's treatment of the general area of trade are applicable to the field of tourism. Essentially, his criticism is based on two particular features of the Agenda: first, it endorses the General Agreement on Tariffs and Trade (GATT) rules which encourage the externalisation of environmental costs; and second, it endorses the idea that only trade liberalisation will bring about sustainable development.

The first of these endorsements stems from GATT's agreement that the degree to which a country internalises its costs is left to choice. This effectively fixes the externalisation of environmental costs as the norm and makes clear that those countries which deviate from this will lose short-term competitiveness.

The second endorsement on trade liberalisation implicitly depends on:

the 'trickle down' mechanism to solve environmental problems – free trade leads to increases in per capita income through the economic growth it engenders, which in turn creates wealth, some part of which can then be invested in environmental protection. . . . The argument essentially says that you must first dirty your own backyard to generate the wealth to clean it up. . . . [This] ignores the facts that:

a) there is no automatic mechanism which guarantees that 'trickle-down' wealth is invested in the environment;

b) environmental damage is cheaper to prevent than cure, and in many cases is irreversible.

The flaws in this argument are being learned painfully around the world, but most notably in developing countries.

(Arden-Clarke, 1992: 14)

Arden-Clarke's critique highlights the ideological values which underpin the priorities of Agenda 21 and reinforces the arguments about the importance of relationships of power made in Chapter 3. The principles of sustainability are not absolute and immutable. In any tourism analysis there is a need to examine the questions of who is stating the principles, priorities and policies, who will benefit from related action and who will lose. And as for the principles, the power of those who use the tools of sustainability, covered in the following section, is an essential feature of an analysis of tourism.

THE TOOLS OF SUSTAINABILITY IN TOURISM

The last section made mention of the tools and techniques available for use in assessing or measuring various aspects of sustainability. Box 4.11 lists these techniques under eight major groupings.

Area protection

As applied to the field of tourism and for the purposes of this chapter, we use the term 'tools or techniques of sustainability' in a general sense. Even the designation of an area of land as a national park or as some other category of protected area can be seen as a tool of sustainable tourism. Those countries with high proportions of their land area under some form of legislated protection might be considered as practising more sustainable tourism than those with low proportions of their land protected. This assumption can of course be questioned. Some governments, for instance, have designated large areas of land as national parks or wildlife reserves but have failed to provide the resources required to afford an appropriate level of protection on the ground. Guatemala and Brazil may be cited as examples here, but they are not alone. It is difficult to blame such governments – they simply do not have the capital resources to pay for land protection, which after all has become a fashionable policy to pursue only in recent years.

115

Box 4.11   The tools of sustainability

The following listing of techniques is not exhaustive. Each group of techniques is briefly discussed in the main text.

1 *Area protection*
Varying categories of status of protected areas:
- national parks
- wildlife refuges/reserves
- biosphere reserves
- country parks
- biological reserves
- areas of outstanding natural beauty (AONBs)
- sites of special scientific interest (SSSIs)

2 *Industry regulation*
- government legislation
- professional association regulations
- international regulation and control
- voluntary self-regulation

3 *Visitor management techniques*
- zoning
- honeypots
- visitor dispersion
- channelled visitor flows
- restricted entry
- vehicle restriction
- differential pricing structures

4 *Environmental impact assessment (EIA)*
- overlays
- matrices
- mathematical models
- cost-benefit analysis (COBA)
- the materials balance model
- the planning balance sheet
- rapid rural appraisal
- geographic information system (GIS)
- environmental auditing

5 *Carrying capacity calculations*
- physical carrying capacity
- ecological carrying capacity
- social carrying capacity
- environmental carrying capacity
- real carrying capacity
- effective or permissible carrying capacity
- limits of acceptable change (LACs)

6 *Consultation/participation techniques*
- meetings!
- public attitude surveys
- stated preference surveys
- contingent valuation method
- the Delphi technique

7 *Codes of conduct*
- for the tourist
- for the industry
- for the hosts
  - host governments
  - host communities

8 *Sustainability indicators*
- resource use
- waste
- pollution
- local production
- access to basic human needs
- access to facilities
- freedom from violence and oppression
- access to the decision-making process
- diversity of natural and cultural life

Indeed, the very idea of protected areas begs the questions of who is protecting the area for whom and from whom. Many such areas have been so designated as a result of the tide of environmental consciousness that has been promoted, especially by environmentalists and conservationists, over the last thirty years (see Figure 6.3, p. 177). In 1994, for instance, WWF International began a fund-raising and recruitment campaign with the patronising slogan: 'He's destroying his own rainforest. To stop him, do you send in the army or an anthropologist?' The advertisement that followed was, as Survival International observed, 'glibly "pro-nature" and implicitly anti-people' (Survival International, 1996: 1). This example is detailed further in Box 6.2 (p. 174).

This consciousness portrays areas of natural beauty as wilderness areas, unspoiled by contact with humans, and reserved for visits by the 'discerning' and appreciative urban dweller in need of rest and recuperation. This view conveniently ignores both the indigenous inhabitants of such areas and the proportion of the national population in search of and in need of land for survival.

Lorenzo Cardenal, a Nicaraguan environmentalist, has characterised this approach as 'parquismo' (Cardenal, 1991). He suggests that a progressive, integrated approach should replace it, referring to the integration of humanity and nature rather than their separation or compartmentalisation as typified by parquismo. These issues are discussed again in Chapters 5–9, and especially in Chapters 6 and 8.

Industry regulation

Regulation of the tourism industry can come from local governments in the form of planning restrictions, national governments in the form of laws relating to business practice, professional associations in the form of articles of affiliation, and international bodies in the form of international agreements and guidelines to governments.

It is axiomatic that government legislation is intrinsically political in multi-party democracies. International agreements may also be explicitly or implicitly political, especially when they stem from a body such as the World Tourism Organisation (WTO) whose 'overall goal is the promotion and development of travel and tourism as a means of stimulating business and economic development' (WTO, 1991). Other international agreements and guidelines, especially those stemming from the work of the scientific community, such as agreements to reduce carbon dioxide emissions, may suffer from a lack of commitment without statutory legislation on the part of national governments and a difficulty in enforcement.

Regulation imposed on the industry by industry associations is normally promoted as a more effective way of preventing unethical or illegal activity than is government legislation. In 1986 the American Society of Travel Agents (ASTA) produced a set of Principles of Professional Conduct and Ethics. They added weight to these with the threat of disciplinary action against those failing to live up to the responsibilities embodied in the set of Principles.

With this type of discipline, the industry tries to promote voluntary self-regulation and to fend off what it sees as restrictive government legislation. On the one hand, it seems to be an intrinsic part of the doctrine of monetarism that any form of regulation should be voluntary and conducted by the industry itself. This accompanies other planks of the doctrine, such as the right to corporate privacy, reduction in public expenditure, transfer of national assets from public to private hands, deregulation of industry, and wholesale support for the notion of 'free trade' which is currently sweeping the globe 'as the key to planetary prosperity and environmental protection' (Carothers, 1993: 15). On the other hand:

> It has to be appreciated that tourism is an industry and, as such, is much like any other industry. . . . There is no more reason to expect tourism, on its own accord, to be 'responsible', than there is to expect the beer industry to discourage drinking or the tobacco industry to discourage smoking – even though many agree that such steps would be socially desirable.
>
> (Butler, 1991: 208)

The tool of regulation is clearly one which allows specific groups to take control of the industry. The argument around regulation represents a power struggle between different interest groups. So should the industry be regulated, presumably by a branch of government? Or should it be left to regulate itself voluntarily? This issue is addressed again in more detail in Chapter 7.

Visitor management techniques

A range of visitor management techniques exists for use by those who cater for and control the movements of tourists. Some of these are listed in Box 4.11. There are several texts which outline these in depth (Lavery, 1971; Elkington and Hailes, 1992; Lindberg and Hawkins, 1993; Witt and Moutinho, 1994).

Worthy of particular note is the current trend towards the restriction of motorised vehicles in areas normally attractive to lovers of the beauty of nature. On the premise that the motor car as currently run is inherently unsustainable, this trend would seem like a move which the scientific community, the hosts, and the planners could all agree works towards the goal of sustainability.[1] This particular issue is currently of topical concern in countries like the UK and USA where levels of car ownership and use are high. It is also a topical issue in many cities in Third World countries, although in national parks and protected areas in the Third World the problems have generally not yet prompted the same level of concern as those of the national parks in the developed countries. There are exceptions to this, as for instance in the case of the highway currently being built through the Metropolitan National Park in Panama City, the largest area of tropical rainforest within the boundaries of a city. Wildlife safari vehicles in East Africa have also created problems sufficient to be widely noticed and publicised in recent years.

Another interesting visitor management technique is that of differential charging for foreign and national visitors. Such a policy is not always understood by the visiting professionals from the north (see Chapter 5), but makes explicit the condition of local participation as an inherent aspect of sustainability.

Environmental impact assessment (EIA)

A technique which has attained fashionability and respect relatively recently is that of environmental impact assessment (EIA). It has been described as 'Among the foremost tools available to national decision makers in their efforts to prevent further environmental deterioration' (Sniffen, 1995). But it can be used at more than just the national scale: 'the EIA process is seen as a means not only of identifying potential impacts, but also of enabling the integration of the environment and development' (Green and Hunter, 1992: 36).

According to Tourism Concern, 'For the tourism industry to develop and survive in a sustainable and responsible manner, an anticipatory approach is essential' (1992: 34–5). This is eminently sensible, as is Goodall's pronouncement that 'Only where the result of the EIA clearly demonstrates that the development will be environmentally responsible and sustain the destination's primary tourism resources should planning permission to proceed normally be granted' (1992: 62).

But EIAs can be manipulated like most other techniques. Their results are responsive to those factors used as inputs. The choice of input, then, is crucial, and it is vital that we recognise that 'If we are to account for environment, . . . then the idea of a politically neutral social science has to be dropped' (Mulberg, 1993: 110). Mulberg's statement is a reference to the externalisation of unquantifiable factors by the practitioners (accountants) of capitalist economics. For 'externalisation' we could read 'ignoring', a practice which has indirectly led to many of the world's worst environmental catastrophes.

The techniques of rapid rural appraisal and EIA will be illustrated and discussed in Chapters 7 and 8.[2]

Carrying capacity calculations

Carrying capacity calculations have already been briefly discussed in this chapter (pp. 106–9), and an example given in Box 4.8. It is worth adding, however, that Mulberg's point about political and social neutrality, which can be extended to include commercial neutrality, is as applicable to carrying capacity calculations as it is to EIAs. Calculations can be manipulated by, for instance, tour operators, protected area officials, officers of conservation organisations, or government officers, to promote either a destination's exclusivity (a low carrying capacity) or its ability and potential to absorb more visitors (a high carrying capacity). Both these strategies might be seen as in the interests of different parties under different circumstances. And it is interesting to note how upper thresholds of visitors, arrived at as a result of carrying capacity calculations, have increased as time progresses. Again, examples are discussed later, specifically in Chapter 8, which also examines a measure becoming more fashionable than carrying capacity, namely, the limits of acceptable change.

It should also be remarked that the notion of carrying capacity reflects the prevailing relationships of dominance between First World and Third World. Management of the carrying capacity of a particular national park or other protected area gives considerable power to those who have that control. And control of the technique itself offers academics their own degree of power in the debate.

Consultation/participation techniques

Stewart and Hams (1991) argue that 'Sustainable development must be built by, through and with the commitment of local communities. The requirements of sustainable development cannot merely be imposed; active participation by local communities is needed.' In the field of tourism, those who speak of sustainable development almost always include the participation of the host communities as one essential element or principle of that sustainability. For this reason, techniques for promoting public participation and involvement in development projects are included in Box 4.11 as one type of tool available for the measurement of sustainability. Furthermore, techniques for measuring public perceptions, attitudes and values (such as contingent valuation techniques and stated preference surveys) are seen here as a necessary stage in the measurement of sustainability. They too are included here, although for details of the application of these techniques, see the sources listed in Appendix 2.

Techniques which allow for consultation and participation (of those people affected) are still young in their development and flawed. They are vulnerable to the type of distortion and bias which is introduced in the selection of inputs. They can also be hijacked to give an appearance of consultation with local people while in reality there is only consultation with so-called 'experts'. Such difficulties are illustrated with the use of the Delphi technique in Chapter 8. Additionally, the point should be made that talk of consultation with local people has become fashionable, especially among conservationists; but as Survival International point out in relation to the role of indigenous people in managing protected areas, 'This looks good on paper, but they are hardly an adequate substitute for land ownership rights and self determination' (1996: 2).

Attempts to value social costs and benefits include surveys of public perceptions of, expectations of and attitudes towards a range of social problems and manifestations such as shopping opportunities, access to recreational facilities, noise levels, litter, standard of living, and vandalism. Measurements of perceptions are placed on a scale, generally extending from highly positive (strongly in favour of) to deeply negative (strongly opposed to). Perceptions, attitudes, expectations and values, however, vary from person to person and from group to group. Difficulties therefore arise with the interpretation of results, which often appear weak and ambiguous, but which are nevertheless often used and excused as participation and consultation. These techniques and difficulties are discussed again in Chapter 8.

Codes of conduct

Recent years have seen a rising tide of codes of conduct for use in the tourist industry. Their design, promotion, contents, relevance, uptake, effectiveness and monitoring have become important features of the industry and are all worthy of attention.

There are two general points that can be made about almost all codes. First, they attempt to influence attitudes and modify behaviour. Second, almost all codes are voluntary; statutory codes, backed by law, are very rare (Mason and Mowforth, 1995).

Many codes of conduct are very impressive in their range of issues and in their depth of discussion and information. But they can be abused by the industry as marketing ploys or as veils extending over many of its impacts. There exist a number of problems associated with the use of codes of conduct which can be summarised under the following descriptions: the monitoring and evaluation of codes of conduct; the conflict between codes as a form of marketing and codes as genuine attempts to improve the practice of tourism; regulation or voluntary self-regulation of the industry; and the variability between codes and the need for coordination.

These issues are discussed at various points throughout Chapters 5–9.

Sustainability indicators

The youngest of all the tools of sustainability are those now described as sustainability indicators, the development of which arose from the Rio Summit of 1992 (see Box 2.3 and pp. 22–5 and 113–14). It is now commonly accepted that conventional indicators of 'well-being' (such as gross national product – GNP) give a restricted, partial and one-sided view of development. It is the search for indicators which show the linkages between economic, social and environmental issues and the power relationships behind them which has given rise to the development of so-called 'sustainability indicators'. Thus far, such indicators have been developed as trials and are currently applied only at local authority level.

One important aspect that has been built into these indicators from their inception has been the participation of local community members in their formulation. There is no doubting here the genuine and diligent attempt to promote such participation as part of the development of sustainability indicators. There is also no doubting that it is precisely this participation which has led to the use of indicators which are much less remote and much more comprehensible to people than are nationally and internationally derived measures such as GNP, gross domestic investment, and the like.

But their acceptance will face an uphill struggle. The measures most frequently used at the level of the national economy relate precisely to that: the economy. Other relevant factors are externalised (that is, ignored). Moreover, their use is well entrenched and perpetuated by conservative media which accept new ideas with great reluctance unless they are forced to do so by a public that has already moved ahead. The need to include the social, cultural, environmental and aesthetic factors which our commercial world and controllers normally externalise has not led to a quick redress for such factors, despite public debate of the issue.

Furthermore, it has yet to be proved that these more locally accountable, more relevant and less remote indicators are less likely to be subject to bias and manipulation to suit the ends of those who use them. Discussion of them is taken further in Chapter 8. (See also Appendix 2.)

WHITHER SUSTAINABILITY IN TOURISM?

It is worth restating here the point that the notion of sustainability has to be taken beyond its current bland usage and interpretation, as best illustrated by politicians and daily media pundits. If it remains a 'buzzword' which can be so widely interpreted that people of very different outlooks on a given issue can use it to support their cause, then it will suffer the same distortions to which older-established words such as 'freedom' and 'democracy' are subjected. Both of these terms, and others, are frequently and regularly distorted by most of our politicians – see Chomsky (1989), Postman (1985), and Curran *et al.* (1986) for analyses of such distortions.

We move on, then, to the question of how sustainability should be taken beyond its current usage and how it should be given a substantial, tangible and unequivocal meaning. This further leads us on to the question of whether we should be promoting the principles and tools of sustainability, as outlined above, towards this end. Both of these questions are addressed in the conclusions given in Chapter 10.

It has already been pointed out that the principles of sustainable tourism are open to manipulation in the service of operators and others in the industry. That is not to say that the principles are not worthy of attention by all those involved in the industry; but it does suggest that the motives of those who apply them should also be scrutinised.

On the assumption that the use of the techniques of measurement and description will help a move towards a clearer, workable and meaningful analysis of sustainability, awareness of the limitations and immaturity of the techniques is also necessary. This means that they are susceptible to manipulation for partisan purposes. In turn, this raises the need to politicise the tourism industry in order to promote its movement towards sustainability and away from its tendency to dominate, corrupt and transform nature, culture and society. The politicisation of the tourism industry would require a clarification and emphasis of the associations between the prevailing power structures and the control of tourism developments, and a clear linking of the goal of reducing uneven and unequal development with the policies pursued by the tourism industry and the governments and international institutions which promote it.

Without this politicisation, sustainability will continue to be hijacked by the prevailing model of development, capitalism, and will increasingly fall into the service of the controllers of capital, the boards of directors of major transnational companies and other organisations which manage the industry. This tendency has already become apparent. And as the new forms of tourism gather ground and increase their share of the tourism market, as seems likely, the current power structure and the processes by which power is held and retained will attempt to subsume them, as has already been shown.

Concurrent with this trend in many areas of the Third World, however, is a grassroots groundswell to take control of and exploit tourist opportunities at the community level. Currently this tendency seems to assume automatically that 'sustainability' is their prerogative, and use of the term is as loose as it is in other tendencies. Automatic assumptions are often used to cover over awkward questions. The existence of these different tendencies highlights the debate between a tourism that is industry-controlled and one that is community-controlled, a debate which is discussed further in later chapters, especially Chapter 8.

The obstacles to change towards sustainability briefly mentioned above and the struggle for power over the definition of the concept are themes which run throughout Chapters 5 to 9. The discussions on global changes, power relationships, dependency theory and sustainability in the first part of the book inform and fuel the analyses of the roles of different sectors of tourism that follow in the second. Specifically, the key themes (uneven and unequal development, globalisation, relationships of power) and key words (new tourism, sustainability and the Third World) of the book will reappear as the basis of discussion in the following chapters.

In Chapter 5 the role of the tourist, particularly the new tourist class of relatively wealthy northerners who view the globe as their village, is considered. In Chapter 6 the high moral ground of the socio-environmentalist organisations which attempt to influence the full range of human development, not simply that of the tourist industry, is examined. Chapter 7 looks at the often criticised operations of the agents of the tourist industry, the tour operators, travel agents and airlines, and asks how they are adapting themselves to the new forms of tourism and the principles which these embrace. Chapter 8 views the concept of sustainability from the point of view of the populations at the receiving end of tourism, especially at the community level, and makes use of a range of levels of participation. Chapter 9 seeks to explain the response of governments to the myriad of factors affecting the development of the tourist industry, its relationship with sustainability, and the power and influence of supranational institutions such as the World Bank and IMF.

Finally, the issues and debates briefly alluded to in this last subsection are discussed again in Chapter 10, in the book's conclusions.

5

A NEW TOURIST CLASS
Trendies on the trail

As Crick (1989) argues, 'human beings', the tourists themselves, are only infrequently the object of consideration in much tourism research. And yet it is upon tourists that the industry is founded. In part this has resulted from academic prejudices of analysing production rather than consumption (Lash and Urry, 1994) and the tendency to focus upon the impacts (economic, environmental, cultural) of tourism where tourists are at best a homogeneous and undifferentiated group and at worst are deemed immaterial.

It has been suggested at various points in earlier chapters that tourists need to be taken more seriously. To summarise, a number of key factors have been stressed. First, Chapter 2 sought to demonstrate the significant economic changes that have resulted in post-Fordist modes of both producing and consuming goods and services. In part, these changes can be traced through to the emergence of new and varied forms of tourism in the Third World. It was also argued that these changes have tended to invest more power in us, the consumers, or the tourists: we have more choice. Second, it has been suggested that the influence of the new middle classes can be linked to the emergence of what may be referred to as postmodern cultural forms. The importance of these social groups in both producing new forms of tourism in the Third World and, of course, forming a significant role in taking such holidays was highlighted. It may be worth taking another look at Figure 2.3, Table 2.4 and Box 3.5 to recall this framework.

Both points stress the increasing significance of tourists, with the second factor highlighting the importance of social class. Of course this is not to say that class is the only factor in studying tourism; far from it. But it is a significant factor and is especially important in the analysis of new forms of tourism. As Crick observes, the 'world of tourism is rife with the class distinction in our everyday world' (Crick, 1989: 334). And yet an analysis of the significance of class and tourism is only weakly developed. It was argued in Chapter 2 that analysis of tourists has centred around either classifying tourists or carrying out motivation and attitude surveys. Although such approaches are of interest, they have tended to limit the scope of tourism analysis.

This chapter explores what use an analysis of class offers in developing a critique of Third World tourism. As elsewhere in the book, this is not a definitive statement of how tourists should be analysed and understood, but an attempt to broaden our thinking and approach to the field of Third World tourism. Initially, the chapter discusses the importance of social class as a concept and how it is reflected in travel and tourism. In particular, the relationships of contemporary global change to the way classes are formed will be considered. The second half of the chapter identifies a number of crucial factors evident in the formation of new middle-class tourism. It will be argued that travel and tourism has an increasingly important symbolic role to play as social classes seek to define and distinguish themselves from other social classes. The final sections of the chapter indicate why social class analysis is of importance in considering the impacts of Third World tourism.[1]

CLASS, CAPITAL AND TRAVEL

Of all social science concepts, arguably, it is class that has been subjected to the most thoroughgoing marginalisation since the 1980s. Politicians began to talk in terms of classless societies, academics became preoccupied with the fragmentation of traditional class lines (working class/middle class/upper class) and more and more people have begun to think in terms of their individual, as opposed to class, status. This is a lure of individualism that led Crompton (1993) to conclude that the 'retreat from class . . . is becoming the sociological equivalent of the new individualism' (1993: 8).

However, a recognition of the role of social class in contemporary capitalism has begun to re-emerge. A number of authors have begun to illustrate the relationships between cultural changes and the development of new middle classes (Featherstone, 1991; Betz, 1992). In common, these authors agree that these significant and numerically expanding class fractions have an instrumental part to play in new cultural forms. Indeed, Lash and Urry refer to postmodernism as a 'hegemonising mission' for the new middle classes and elsewhere they are considered as both producers and consumers of postmodernism – *par excellence* (Featherstone, 1991). In Box 3.3 (p. 53) we identified a number of key characteristics of postmodernism and highlighted their relevance to the analysis of Third World tourism. Briefly these involved: the emergence of specialist agents and tour operators (and its adjunct, more individually centred and flexible holidays); the de-differentiation of tourism as it becomes associated with other activities; and the growth of interest in *other* cultures, environments and their association with the emergence of new social movements.

Nevertheless, the different ways in which different social classes consume tourism is vastly under-researched. This is significant in that we have argued above that much new tourism is based on the notion of individual travel. As Figure 5.1 conveys, a culture of travel (as opposed to tourism) that is shamelessly hedonistic has emerged: chill out, the world's a breeze, so experience it.

Figure 5.1 The culture of travel

Box 5.1   Tourists, tourism and pollution

The fundamentally conservative discourse of sustainability transforms travel into an altogether differently conceived activity that is underpinned by an almost missionary zeal (Kutay, 1989; Munt, 1994a).

'The Barbarian of yesterday is the tourist of today', (N. Mitford, 'The tourist' 1959, quoted in Urry 1990a).

'With package tourists invading the old hippie hang-outs, the desire to find new destinations becomes intense.' (Calder, 1994a: 35)

'I'm an adventurer. . . . The Africa I experienced was literally that of the 'Voyage au Bout de la Nuit.' It was a time when people were travellers, not tourists. . . . I travelled on a shoestring. Regimented tourism is a contagious form of pollution.' (Cartier Bresson, quoted in Guerrin, 1991)

'Perhaps we need to accept that some areas will remain "sacrificial resorts" which continue to attract tourists who do not seek an experience of "deeper" quality.' (Rosenthal, 1991: 2)

'With the days of the package holiday . . . drawing to a close, we are about to discover very special holidays for very special people: like you.' (Sherpa Expeditions 1992 brochure)

But most of the existing commentary amounts to little more than the rhetorical and polemical ranting of middle-class writers bemoaning the scale, effects and popularity of mass tourism (see Box 5.1). The analogy of tourism to pollution is commonplace (Brooks, 1990; Guerrin, 1991; Turner and Ash, 1975) and is often expressed in socio-cultural terms where it is the swarms or tidal waves of tourists that constitute a problem that must be contained (Box 5.1).

Equally limiting is the fact that the role of geography, or the spatial dimension, is all but ignored. But this is fundamental both to understanding tourism as a process and how social classes are constituted as part of that process. As Thurot and Thurot note, the ability of the middle classes to maintain their 'social distance' from others has become more difficult because 'great distance no longer offers substitutable destinations' (1983: 182). And hence the references in Box 5.1 to 'invading', 'sacrificial resorts', and so on. As Box 4.1 suggested, the circulation of tourist areas has tended to focus either on the deterministic cyclical nature of tourist development or the cyclical character of tourists themselves. In the former approach tourist areas are modelled as moving from discovery to overcrowding (Butler, 1980; Doxey, 1976). In the latter approach, a taxonomy of visitor characteristics are noted (Cohen, 1974, 1979a; Smith, 1989). Although this approach offers greater scope for thinking about the relationships between social class and tourism, the links remain weakly developed.

Bourdieu and 'habitus'

By acknowledging the importance of cultural consumption in the study of social classes, the French sociologist, Pierre Bourdieu (1984), has provided a very productive way for thinking about the relationships of class and tourism. A number of authors have noted the applicability of his influential work to tourism (Bruner, 1989; Errington and Gewertz, 1989; Lash and Urry, 1994; Urry, 1990b).

Briefly, Bourdieu (1984) argues that social classes wage 'classificatory struggles' seeking to distinguish themselves from each other by education, occupation, residence, and so on, and, of course, through commodities, which is taken to include both objects (cars, furniture, and so on) and experiences, such as holidays. They achieve this, Bourdieu (1984) argues, by constructing 'lifestyles'. We have already alluded to this notion of lifestyle ('Sustaining lifestyles', p. 31) as a useful way of considering individuals' uses of a range of objects, experiences, hobbies and beliefs to, sociologically speaking, 'mark their territories'.[2] These lifestyles, Bourdieu concludes, are the products of what he terms *habitus* and Box 5.2 presents the usefulness of the concept in the study of tourism. Box 5.3 provides an application of the concept.

Two important points emerge from Bourdieu's work. First, social classes are in constant struggle to ensure that differentiation from class fractions above and below are maintained. These battles are an example of everyday hegemony in practice (as discussed in Chapter 2). Of course, this also suggests that social classes are themselves constantly formed and hence the term class fractions to distinguish from the 'once-and-for-all' idea of *the* working class, or *the* middle class. Social class formation is, therefore, a dynamic process, 'a flame whose edges are in constant movement, oscillating around a line or surface' (Bourdieu, 1987: 13). Second, the struggles inherent in social classes have both a social element (ecotourism is better than package tourism) and spatial element (the Brazilian rainforest has more kudos than a Gambian beach). In other words, tourism can be used to differentiate both socially and spatially.

So exactly how are these struggles waged and differentiation achieved? An important part of the answer lies in the accumulation of capital. But in this case it is not the finance capital necessary to sustain profits, but cultural and symbolic capital linked to certain kinds of tourism.

Cultural and symbolic capital

As a number of authors have contended, consumption has become more skilful in recent years (Lash and Urry, 1994: 260). Not only do consumers ask more questions of the commodities they buy (What is the least polluting car to buy, or holiday to take?), they are also more sensitive to the symbols they are consuming. The consumption of holidays has assumed an increasingly significant role in the cultural differentiation characteristic of other forms of consumption

Box 5.2 Habitus

What is habitus?

Habitus represents the ability and disposition of individuals and social classes to appropriate objects and practices that differentiate them from others. A knowledge of 'foreign' food, good wine, classic literature or Latin American film, for example, may all assist in differentiating from others without such knowledge or appreciation. Habitus is, therefore, a 'cognitive structure' (Jackson, 1989) which 'gives people a sense of their place in the world' (King, 1995: 28). And for Bourdieu it is the necessary starting point for a theory of *strategies* that 'aims to account for the logic according to which groups, or classes, form and break up.' (1986: 119)

How is it reflected in tourism?

Travel has always had an important role to play in this process of differentiation, and as tourism has become more widespread the struggle has intensified. It is difficult to deny that Marbella, Kos, Phuket and 18–30 provide a very different set of meanings and representations from Tikal, Chiang Mai, Kilimanjaro and Explore. So different places, the different experiences to be had, and even different operators, add up to what we could describe as a 'symbolic system'; it is the way in which we represent objects and experiences and then communicate this to others.

The taste of social class!

And we can also argue that these different components 'each embody particular class "tastes"' (Allen and Hamnett, 1995: 240). Habitus, therefore, represents a certain 'class culture and milieu' (Zukin, 1987: 131) and provides the basis for class reproduction and differentiation. The traveller/tourist distinction, although highly stereotyped, is reflective of a wider debate over social class differentiation (a point to which we return below). Box 5.3 presents an example of habitus in 'Ecotourists: a personal profile'.

Exploitation and domination

A discussion of class also suggests the importance that exploitation (of one class by another) should play in the analysis; in 'order for a social collectivity to be regarded as a social class it has to have its roots in a process of exploitation' (Savage *et al.*, 1992: 5). Moreover, an exploitation-centred analysis of class provides for a more coherent strategy for considering the characteristics of the middle classes in contemporary capitalism (Wright, 1985: 131); hence the straightforward observation 'one person's welfare is obtained at the expense of another' (1985: 65). Traditionally, class analysis has been undertaken with a production-side bias with the focus of the exploitation of the workforce by capitalists (Wright, 1985), and a principal focus on economic measures (occupation, income, and so on) as indicators of class position. The critical part of Bourdieu's work is that 'domination remains, but it must be reconceptualised in

Box 5.2
continued

a world of consumption. Domination is now mediated by taste' (Frank, 1991: 66). Reflecting this power, Eagleton refers to habitus as a 'practical ideology', or what we might call an 'everyday hegemony', whereby the need to dominate is translated into routine social behaviour.

Box 5.3   Ecotourists: a personal profile

Age 58, retired World Bank agricultural economist

Meet a member of the new 'whoopie' club – wealthy, healthy, older people. Like many ecotourists, they are from the rich industrialised countries, aged between 44 and 64 years of age, and this growing élitist club has 'been there and done that'.

   Not for them the packed beaches of the mass tourist resort. According to a survey carried out by the US-based financial company American Express, these travellers have exhausted traditional destinations and are now in search of original, preferably pristine destinations. As the gap widens between retirement and death, and an ageing population grows, these rich people are destined to make up the core of future eco-travellers.

Age 33, teacher, partner also in work, vegetarian and eco-friendly, donates to Third World charities

Meet the North's sensitive soul. A member of the liberal middle classes who have grown more environmentally aware over the eighties. For these people various eco-travel operators seem to offer the exotic destination, but in a way that can help the local economy and not be too disruptive to the environment. And in a busy, do-gooding life, a laid-on tour is convenient.

Age 20, student

Meet what some observers call the 'curriculum vitae builders', or 'ego-tourists'. Young northerners who may be more mindful of environmental issues and who take off for longer holidays in the Third World, searching for a style of travel which reflects an 'alternative' lifestyle. Many will backpack, others will join tours that explore the last great wildernesses. What they have in common with all other ecotourists is the necessary capital to embark on the adventure.

Age 36, director of southern-based NGO

Meet the new affluent southerner. Works for an organisation which receives most of its funds from individual and corporate donations and grants from the north. Intensely concerned with his work, which allows him to travel to other areas in the Third World. Is keen to network with northern environmentalists. Spends holidays with family visiting national parks in own country and neighbouring countries.

(Panos Institute, 1995)

and the Third World has an important role within this and represents a sizeable cultural asset. It is a way of investing in and accumulating what Bourdieu refers to as cultural capital.

Unlike economic capital, cultural capital is not something that can be bought, but instead relies on the ability of individuals to join the game of being able to 'know' and 'appreciate' what to eat, drink, wear, watch and what types of holiday to take. In other words, it requires the skill of reading the cultural significance of certain types of consumption; for example, the significance of 'original features' in the process of gentrification (Jager, 1986) or the cultural significance of certain types of holidays, whether ecotourism in Kenya or backpacking in South-East Asia. In the former case, 'it is not simply that a desire to see ecological attractions in far-off destinations conveys a sense of superior status; rather it is the actual ability to appreciate ecology which indicates cultural competence' (Allen and Hamnett, 1995: 240). In the latter case, it will be argued that cultural capital is also accumulated through journeys to remote and hazardous regions (recalling the discussion of fetishism and aestheticisation in Chapter 3). It is these young travellers who often emerge as 'figures of admiration' (Jardine, 1994: A). As Bob Samuel, information manager at Trailfinders, observes: 'There is a lot of kudos in coming back and telling your friends you have travelled overland in Cambodia. It gives you a certain cachet' (quoted in Jardine, 1994).

If tourism has increasingly taken on a symbolic meaning (in part due to the dramatic increase in tourism), it has also meant that tourism as a commodity embodies 'sign-value', an important means of stocking up on cultural capital. It is symbolic in the way in which travel and tourism embody certain attributes: personal qualities in the individual, such as strength of character, adaptability, resourcefulness, sensitivity or even 'worldliness'. In a somewhat traditional sense, travel is widely regarded as 'character building'.

In this way tourism is not only something to be enjoyed but represents a strategy for building a reputation that can be converted into economic capital. In other words, it transforms travel into a commodity with exchange-value and can be traded in, so to speak, for material benefits; and hence the phrase 'curriculum vitae builders' in Box 5.3. Trailfinders' 'long haul travel consultants' must be aged 22–30 and 'have travelled extensively'. But of course it is not only the travel industry that recognises the 'benefits' of travelling and there is a number of occupations where overseas travel is an increasingly significant component or, indeed, prerequisite. How are the sign and exchange values of these travel commodities maintained or enhanced? Before trying to answer that question, consideration will be given to how class fractions may be conceived in relation to tourism.

Ecotourists

The first principal grouping Bourdieu terms the new bourgeoisie. They are located firmly within the service sector with jobs involving finance, marketing and purchasing, for example. They are also class fractions that earn a lot and are correspondingly high in terms of economic capital (finance). They can afford expensive holidays that are exclusive in terms of price affordability and the numbers of tourists permitted: private game reserves with luxury accommodation and limited capacity, for example. They are class fractions, therefore, which can take holidays in environments which, by virtue of price, are exclusive. For shorthand, we shall refer to this group as ecotourists, older and professionally successful tourists (Errington and Gewertz, 1989) or as Wood refers to them, the 'cream of Manhattan and the City' (Wood, 1991: 65). 'Ecotourist' has a double meaning, however, for not only does it signal an interest and focus of this type of tourist on the 'environment' (*eco*logy), it also indicates the ability to pay the high prices that such holidays command (*eco*nomic capital).

It is not a coincidence that the 'taste' of which Bourdieu speaks is actually translated into culinary metaphors for the holidays aimed at this social class, for luxurious eating is another of this fraction's cultural characteristics. Thomson Holidays' 'Travel à la Carte' urges you to 'choose from the menu or dream up a dish of your own' and Barrett (1989) refers to 'delicatessen' travel agents. It embraces the delicate, discriminating and generally luxurious holiday to an off-the-beaten-track, out-of-the-way place. And there are apparently no expenses spared on the level of service and culinary provision. As Africa Exclusive, a tour company specialising in East African safaris, put it, 'gastronomy is taken as seriously as zoology'. It also carries a level of cultural capital as well. It is a way of buying into a shared meaning that the environment has an acknowledged value: 'ECO-TOURISM. . . . It's not in any dictionary yet, but if you want to impress at your next dinner party, it's a dead cert' (Wood, 1991: 65). So, not only are the new bourgeoisie high on economic capital, they are also high in terms of their ability to accumulate cultural capital.

Ego-tourists

The second important class fraction that Bourdieu identifies, and one that is potentially more important, is the 'new petit bourgeoisie'. This social class is characterised by the increasing numbers working in employment involving 'presentation and representation' (Bourdieu, 1984) such as media and advertising, but also encompasses a wide range of service jobs (such as the guides working for small tour operators) and the growth in the number of people working for charities, for example.

You may recall that some commentators have emphasised the significance of this social class in interpretation and representation. This includes not only opinion-forming in terms of the latest film to see, clothes to wear or designer alcohol to drink, but also how we should perceive issues such as homelessness or overseas development: in other words, a guide to the things we should be 'into' (consider the profiling of tour leaders in Box 5.4).[3] Chapter 2 gave references to this group as taste-makers, trend-setters, new cultural intermediaries or cultural

---

Box 5.4   The Explore Team: tour leader profiles

Becca Hardy

Becca leads our groups in countries as diverse as Egypt and Bhutan. Her hobbies include drag-racing and cooking! This came in very handy when she led our Scandinavian camping trip a few years ago.

Lucian De Silva

Lucian is our local leader in Sri Lanka. A registered guide since 1983, his welcoming nature, enthusiasm and boundless knowledge on almost every aspect of Sri Lanka, make him one of our most valued leaders.

Suzi Poole

Suzi led tours for us all over the world before joining our Tour Planning department in 1993. She is also responsible for implementing and monitoring our environmental policy. As well as being fluent in French and Spanish, Suzi has extensive travel experience in most of Asia, South America, Australasia and South-East Asia.

Gerard Heelan

After leading tours in India, Spain, Turkey and throughout the Middle East, he has recently joined the Marketing Department and presents many of our Slideshow evenings. His previous occupations vary from teaching Economics in a college, to driving London buses. Interests include rugby, chess and theatre.

(Explore brochure, 1996/97)

---

brokers. These are terms which seek to capture the structural significance of such groups in postmodern culture.

Unlike the new bourgeoisie who are high on both economic and cultural capital, however, the new petit bourgeoisie are not so economically well-endowed and must seek out cultural capital in order to establish their social class identity. In this way the burden of class differentiation weighs most heavily on this social class who must differentiate themselves from the working classes below (mass packaged tourists) and the high spending bourgeois middle classes above (ecotourists). Part of the reaction has been for the new petit bourgeoisie to deem themselves unclassifiable, 'excluded', 'dropped out' or perhaps, in popular tourism discourse, 'alternative'; anything other than categorisation and assignment to a class (Bourdieu, 1984). And yet, as Bourdieu concludes, the A to Z of their practices, from 'aikido' to 'Zen' (or from alternative tourism to wildlife tourism – Box 4.6, p. 100) exudes classification, being nothing other than, 'An inventory of thinly disguised expressions of a sort of dream of social flying, a desperate effort to defy the gravity of the social field' (1984: 370).

This slide into individualism is an expression of what Smith (1989) refers to as a new ideology founded upon the 'pursuit of difference, diversity and distinction'. Indeed, the pursuit of difference is of critical significance in understanding the expansion of tourism in the Third World. Contradictorily, while the assertion of individualism becomes a frame of action of the new middle classes, it also reaffirms their class status and position. And, as already argued, the struggle, and by inference, exploitation, is 'quasi-inevitable' (Bourdieu and Eagleton, 1994).

The idea of 'classlessness' or 'individualism' is an indelible mark among these class fractions, resulting, some argue, in the arrival of ego-tourists (Wheeller, 1992, 1993a, 1993b; Munt, 1994a) or, as Box 5.3 suggests, 'curriculum vitae builders'. Ego-tourism is more characteristic of less formalised forms of travel, such as backpacking, overland trucking (Truck Africa refer to their customers as 'would-be backpackers') or small group travel, which often involve longer holidays overseas, especially in Third World regions. Indeed, 'going-round-the-world' or 'taking a year off to travel' has become a latter-day equivalent of the Grand Tour and an important component of the culture of travel. Ego-tourists must search for a style of travel which is both reflective of an 'alternative' lifestyle and which is capable of maintaining and enhancing their cultural capital. Furthermore, it is a class fraction that attempts to compensate for insufficient economic capital, with an obsessional quest for the authentication of experience (Cohen, 1979b, 1979c; Errington and Gewertz, 1989). MacCannell (1976) portrayed the middle classes as systematically scavenging the earth in search of new experiences and Errington and Gewertz note, 'For travellers, the encounter with what was seen as "primitive" – the exotic, the whole, the fundamentally human – contributed to their own individuality, integration and authenticity' (1989: 42). Stressing the link of ego-tourists to environmental issues, Gordon (1990) associates tourists he observed in Namibia to 'an emergent, urban-based, alienated petit bourgeoisie'.

Ultimately it is a competition for uniqueness (Cohen, 1979b, 1989) with which ego-tourists engage. This is bound up with the knowledge, difference and distinction that certain forms of tourism imply – a struggle, so to speak, to stamp the hallmark of individualism in the traveller's passport. While individualism has always been a key characteristic of bourgeois culture, which regards travel as an 'activity for the stimulation and development of character' (Rojek, 1993: 114), the rate at which individualism is sought now has significant consequences and impacts for places, especially in the Third World.

Individualism also underscores why ego-tourists feel they are exempt from the criticism levelled at much Third World tourism. First, as Sherpa Expeditions contends of its customers, 'They are involved and interested in the world around them, and in their turn are interesting people.' And second, individualism provides the foundation for claiming the moral high ground in the battle over 'what kind of tourism'? (see Figure 3.3, p. 72). These are people travelling on a shoestring, and as individuals their practices are not harmful to environment or society. Ego-tourists conveniently disaggregate their actions – they are individuals and their actions therefore have no significant impact. In other words, they refuse to acknowledge their part in a larger entity or mass. Indeed, by contrast some would argue their travels are actually beneficial (Kutay, 1989; Seabrook, 1995, for example). As we shall see below, however, moral claims are by no means sufficient to secure victory in this battle. The new middle classes have adopted strategies to remedy this by, for example, creating jobs and professionalising not just occupations, but consumption practices as well.

In effect, the boundaries between these new middle classes – between ecotourists and ego-tourists – are blurred: they share their ability to make relatively expensive journeys overseas and their willingness to wage hegemonic struggles to define and differentiate their forms of tourism. In this way, tourism has become part of that critical assemblage of goods, practices and experiences that are taken up as social 'bridges and doors', so important in bonding some and excluding others (Featherstone, 1991: 111).

A NEW CLASS OF TOURIST?

Tourism cannot be understood as just a means of having some enjoyment and a break from the routine of everyday, an entirely innocent affair where there are some unfortunate, incidental impacts. So far this chapter has argued that we need a deeper understanding of tourism fully to appreciate its content and expression and, of course, its potential impact. There are two principal components to our argument. First, tourism must be understood as a commodity with both a symbolic or sign value and an exchange value. Second, much so-called new tourism is an expression of the new middle classes' hegemonic struggle for cultural and class superiority (Stauth and Turner, 1988).

Of course travel has always been an expression of taste and a way of establishing class status (Adler, 1989). But, with the rapid growth in the numbers of people taking holidays, it has never been so widely used as at present. Put simply, the democratisation of tourism has created a social headache when it comes to classes attempting to differentiate themselves from one another.

Two fundamental questions arise from the discussion. How are the sign and exchange values of these travel commodities maintained and enhanced? And how do social classes pursue differentiation and to what effect? In answering the first question, we shall concentrate on three key methods that are used to define and legitimise contemporary tourism: the appeals to intellectualism and professionalism, together with the discourse that has arisen to describe travel today. In addressing the second question we shall identify what we have labelled 'hegemonic spatial struggles' pursued by the new middle classes. Underlying this discussion is the acknowledgement that much new tourism focuses on the Third World.

The new intellectuals

It has been argued that the new petit bourgeoisie have sought to intellectualise new areas of activity and expertise (Bourdieu, 1984; Featherstone, 1987). These 'new intellectuals' attempt to ape and popularise intellectual lifestyles and transmit intellectuals' ideas to a wider audience – arguably the ultimate in a cultivated, even scholastic and 'romantic flight from the social world' (Bourdieu, 1984: 371). Tourism is a particularly important component in this respect.

In Box 4.6 (p. 100) the 'de-differentiated' characteristic of much contemporary Third World tourism was noted, in which academic prefixes are increasingly apparent. Holidays have moved beyond sheer relaxation towards the opportunity to study and learn, to experience the world through a pseudo-intellectual frame. With the acknowledgement of tourism as an environmentally, socially and culturally problematic activity among certain fractions of the new middle classes, tourism and tour companies catering for the intellectual demands of these class fractions are of increasing importance in the legitimation of travel. Thus we find companies such as Academic Travel (part of STA, an ostensibly student-oriented travel agency), Adult Education Study Tours (established in 1982), who offer 'special interest travel for thinking people', and Temple World, offering 'exclusive journeys into wildlife, archaeology, natural history and ecology', whose guest leaders are academics such as the geographer, Andrew Goudie.

Moreover, the intellectualisation of travel has been enhanced by interlinkage between academia and tour companies. It is convenient for companies such as Himalayan Kingdoms (1992–3) to prioritise places on their Bhutan visit (a country that has imposed severe restrictions on the number of tourists each year) to Fellows of the Royal Geographical Society, and for Temple World to run a series of 'Royal Geographical Society African Tours'. At the height of what we might term 'intellectual mimicry', the travel operator Voyages Jules Verne offers a tour brochure that is indistinguishable from a weekend broadsheet newspaper supplement and is appropriately entitled *Travel Review*. Intellectual legitimation is sought in this instance by interspersing descriptions of holidays and itineraries with other considered features – 'Islam', 'The Turkish inner man' and the 'Shepherds of Shanghai', for example.

Among the younger new petit bourgeois tourists, with less economic capital, intellectualisation is sought in other ways. Most frequently this involves the claim of cultural superiority, of true and real contact with indigenous peoples, pursued through organised tours, such as 'overlanding', or 'individual' travel informed by travel guides such as the *Rough Guide* or *Lonely Planet Guide* series. As Guerba put it, this means more than 'scratch[ing] the surface', it requires getting participants 'personally involved in the country, its people and its wildlife' (1992). In addition, as noted earlier, longer periods of travel overseas have in themselves evoked the spirit of intellectualism and are increasingly referred to as sabbaticals among the new middle classes. But intellectualising is insufficient in itself, and has necessitated the use of other forms of classification and legitimation.

Travel professionals

In the context of socio-economic and cultural change discussed earlier (Chapter 2) coupled with the associated growth of the new middle classes, there have been two notable processes at work. First, professions have become relatively more open following the educational 'revolution' of the 1960s, and the aura (medicine, law, the City and so on) with which they had been traditionally conceived has, at least in part, been dismantled. Second, however, with still relatively limited access to the established professions, the new middle classes have been busy legitimising new licences and certificates (Featherstone, 1991; Lash and Urry, 1987): a classic process of establishing new professions (with professional education and qualification periods, codes of ethics and accrediting bodies or institutes), or professionalisation, as Bourdieu (1984) notes. This dual process may indicate that the infallibility of professional distinction is beginning to falter and surprisingly little analytical consideration appears to have been focused on a logical extension (or migration) of the struggle for distinction through professionalisation, into consumption practices.

Professionalism is a key characteristic assigned to company tour leaders, who in many cases are introduced personally: 'dedicated professionalism' as one brochure says (Ambercrombie and Kent, 1992), indicating that a tour manager not only has a 'deep love of and insight into the areas visited', but also has 'academic credentials and expertise in a specialised field'. This citation of 'experience' and academic qualifications is commonplace. The following pages cite and quote many tour company brochures dating from 1992 to 1996. In many cases the quoted passages are repeated from year to year.

In tour companies specialising in overland trucking and travelling for ego-tourists, this also means that tour leaders are 'experienced international travellers in their own right' (Hann Overland, 1992). At Journey Latin America ('Small Group Escorted Trips 1991–2') even the reservation staff are 'South American experts' and have 'travelled extensively in Latin America'; the tour leaders are all graduates (some in Latin American studies), a portfolio similar to Trailfinders. Tour leaders in these companies also embody, and are ambassadors of, new middle-class lifestyles, as the tour leader profiles in Box 5.4 testify.

In addition, the spirit of professional dedication is also widely cited, where work in this part of the tourism industry is more akin to a vocation. Encounter Overland refer to their trips as projects and talk of their leaders in glowing terms: 'ordinary men and women – often previous trip members – who have elected to put promising careers on hold and devote half a dozen years or more to what they like doing best' (1995: 10). And, of course, with greater flexibility in the service sector a higher proportion of new middle-class employees are able to take longer periods off between jobs or contracts, or are able to negotiate relatively long periods of absence. All in all a more dedicated, professional and avaricious tourist class has emerged.

The point anticipated earlier is that 'travelling' has emerged as an important informal qualification, with the number and range of stamps in a passport acting, so to speak, as a professional certificate; a record of achievement and experience. Not only is travel a professional prerequisite for employment in parts of the tourism industry, but it is an important attribute in many new professions, such as overseas development work.

There are also indications that professionals working in other disciplines have begun to diversify into travel. Many of the occasional tour leaders, for instance, are professionals in other fields. For example, the two directors of Papyrus Tours, a company established in 1984 with an aim to provide tours to East Africa, who were supportive of conservation efforts are also a senior officer in a large local authority and a self-employed professional landscape architect, respectively. Of course, we should also note that Third World tourism has created a huge range of job opportunities for the new middle classes in the First World. It is not just operators in the industry but consultants, journalists, tourism commentators, academics and charities focusing on tourism issues – take the example of The Ecotourism Society (TES) given in Box 5.5.

Box 5.5   The Ecotourism Society (TES)

TES has a professional membership of 745 . . . the people in this directory are inspiring. There is breadth of discipline, profession and of geographic distribution. These listings are solid evidence of ecotourism's international appeal and its multidisciplinary tactics. The quality of the individuals and institutions who are members speaks volumes about the organisation, and why it maintains intellectual leadership in the field . . .

Dr David Western, . . . produce[d] a one sentence definition of ecotourism. Ecotourism is, according to the definition that emerged . . . *responsible travel to natural areas that conserves the environment and sustains the well-being of local people* . . .

It was decided that The Ecotourism Society should be an organisation for professionals. It was clear in 1991 that a broad array of professionals from a variety of disciplines was needed to make ecotourism a genuine tool for conservation and sustainable development . . .

The members listed here are the pioneers – they are forging a path that many others will learn from . . .

(Epler Wood, 1994)

It is these professionals who are the opinion formers, the teachers, the advisers, even the ones who take decisions. In Tensie Whelan's book, the 'second section provides more technical information on how to do ecotourism "right"' (1991: 4). But surely these superficially 'benevolent' organisations are not somehow value free? It is clear who these professionals are: in TES's case, middle-class Americans, many of whom represent other powerful NGO interests (WWF, Conservation International, IUCN and so on – see listing of boards of directors and advisers). We must ask what vision of the world they are pursuing and how such visions are imposed from the First World on the Third World, a question considered in the context of the activities of these organisations in Chapter 6.

The commencement of professionalisation processes in consumption are also detectable. As noted earlier, consumption has become more skilled. First, there are clear signals that the distinctions between occupational professionalism, and consumption and leisure, are beginning to blur. Illustrative of this is the growth of outdoor management and 'team' building exercises, especially popular among the new middle classes. High Places, for example, which offer mountain holidays in a number of Third World countries, also offer 'another type of HOLIDAY! . . . training programmes for people at work in industry, commerce and the public sector'. This is an operation appropriately called 'HIGH PROFILE', an experience for people like 'YOU, the sort of people who come on our holidays', described elsewhere as 'intelligent and discerning people who wish to retain a taste of independent travel'.

Second, and more significant, tourists themselves are attempting to professionalise travel, a process encouraged, it would appear, by tour operators and environmental organisations. While the emergence of a formal travel qualification, such as the need to support an application for a Himalayan Kingdom expedition 'with your climbing CV', remains the exception rather than the rule, for the time being anyway, the number of tourist codes addressing the ethics and conduct of travel have exploded (see Chapter 7). A tourist 'Code of Conduct' established by the Ecumenical Coalition on Third World Tourism (ECTWT), has been reproduced in many places, especially by the network of organisations concerned with the effects of tourism, but also, for example, among travel agency associations and tour operators who have formulated their own versions. Area-specific codes with emphasis on ecological and cultural issues such as the 'Himalayan Tourist Code' formulated by Tourism Concern (in the UK) have also appeared. Organisations such as these have attracted increasing support from members of the new middle classes and from new independent tour operators like High Places, which reproduces the Himalayan Code in full and which claims to strive to 'adhere to the ethics' of Tourism Concern. It is the emergence of an ethical, if not professional, approach to tourism reminiscent of Kutay's observed 'Peace Corps-type travellers looking for a more meaningful vacation' (1989: 35).

These codes of ethics form the backbone of a hegemony of travel (or ecotravel as it is now known in North America), which is advanced by the new middle classes, the small independent tour operators and, the vanguard of this hegemony, tourism organisations. Together they begin to represent a new tourism social movement with organisations such as Tourism Concern, Green Flag International (a 'conservation company'), the Campaign for Environmentally Responsible Tourism (CERT) and the US-based Ecotourism Society emerging as symbolic 'institutes of travel'. These institutes now provide ethical yardsticks against which the activities of operators and tourists can be measured and classified. With an overwhelming concern for environmental ethics (such as the

US National Audubon Society's 'Travel Ethic for Environmentally Responsible Travel'), it is in the ecological heartlands of the Third World that these hegemonic struggles are most readily detected and played out.

The discourse of 'new wave tourism'

'Out goes the "'ere we go, 'ere we go" Spanish Costa-style, in comes a more thoughtful middle class approach' (Barrett, 1990: 2). This was how Frank Barrett of the *Independent* newspaper, the UK's champion of new middle-class values, described the shift to 'new wave tourism' (Wheeller, 1992).[4] There can be little disputing that the attributes and qualities of the supposedly more alternative, individualised and sensitive forms of travel are generally unspoken but appear to be deep within the new middle-class psyche. As The Travel Alternative, a specialist company offering craft and textile holidays, comments: 'Dear Friends, On a recent tour a passenger said "You can either be a tourist or a traveller." We would like to think that our tours allow everyone the opportunity to be travellers' (*Newsletter* 7, 1994).[5]

Of course, it is not only through the classificatory distinction between tourist and traveller that differentiation is pursued through discourse. But it is highly illustrative of the battle being waged to put into words exactly what kind of holidaymaker we are and what we stand for; and, conversely, state exactly what we are not, for, as Barrett argues, 'tourist' is the 'worst kind of insult' (Barrett, 1990: 3).

Some observers have momentarily reflected on the highbrow nature of traveller and the intellectual snobbishness within which the whole debate is framed and have attempted to rescue 'tourist' from middle-class derision. Tourist has become prefixed by benevolent adjectives such as the 'The Good Tourist' (Wood and House, 1991) or 'The Reluctant Tourist', title of a series run by the *Independent* which culminated in *The Reluctant Tourist's Handbook*. There are also references to ethical tourists, alternative tourists, and so on. Arguably, though, this is further confirmation that contemporary tourism is charged with the classificatory struggles in and between class fractions (Walter, 1982). So, as Culler observes, 'Once one recognises that wanting to be less touristy than other tourists is part of being a tourist, one can recognise the superficiality of most discussions of tourism, especially those that stress the superficiality of tourists' (1988: 158).

By contrast, the term 'traveller' assumes that it is no longer a process of tourism with which the individual is engaged, but a considerably more de-differentiated, esoteric and individualised form of activity; travel is to tourism, as individual is to class. It is a strategy for seeking differentiation or a 'paper-chase aimed at ensuring constant distinctive gaps' (Bourdieu, 1984: 481). Most importantly, the discourse adopted by these class fractions is a further reminder that consideration of the way in which power is expressed and imposed must lie at the heart of the analysis of tourism (Munt, 1994a; Wheeller, 1993a; Urry, 1990a).

Many small tour companies and individual travellers have attempted to maximise the distinction. The practices of travellers are perhaps best conceived as part of the 'cult of individualism' (Pels and Crebas, 1991), though it is deeply ironic that they are largely indistinguishable from each other by virtue of their discourse, dress codes and the informal 'packages' they follow through travel guides. Whole regions have become travel circuits (in popular travel discourse 'doing' South-East Asia, Central America, and so on). Errington and Gewertz (1989) reproduce excerpts from a Papua New Guinea guesthouse visitor's book to illustrate the fundamental traveller distinction, with two US travellers advising: 'Explain difference between tourist and traveller. . . . Be a traveller, not a tourist. It makes a big difference' (1989: 40). Travellers, therefore, attempt to create their own aura, while attempting to prevent the encroachment of 'tourists' in their quest for authentication:

> Momentarily alone in one of the wildest places on earth, you feel exhilarated, exhausted and scared. Unfortunately the feeling is unlikely to last more than a few moments. If you are lucky, all that will happen is another traveller will appear on the ridge to exchange pleasantries and wonderment. If you are unlucky . . . your reverie will be interrupted by half a dozen Londoners swearing and shouting.
>
> (Edwards, 1992: 19)

As Box 5.6 suggests, mass tourism and tourists have also become the target of independent tour operators. Intellectualism oozes through, with package tourists 'invading' backpackers' discoveries and the celebration of areas where the invasion has been forestalled: at the height of aestheticisation, Cambodia is lucky for its 'fascinating culture bruised by war' and even luckier with a culture 'unscathed by tourism' (Calder, 1994). Companies refer to the 'anonymity and inflexibility' (J. & C. Voyageurs), of 'run of the mill' (Ambercrombie and Kent), 'conventional' (Explore) and 'impersonalised' (Africa Exclusive) mass packaged tours, and celebrate their demise (Magic of the Orient). Ultimately, it is an appeal to the idea that travel is 'individual enough to be a sustainable alternative to the juggernaut of mass tourism' (Explore).

In its place, and by contrast, we have the emergence of Zimbabwe, where 'nothing is packaged' (Africa Exclusive), and the 'Unpackaged Caribbean' (Kestours). For luxury tour operators and their new bourgeois clientele (eco-tourists), exclusiveness is sufficient enough a distinction, with Cox and Kings taking 'a select few to Africa', and operators speaking of 'a small and select clientele' (Ambercrombie and Kent) and 'limited edition holidays' (EcoSafari).

For tour operators catering more for the young and adventurous ego-tourists where economic capital is clearly insufficient to confer taste, it is the invidious distinction of participant travellers from tourists that has become critical. It advances the distinctions beyond a mere reference (or 'charge') to packaging and instead focuses on the qualities and practices of travel and contrasts these to tourism. This is particularly notable of overlanding operators such as Dragoman

---

Box 5.6   On tourism and travelling

Tourism	Travelling
Invasion, rape, poisoning, tidal wave, pollution, swarms, juggernaut	Discovery, exploration, understanding, peaceful contact
Benidorm, Torremolinos, Kos	Tikal, Phnom Penh, Zanzibar
Sun, Sand, Sea and Sex	Sensible, Sensitive, Sophisticated, Sustainable and Superficial (Wheeller, 1993a: 122); or Intelligent, Inquisitive, Independent and Idealistic
Unadventurous, narrow-minded, undiscerning, unenergetic, inexperienced, unimaginative, unintelligent, boring, unreal, false	Adventurous, broad-minded, intelligent, discerning, energetic, experienced, keen, imaginative, independent, intrepid, real, true

---

specialising in trucking for young travellers, a world 'shunned by the masses who prefer the resorts and beaches' or, as Explore put it, travel for 'people who want more out of their holiday than buckets of cheap wine and a suntan', a stereotype also used by Exodus Adventure. In similar vein Journey Latin America contrast their ('Small Group Escorted Tours') participants to those who prefer 'two weeks in Torremolinos' and where preference is for a 'cold beer after a three hour walk in the jungle, not nightly booze-ups'. As Tours to Remember warn, if you are looking for 'two weeks in the eastern equivalent of Benidorm', look elsewhere.

Logically, tour companies also promise 'tourist-free' zones (Naturetrek), 'few lager and litter louts' (Cara Spencer) and avoidance of 'Tourist haunts' (High Places). In a more 'positive' sense, Encounter Overland celebrates the traveller-cum-wayfarer: 'today's custodian of the ancient relationship between traveller and native which throughout the world has been the historic basis of peaceful contact'.

So what of the qualities and practices that are claimed to constitute the distinction and embody the traveller? The simple listing of adjectives applied to travellers in travel brochures (Box 5.6) begins to map the key coordinates. By implication, if not explicitly as noted above, the counter-distinctions constitute tourists. If, for any reason, the citation of these *qualities* proves insufficient, however, there are other prerequisites which help define travellers. High Places tours of India, for example, demand 'patience, stamina, humour and adaptability', and Journey Latin America cite 'essential qualifications for all trips' as 'emotional balance, maturity, a spirit of adventure, and a desire for good companionship'.

There are two other notable features of travelling and travellers. First, with an overwhelming emphasis on a mixture of often physical activities, especially among ego-tourists, bodily fitness is important. High Places offers the symbolically new middle-class holiday: mountain biking. Second, and most important, there is need for the fellowship of other like-minded travellers, whether it be small intimate group tours (with generally between six and twenty participants) or individual backpackers meeting at predestined travel guide-recommended hotels, restaurants or sites. In both cases, albeit in rather different ways, the emphasis is on participation, on 'action and involvement' (High Places) and, ultimately, on accomplishment (a recurrent theme in tour brochures).

Not only is there an intense struggle in classifying and legitimising this notional traveller/tourist distinction, but ego-tourists vie among themselves over what actually constitutes a traveller in the first place. As was seen in Chapter 3, Borzello (1991) considers that 'Truck travellers are not travellers but a very peculiar sort of tourist.' And it is an intense classificatory struggle that has pronounced spatial reflections, a point to which we turn below.

Tales of the unexpected

Both explorers and dragomen journeyed through far-flung lands in search of long-forgotten civilisations and empires. They explored hostile deserts to find nomadic tribes and ancient cities. They went in search of legendary mountains deep in the heart of jungles and brought back stories of fabulous wildlife.

(Dragoman, 1995: 1)

These fabulous stories or 'travellers' tales' (Hall, 1992b; Massey, 1995) have a deep history. There are noted discussions and a resurgence of interest in ethnographic travelogues and travel writings (see, for example, Fussell, 1980; Blunt, 1994; Pratt, 1992). But debates around the contemporary nature of tales, or myths as referred to by some (Borzello, 1994; Calder, 1994b; Selwyn, 1994, 1995), have remained sketchy. This is of particular significance for, as Massey (1995) argues, tales are a culmination of the ongoing process of travel which fill out our geographical imaginations. Travellers' tales are a modern (or perhaps postmodern) continuation of the stories of colonial encounters. They are a useful way of demonstrating how discourse is used to represent or recreate a reality (the reality that First World tourists experience) and impose meaning, in much the same way as the fantastic tales retold by colonialists in the nineteenth century. (In this way discourse can be conceived as part of that everyday hegemony.)

Latter-day travellers' tales are a way of sustaining the aura and mystique of the Third World (reminiscent of Said's *Orientalism*) and in turn sustaining the value of travel as a commodity. Their construction is often complex and draws upon a range of images and representations, including film (as suggested in

'Weekending in El Salvador', Box 3.12, p. 81), travel writing and colonial explorations (Adams, 1984, 1991).

Tales also have an important part to play in the reproduction of social class. It has already been suggested that tales have an important symbolic currency as a form of cultural capital; the example used suggested that travelling on a 'shoestring' simultaneously denies the existence and need for substantial lumps of economic capital and confers status, uniqueness, worldliness and resourceful-ness. As Calder records of travelling in rebel-held Nicaraguan territory, by the time you reached the 'safety' of neighbouring Costa Rica 'all that backpack-ing bravado returns and you have a couple of cracking stories with which to trump fellow travellers' tales' (1994a: 35). Travel and the construction of tales, therefore, represent an informal credentialism.

The construction of tales is also intimately bound up with the quest for authenticity (see Box 3.6, p. 58). It is a way of creating an aura in an effort to remove this experience from the tourist's world. These are feelings that cannot be 'snapped'. Take this example from High Places:

> The trip begins with a short drive from Nairobi down into the Great Rift Valley, to our camp set among the trees, overlooking grassy plains where zebra and giraffe graze. The only sound is the breeze and a kettle boiling on the stove. Time to relax as the trip unfolds. . . . Walking through the Loita Hills, in Maasai territory, following little-used trails and using local warriors as guides; . . . watching a majestic herd of elephants cross the Mara plains, or a cheetah stalk its prey; sitting round the fire, swopping stories with Maasai warriors; relaxing, as the sunset turns the empty plains red.
>
> (High Places brochure)

Of all modes of travel, it is trekking in areas of solitude that most gives rise to a romantic gaze and the construction of tales almost spiritual in their content. The phenomenal growth of trekking in South-East Asia, Latin America and Africa (Brockelman and Dearden, 1990; see Chapter 8) underlines its impor-tance in new middle-class travel. It is the authenticity of this form of alternative tourism that lies at the heart of trekking (Cohen, 1989), and the accumulation of tales relies upon authenticity to verify its uniqueness and currency: as High Places promote, 'Our trips are unpretentious, authentic and adventurous', and Trailfinders contrast 'staged tourist "shows"' with the experience of 'untouched' and 'traditional' villages in the highlands of Thailand.

In this case authenticity is expressed in a somewhat different manner, conveyed through an emotional response that is either unique or intensely difficult to achieve. It is an exclusionary experience, that even if the struggle to dominate and control a spatially discrete area is lost, tourists are still unable to share the aura. Frank McCready (Sherpa Expeditions Managing Director) comments on the realisation of dreams: 'When the ordinary person reaches a place like Annapurna Base Camp it changes their whole view of life. . . . Some of the places we take people are so fantastic I've seen grown men entirely overcome with emotion.' It is this experience of exhilaration and exhaustion (Edwards, 1992) that culminates in the eulogising of trekking. Himalayan Kingdoms comments: 'The blend of grandeur and timelessness defeats all superlatives and such a place should be visited at least once every lifetime' (1).

To return briefly to the theme of aestheticisation: in order to legitimise and authenticate the travel experience, tales must be aestheticised in two important respects. In the first place, there is the dual requirement to make trips both purposeful and distinguishable from those of the average tourist or, indeed, traveller. Travel is purposeful, tourism is not. For example, one travel writer notes of a visit to Nepal, 'We had tea and increasingly battered chocolate to deliver to a friend of a friend working in a health clinic there' (Norris, 1994). Sabbaticals, 'research' work, the presence of friends, colleagues and relatives or the emergence of work-brigades and a range of work- and activity-centred holidays, are all used by the new middle classes to signal that *this is more than just a holiday* (see Boxes 7.13 and 7.14, pp. 231, 232). It must be sufficient in distancing itself from supposedly inactive or inert forms of tourism. Exodus ('Overland Newsletter', summer 1994, 'What the customers say' section) notes: 'I think that was the best trip I've ever done, partly because of, rather than despite the difficulties and challenges we faced.'

Second, as suggested in Chapter 3, there has always been a nagging inadequacy around the assertions that 'one cannot sell poverty, but one can sell paradise' (Crick, 1989; Rojek, 1993). The quest for authenticity, like travel itself, has had to seek out new experiences. No longer can it rest upon the cultural preservation and discovery of highland tribes and deep forest natives. It has migrated to the quest for danger and risk and the emergence of contemporary travelogues, diaries: '"I'm keeping a journal; maybe I can make it into a book," says one traveller to me as we sit in a bar in a small village' (Gordon-Walker, 1993: 19). Tales legitimise travel by sensationalising it, making it appear daring and risky, while at the same time peripheralising danger (Jardine, 1994). Of Vang Tau (Vietnam) the *STA Travel Guide* notes: 'a two hour boat ride from Ho Chi Minh City and a million miles away from conflict' (5). Similarly, Encounter Overland's philosophy remains: 'With true adventure there is always an element of chance and of risk. This fact is not regrettable. It is often the fact upon which the best travel experiences are based.' Suffering can often enrich the travel experience, as Box 5.7 makes clear.

Most invidious, however, is to utilise tales to aestheticise risk and boost the cultural capital accumulated in travel. As suggested earlier (in Chapter 3), while the hazards of travelling in particular regions may act as a warning, it simultaneously signals the degree of legitimacy or coolness to be attributed to particular destinations. As Gordon-Walker notes of Peru, 'travellers who are disappointed not to have captured a Sendero Luminoso flag raise their spirits . . . and swap stories about friends of friends who have been robbed' (1993: 19). Similar experiences were recounted in 'Where thieves prosper' and 'Weekending in El Salvador'

Box 5.7 Tales of the unexpected

A glass of fiery rice liquor was thrust into my hands, and I sat down with them and joined in their infectious laughter, without a single word of common language between us . . .

Seven years on I'm none the wiser about the purpose or origins of that strange harvest festival in northern Thailand . . .

As it was I could find no mention of the festival or town in anything I read, which probably accounted for the fact that I was the only westerner on the scene. But it wasn't this smug one-upmanship that made it so special, but rather the very act of discovery. For one impetuous moment I dared to stray from the charted territory . . . and it was bliss . . .

To me, surprise is the primary reason to travel. All those other time-worn clichés about broadening the mind and stimulating an appreciation of the world are merely justifications, not motivations. . . . Think about those special moments from your last few trips, the ones that you bore your friends rigid with at parties. Are they accurate reports of the splendour of the temples or the excellent cuisine? I hope not. No, it's those unexpected events when your plans are interrupted by that out-of-the-blue experience . . .

To my mind, if there's one thing worse than a holiday that fails to meet your expectations it's one that *does*. . . . And let's be brutally honest here, if in the second week of your big adventure you haven't had at least one near-death experience you're going to want your money back. Right? . . .

. . . do we really want a boil-in-the-bag cultural experience? No we bloody well don't – we want an unpolished, unbleached wholegrain adventure with bits of stone and fingernail in the bottom!!!

(Morrison, 1996)

(Boxes 3.11 and 3.12, pp. 80, 81). Travelling in potentially dangerous regions, being hoisted from a bus and frisked at midnight or braving certain urban areas are experiences to be enjoyed and admired by other travellers. Risk titillates, even eroticises, adventures in the Third World.[6] And it is this genre of tale-telling that has come to the fore in discussion among fellow ego-tourists, in travel guides and reviews and within television coverage. In short, it is now widely adopted discourse focused toward social differentiation and the insurance, or hope, that certain places will stay unexplored by other groups of tourists.

THE SCRAMBLE FOR THIRD WORLD DESTINATIONS

We have argued that a pseudo-ethical and moral infrastructure underpins the growth of Third World tourism (Lea, 1993), with new middle-class tourists (or travellers) contrasting their morally justifiable means of travel with the morally reprehensible practices of tourists. There is a temptation to conclude from this, as Gordon-Walker does, that you 'cannot help suspecting that the campaign for

147

"sustainable tourism" is little more than a rationalised desire to keep the Third World a cheap place to visit' (1993: 19). But such a conclusion would result in underselling the importance of how First World tourists impose their desires on Third World destinations. If the new middle classes are forced to wage this hegemonic classificatory struggle founded on taste, then it might be reasonable to expect that such struggles will also be reflected spatially. So at the very heart of this campaign lies, not just the preservation of a nice cheap place to visit, but the playing out of the battle of social class differentiation.

The professionalisation and intellectualisation of travel, together with its associated discourse, have been insufficient in themselves to ensure social differentiation and, more importantly, the creation of physical distance between mass tourists and travellers. The new middle classes must adopt strategies of exclusion, to seek and protect the new travel commodities. Extracts from Martha Gellhorn's piece, 'Too good for tourists' (Box 5.8) illustrate the process by which this is achieved. It is an attempt to persuade and to impose the idea that this First World view is correct.

As some authors contend, the near-absence of other tourists to which Gellhorn alludes is most efficiently achieved by the consumption of 'positional' goods (Hirsch, 1976), with dominant class fractions attempting to impose scarcity (Featherstone, 1987; Leiss, 1983). At the simplest level this involves claiming

---

Box 5.8   Too good for tourists

The airport terminal at Belize City is very encouraging. It is an overgrown shack. . . . After each passport has been examined as if passports were new-fangled inventions, you enter a square, wooden room where the baggage has been dumped in the middle of the floor. A free-for-all follows. . . . My heart rose like a lark. This is how it used to be before the Caribbean was homogenised by tourism. Tourism, even . . . minor, modest tourism, corrupts.

I cannot exaggerate the pure physical pleasure of underpopulation and empty space. . . . Half a century ago I thought this ghastliness of packed humanity was the peculiar doom of China: now much of the world I knew and loved is ruined because there are too many of us and we move everywhere. For my taste, Belize is ideal.

A young American in the shortest possible frayed shorts was questioning the American Rum Point owners about the effects of tourism on the population and the environment, his thesis for the University of California at Berkeley. We agreed that tourism is a destroyer and Belize is far too good for it. None of the inn owners I had met, English, French, American, Belizean, Irish, want to expand, though, between them, they could take care of less than a hundred perceptive guests. They love the country as it is. How much better if oil is found in viable quantities in the north, and tourists go somewhere else. It is astonishing at my age to stumble on a new country. I feel astonished.

(Gellhorn, 1990)

that certain areas 'receive very few tourists' (Detours, 1992), are 'tourist free' (such as national parks visited by Naturetrek), or that even whole countries are for the 'traveller rather than the tourist', as Meon describes Ecuador (1992).

But this amounts to little more than marketing, and alternative strategies and claims must be pursued, such as an active search for 'off-the-beaten-track' (or, as Asian Affairs claims, 'even further off-the-beaten-track') and 'lesser visited areas', areas 'rarely visited' (Himalayan Kingdom) or just 'secrets' (Okavango Tours and Safaris). Exclusiveness, uniqueness, romanticism and relative solitude are central to the philosophy of these tour operators.

This philosophy both embodies the contradictions in contemporary travel to the Third World and reveals the protracted and increasing difficulties which companies and travellers have in spatially defining separated practices from other like-minded travellers. As tourism spreads to more and more destinations in the Third World, the distinction of a mass packaged tourist from the individual character of travel is spurious. Ego-tourists crowd into cheap guest-houses, basic bus stations and out-of-the-way beach resorts, while ecotourists crowd into game reserves and national parks. Travel is little more than a figment of wishful middle-class thinking. It is worth considering the views of a guide from Bali (see Box 5.9). The example of the island of Bali is taken further in Box 5.10, which alludes to this blurring of tourist distinctions. Figure 5.2 (p. 152) provides some background to this example, which is taken further in Chapter 9 with an examination of the effects of policy on the island.

It is a context, therefore, within which new tourism must not only do battle with mass tourism and tourists, but contend within itself both spatially and qualitatively for the most virtuous practice. This is particularly notable in safaris.

Box 5.9 *A view from the Third World*

Interview with Budi, 28, an Indonesian

Q   Do you think there is a difference between a 'tourist' and a 'traveller'?

A   The traveller thinks they know everything about the local people and the country. But it's usually because some other traveller told them before. But they do whatever they like – some travellers are good, but 90 per cent are not, they can be very impolite. With the tourist, everything is organised, so they don't destroy as much. The traveller wants to see something new, and wants it to be cheap and then tells others about it. I prefer tourists . . . , they go to specific places, it is more professional. But the traveller is uncontrolled – they won't go to the places already prepared for them; they want to go to other places and then they spoil it – and don't spend any money! Travellers always talk the same and say: 'Don't go to Kuta because it's spoiled.' Then they go to a new place.

(Wheat, 1994: 9)

Box 5.10 Paradise lost: Bali and the new tourist

Tourism is no recent phenomenon on the Indonesian island of Bali. Colonised by the Dutch, the Balinese held on to a unique blend of Hindu, Buddhist and animist religions that imbued their cultural and social life. Tales of this exotic society living in tune with art and nature attracted foreigners as long as a century ago, and since independence, Bali has been the jewel in the crown of government tourism policy, hosting ever larger numbers of visitors (see Figure 5.2).

Contemporary tourists to Bali display a wide mixture of motivations. The island accommodates seriously wealthy hedonists and conventional package tourists (many on long-haul stopovers or multi-centre holidays), alongside large numbers of young Australians who arrive on charter flights and regard the island as a cheap and cheerful holiday spot and a surfing mecca. There is a huge pool of homestays and guesthouses, known locally as *losmens*, catering for this budget market.

Bali's enduring image as a cultural paradise still draws independent travellers who also patronise the *losmens*, and it features frequently in the brochures of companies appealing to the new tourist as a land where it is possible to get off the beaten track while never straying far from creature comforts: 'the industry has been controlled so that hotels are primarily located in the southern peninsula, leaving the rest of the island essentially unchanged and uniquely Balinese' (Mortlock, 1988).

The village of Kuta, close to Bali's international airport, has been a favourite haunt of budget tourists since the 1970s, when its residents woke up to the lucrative possibilities of turning their homes into *losmens* and offering other services from bike hire and restaurants to massages on the huge crescent-shaped beach, which soon became a haven for novice surfers. *Losmens* began appearing in nearby Sanur, and on the north and west coasts too – for example at Lovina Beach patronised by the backpacking fraternity as a rest-ful retreat from an overdose of culture or socialising in southern Bali: 'It's a good place to meet other travellers, and there's quite an active social scene' (Wheeler and Lyon, 1994).

The inland village of Ubud, renowned for its artists and lush beauty, has also grown rapidly. From a scattering of *losmens*, restaurants and artists' studios attracting backpackers and longstay visitors, Ubud has mushroomed into the alternative inland base for anyone not locked into a standard beach hotel package deal. Nowadays, new tourists wishing to escape the mass market experience still favour Ubud, but may well avoid the commercialised southern coastal resorts, opting instead for accommodation in less developed spots.

However, if the entrepreneurs of outposts such as Candidasa and Lovina Beach are counting on increased custom from this source, they may be disappointed. Like the first visitors to the island, new tourists are specifically interested in Bali's unique harmony of culture and landscape and they still tend

---

Box 5.10
continued

to visit the prime sites on every tourist's itinerary. As many commentators (see for example, Picard (1991) and Noronha (1979)) have remarked, Bali's cultural life has so far been remarkably resilient. But as tourist sites become increasingly crowded and as its distinctive art forms become increasingly commercialised, new tourists may well decide not to bother with Bali at all, taking up instead the Indonesian Tourist Board's invitation to explore the rest of the republic's giant archipelago and leaving this particular 'paradise' to the surfers and conventional package tourists.

(specially prepared for this book by Alison Stancliffe)

For example, Papyrus Tours ensure that they spare you 'any involvement in safari bus "rat runs"', referred to by Africa Exclusive as the 'herds of landrovers that ... plague some other parts of Africa' (but which do not afflict, of course, this tour operator's safaris in Zimbabwe). Even vehicles themselves are a point of differentiation among safari operators, and overland trucking companies (as shown in Box 3.10, pp. 75–6), with Into Africa claiming that 'landrovers do less damage' than larger trucks and are more sensitive to both wildlife and local people.

Similarly, trekkers face increasing difficulties in spatially legitimating their experiences, as Edwards' (1992) lamentations have already indicated (Cohen, 1989). With the sharp increase of trekking in northern Thailand, for example, it is more difficult to find a trek which visits 'untouched' or 'traditional' villages, we are warned by Trailfinder, a company with which it is possible to visit 'more remote villages less exposed to western influences', where 'groups are welcomed as a refreshing diversion to normal village life'. At possibly its most advanced, spatial legitimation involves promises of carefully researched itineraries which avoid the 'overcrowded trek routes, often exploring untravelled or Restricted Areas, or visiting cultural festivals unknown to most westerners' (Himalayan Travel, 1992). Ultimately, legitimation for this company is sought through the pages of the *Geographical Magazine* with the Co-director writing of their trek to Mustang:

An understanding and appreciation of both cultural values and ecological balance is essential when visiting such remote and unspoilt regions. This expedition was the first in a series which it is hoped will help turn the tide of 'tourist pollution' to remote areas of the world.

(Brooks, 1990)

This form of tourism therefore means the need to ensure the absence of other tourists. Some areas are considered especially attractive by virtue of the exclusionary nature of state policies which limit the number of tourists. Box 5.11 shows that Bhutan, for example, allows access to only 5,000 foreign tourists a year. Trekking companies thus enter an intense competition for the most

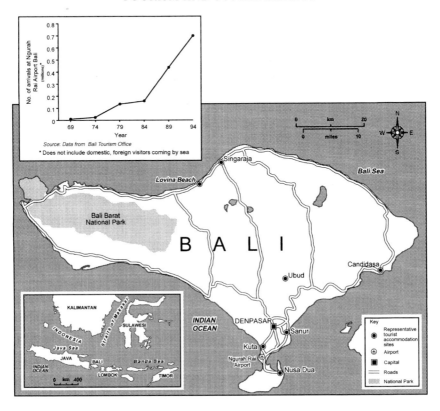

Figure 5.2 The island of Bali and its tourist arrivals

authentic locations and that means destinations with the least tourists, by implication, the most difficult destinations to enter. Take the statements from the UK-based Himalayan Kingdoms in Box 5.12, for example.

But it is with environmental conservation that tour companies and tourists have discovered perhaps the most effective method of exclusion, or 'inclusiveness'. Many tour companies are now supportive of conservation measures and cite membership of both global (World Wildlife Fund, CARE, Friends of the Earth, and so on) and the more localised (such as Elefriends and Mountain Gorilla Project) environmental organisations. Such projects are essential to both ecotourists and ego-tourists in restricting the number of visitors to such areas and in so doing securing some form of exclusiveness, however circumscribed in practice. Beyond this, however, some companies have actively pursued conservation policy. Examples are Papyrus Tours, where the Director is a founding trustee of the Kenya Wildlife Trust (a UK-based trust seeking to support conservation initiatives in Kenya), or Worldwide Journeys and Expeditions (1992), which claims management involvement in the only privately managed national park in Africa, Kasanka (Zambia). Similarly, the certificated purchase of an acre of rain-

---

Box 5.11   *Class versus mass tourism*

The political scientist, Linda Richter, identifies a key issue in tourism planning as *class versus mass tourism* (Richter and Richter, 1985). A number of countries, of which Bhutan is undoubtedly the most unequivocal example, have chosen to concentrate their tourism on high spending, 'classy' tourists.

Bhutan is a small, land-locked, Himalayan mountain kingdom to the north of India. The country allowed entry to its first tourists in 1974, but only at the rate of 1,000 per year. The rate has increased in recent years so that up to 5,000 may now enter per year, although in 1988 the government closed some monasteries to foreigners because of 'growing materialism' among monks who accepted trinkets such as sweets and pencils together with money from visitors (Smith, 1989: 14–15).

A glance through several of the trekking brochures shows that Bhutan is an expensive destination as a result of the weekly charge levied on all foreign tourists. This restrictive entry has been utilised by some 'alternative' travel companies to demonstrate the exclusivity of their holiday destinations (Naturetrek; Himalayan Kingdoms; Coromandel).

In 1986, the study made by the World Tourism Organisation (WTO) praised the Bhutanese system. WTO planner, Edward Inskeep, agreed that 'selection' can be achieved: 'the limited tourism approach can be applied, through selective marketing techniques, to attract tourists from any socio-economic groups who respect and do not abuse the local environment and culture' (Inskeep, 1987).

---

forest is also encouraged by multinational environmental organisations such as Programme for Belize and the Nature Conservancy; these areas are often reserved for the consumption of discerning First World ecotourists.

Sustainability has proved the perfect ally of the new middle classes, with the social construction and application of ecological concepts such as carrying capacity proving the ultimate justification for natural exclusiveness. As already argued, notions of sustainability are hegemonic in their own right and are readily translated spatially. The inflated ranks of academia have been quick to seize upon this, using 'research holidays' to impose their own ethnocentric analysis and recommendations. Of Thailand, Brockelman and Dearden (the latter a former President of the Ecotourism Society) argue that the number of trekkers be kept relatively low, concluding that 'nature trekking should . . . be promoted for special tours and not among general tourists' (1990: 146). The social exclusiveness of their proposals, the class-laden nature of this debate, and the cultural capital that is to be accumulated and invested, are plain to see:

> It should be designed for serious naturalists or Nature enthusiasts and others wishing to experience genuine local cultures. Most such persons

are reasonably affluent and educated and read a variety of Nature, conservation, and botanical or zoological society publications. . . . Such clientele is likely to get the most out of the experience, and probably also leave behind the best impressions.

(1990: 146)

Box 5.12   Keeping tourists at bay

As yet undiscovered by mass tourism, our 'Hindu Kush Trail' provides an unspoilt route to the base of Tirich Mir. (4)

Special permission is needed to enter Chitral . . . so already there will be a noticeable lack of tourists. (5)

This is a trek with plenty of time to interact with local people. . . . The inhabitants have rarely seen Europeans and are friendly and hospitable. (5)

. . . [we] will not see human habitation for two weeks. (5)

In March 1992 the first trekkers were allowed access to the former Kingdom of Lo, and it is now the Government of Nepal's policy to restrict numbers. . . . Our first trek . . . came back with wildly enthusiastic descriptions. (9)

Our own expedition to Ramdung . . . was the first British group to gain entry for twelve years. (10)

[Tibet] has always been a difficult country to enter. (12)

This is little frequented by western trekkers. (13)

. . . but few follow the routes we use . . . (32)

(Himalayan Kingdoms, *Treks 1996* brochure)

CONCLUSION

In this chapter we have attempted to show how the contemporary socio-cultural changes outlined in Chapter 2 are reflected through the growth and popularity of travel in many parts of the Third World. While the discussion does not imply a determinism that social class lies at the heart of all tourism, it has sought to show that it is a central and frequently overlooked factor in attempts to explain why tourism spreads and seeks out new frontiers. In this respect, it is a further demonstration of why the boom-to-bust cycles of tourism, which are most frequently used to demonstrate the circulation of destinations, amount to descriptions rather than explanations.

The chapter has argued that a sizeable proportion of new forms of tourism in the Third World can be related directly to the burgeoning new middle classes of the First World. This group is being joined increasingly by new middle-class

154

tourists from the Third World, most particularly from South-East Asia. This growth was also related to the increased interest in otherness, with a particular concern for ecology and ethnicity, both of which are found in plentiful quantities in the Third World. In addition, it was argued that central to the swelling ranks of the new middle classes is the necessity to differentiate socially from other social classes. This is most readily achieved through the construction of lifestyles, of which holidays are undeniably an important part.

The chapter explored how both ecotourists and ego-tourists are able to accumulate what has been identified as cultural capital and the strategies – professional, intellectual and discursive – that are employed to ensure that the accumulation of such capital is sustained and enhanced. In particular, we have argued that this provides a rather different understanding of the current debate over sustainability. This debate can be recast, in part, as the drive towards sustaining the opportunity and ability to consume authentic and exciting experiences, which simultaneously necessitates the exclusion of other types of tourists. Ultimately, this represents a cultural and social reaction of the new middle classes to the crassness which they perceive in tourism and a craving for social and geographical distance from tourists.

6

SOCIO-ENVIRONMENTAL ORGANISATIONS
Where shall we save next?

Much of the discussion so far has been underlined by the uneven and unequal nature of tourism development. This is most vividly expressed at a macro scale as the inequality of First World and Third World tourism. In particular it has been suggested that questions of power which underlie the way in which tourism is owned and controlled largely from the First World for example, should not stop with the analysis of the tourism industry itself. This is especially true in an analysis of new tourism.

In this chapter we consider the involvement and activities of the socio-environmental movement and the role of socio-environmental organisations; although, as we examine below, these are very broad terms. Chapters 1–4 emphasised that a good deal of the debate concerning the emergence of new forms of tourism has centred upon questions of environmental harm attributed to traditional forms of tourism and has sought ways in which to prevent or mitigate these negative effects. Equally, however, it has been argued that such activity can also be traced to the desire to preserve environments; the areas of so-called wilderness and virgin territories where nature can be experienced by 'discerning' new tourists.

It was suggested in Chapters 1–4 that this also reflects two facets of sustainability: on the one hand, an environmental or ecologically centred meaning of sustainability as protecting and enhancing resources and biodiversity; and, on the other hand, an attempt to sustain cultural products for the benefit of (predominantly) First World new middle-class tourists, or, in other words, retaining places that are free from mass tourism and tourists.

As might be expected, the socio-environmental movement has spearheaded the advocacy and implementation of programmes and policies centred upon sustainability. This movement is, therefore, a key interest in the analysis of new tourism, and the Third World has become a major focus for environmental concerns partly as a result of its spectacular environments. There are two introductory observations that can be made, each of which embodies a paradox.

First, in general, environmental issues and environment-centred organisations have been treated as benevolent causes and have attracted widespread interest and support, even in the light of quite interventionist policies. While this support

156

reached its height during the 1980s and has subsequently waned somewhat, it nevertheless remains a significant pull for the new middle classes.

Second, tourism very much represents a double-edged sword for the socio-environmental movement, in that it is an activity which is both reviled and revered. It has become a focus for both criticism, as a result of its impacts, and promotion, as one means of achieving sustainable development. With the concerns expressed primarily in terms of the environmental and cultural impacts of tourism, the new socio-environmental movement has rounded upon some forms of tourism (that is, mass tourism) and promoted others (alternative, appropriate, sustainable, and so on). Concerns have centred on the need to protect endangered habitats, maintain biodiversity and promote minority rights. An example of the latter is Survival International, an INGO supporting and campaigning on behalf of tribal peoples. Survival and its tourism campaign, 'Danger Tourists', has already been introduced in Chapter 3. This is an interesting contradiction in that it has tended to result in the advocacy of exclusionary policies that subsequently lead to a reduction in visitor numbers and the inevitable outcome of élitism in tourism. Where new tourism projects exist they are often lauded for their vision and sensitivity, while mass tourism stands accused of crassness and environmental genocide.

This chapter attempts to introduce these debates and to reflect the complexity of the positions taken, the range of environmental concerns and how these relate to tourism. Initially it reflects back on the previous discussion to suggest the links between new tourism, the new middle classes and so-called new socio-environmental movements (of which the environmental movement is a key part). It then traces the different streams of the environment and tourism debate through a continuum of environmentalism. In doing so we are seeking to avoid a simplistic reading of environmentalism as an undifferentiated whole. Characteristic of the argument, we also seek to demonstrate how and why power is a fundamental component of tourism analysis.

NEW SOCIO-ENVIRONMENTAL MOVEMENTS

Chapter 2 sketched out the most noted social, economic and political changes in the late twentieth century and suggested ways in which the growth of new forms of tourism were related to these factors. In particular it was argued that there are apparent relationships between the growth of the new middle classes and a concern with 'otherness' which includes an interest in minority cultures, religion, ethnicity and, arguably most significant, a concern with environment and ecology. Notably, the 10–12 per cent of all tourism that is attributed to new tourism activities is similar in proportion to that of the First World populations interested and concerned with these other issues (World Tourism Organisation, 1995b: 28).

The growth of interest in *sustainable lifestyles* was singled out as the most momentous of these movements and an important component of a so-called 'new politics' which stands in contradistinction to 'old' style political parties. The emergence and growth of a 'new social movement' (or what we refer to here as socio-environmental organisations) that lies at the heart of the new politics and which has been responsible for campaigning and politicking on single issues (anti-nuclear and world peace, anti-racism, environment, and so on) has in some ways 'transformed the political scene' (Crompton, 1993: 16). The socio-environmental movement in its many guises has become one of the most enduring images of the last twenty years and has captured the public imagination in a way that has far surpassed other movements. As Eckersley comments in her study of environmental political thought:

> The environmental crisis and popular environmental concern have prompted a considerable transformation in Western politics over the last three decades. . . . Whatever the outcome of this realignment in Western politics, the intractable nature of environmental problems will ensure that environmental politics (or what I shall refer to as 'ecopolitics') is here to stay.
>
> (1992: 7)

Two of the most prominent organisations, both globally and within, for example, the UK, are Friends of the Earth and Greenpeace, and both reflect these significant changes. Figure 6.1 charts the growth in their membership

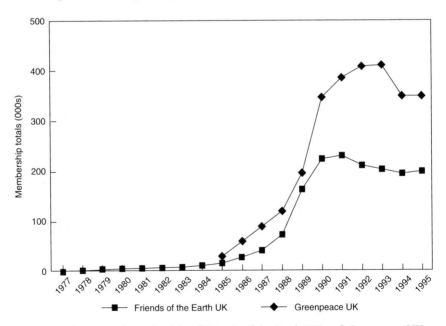

Figure 6.1 Annual membership of Friends of the Earth UK and Greenpeace UK
Source: Friends of the Earth UK and Greenpeace UK (1995 values estimated)
Note: Change of membership categories for Greenpeace UK in 1994

numbers from their early days. In both cases, the most dramatic rise in their popularity came in the second half of the 1980s, reaching a peak in the early 1990s and falling off slightly to the mid-1990s.

This growth accompanied a variety of other factors, including an increasing visibility through the media of environmental disasters and problems, and a concomitant rise in the public perception of these problems. Growth can also be attributed to concern about the increasing importance of the role of supranational organisations such as the World Bank, and the rising intrusion into most ways of life of transnational companies, and possibly a growing disenchantment with the potential for bringing about change through increasingly distant, but powerless, democratically elected representatives.

The recent slight falling-off of membership numbers is a little less easy to link with other factors (perhaps because of the lesser benefit of hindsight), but may well be traced back to the demise of the Soviet Union and its geopolitical counterweight to the First World. This has given rise to both complacent attitudes of 'victory' and equally despair and apathy in opposition, both of which may well have led to a disillusionment with the purpose behind involvement and activism.

A second factor of relevance here is the fact that the environmental lobby, initially perceived as a single-issue movement in itself, has managed to raise its public profile high enough to promote a range of single issues under the general umbrella of the environment. Environmental issues are still widely equated with ecology rather than society or culture, although this broadening of issues under the general term of the environment should be acknowledged. This division of issues has also been associated with a rise in the number of organisations now dedicated to what are more justifiably described as single issues, especially within the field of tourism. Tourism Concern, for instance, was first established in the UK in 1989 with 200 members. By 1995 it had around 1,000 members and had gained a respectable reputation among the relatively highbrow sector of the country's media. It had also become an organisation whose blessing was sought by some of the new and specialised tour operators (which are described in Chapters 3, 5 and particularly 7).

A straightforward listing of non-governmental organisations concerned with the impact of tourism is given in Appendix 3. The list is far from comprehensive, but serves to highlight this recent burgeoning of organisations focused on tourism. Few of the organisations listed have existed for longer than fifteen years. Many have their headquarters in First World countries and see the issues of global tourism or tourism to Third World countries as their major focus.

Another organisation whose remit can be considered to be a single issue is Survival International (an INGO concerned for tribal peoples). It is of particular relevance to the tourism industry partly because it has recently conducted research and produced reports on the effects of tourism on tribal peoples and partly because of the nature of these effects – an issue taken up again later in this chapter and in Chapter 8. Figure 6.2 indicates not only its growth since 1987 but also its

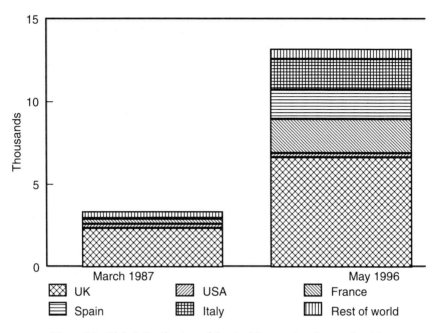

Figure 6.2 Global distribution of Survival International's membership
Source: Survival International

predominantly European membership. It has its base and was founded in the UK, but since 1990 has opened satellite offices in France, Spain and Italy, which accounts for its high growth rate of membership in those countries.

This recent growth in membership of and support for socio-environmental organisations with their bases in First World countries is likely to result in the wider dissemination of their particular interpretation of the role of tourism in the political, social, economic and cultural systems which prevail in Third World countries. As support for such organisations grows, so will their ability to influence the relevant and topical debates. Whether this influence will act as a counterweight to the geopolitical forces currently wielded by transnational companies, supranational institutions and First World government interests or, on the contrary, as an additional weight to these forces is still a matter for debate. There can be little doubt that many organisations representing the socio-environmental movement rail strongly against the activities of supranational institutions and their effects. But in different ways and in differing degrees they also represent a significant hegemonic interest in their own right, in convincing their constituents of the positions they adopt on tourism. The debates also reflect the contested nature of sustainability as competing interests struggle to legitimate their own definitions of the concept. How this debate will evolve is a matter for speculation, but it is already clear that all the interests involved in this struggle represent different facets of First World power.

While new socio-environmental organisations are of special interest in the way in which they defy, as it were, the traditional left/right political ideologies, they also tend to be closely related to class in that their memberships are drawn largely from the middle classes.[1] As Dobson asserts, 'there is plenty of socio-logical evidence to show that the environmental movement is predominantly a middle class affair' (1995: 154) with its ranks drawn from the educated, 'intellectual', and 'socially aware' elements. And as Jonathon Porritt argues:

One must of course acknowledge that the post-industrial revolution is likely to be pioneered by middle-class people. The reasons are simple: such people not only have more chance of working out where their own genuine self-interest lies, but they also have the flexibility and security to act upon such insights.

(Porritt, 1984: 116; quoted in Dobson, 1990)

The reason for the predominance of new middle-class involvement in the new socio-environmental movement (as both supporters, members and employees) and their participation in 'green politics' is a matter of considerable debate and disagreement and one that is reflected in the degree of involvement of the new middle classes in new tourism. As Eckersley explains, on the one hand there are the selfishness and self-interest arguments that posit that 'it is mainly the new class that is involved . . . because it is a means by which it is able to further its own class interest' (1989: 210). On the other hand, there are those who argue that the new middle classes are the vanguard of post-material values, the har-bingers of sustainable lifestyles. Eckersley argues in favour of the importance of education and the new middle-class awareness of the 'scale and depth of the social and ecological problems' (1989: 222). This is what Inglehart (1977, 1981) refers to as the empathy and sympathy that the new middle classes have with such issues and the new socio-environmental organisations that mobilise around them and which produce, he argues, an 'ideology of the new middle classes'.

Although Eckersley strenuously defends the socio-environmental movement against charges of élitism, it is difficult to deny the piety and sacrificial overtones which are apparent in much promotion of new middle-class ideals. They are reminiscent of the 'Peace-Corps type tourists' (Kutay, 1989) and the Explore leaders who put 'promising careers on hold'. As Eckersley suggests:

the assumption that environmentalists will necessarily defend their own level of affluence simply overlooks the fact that many committed environ-mentalists have deliberately forsaken the material lifestyles and career opportunities of their fellow class members (which they argue are wasteful and devoid of purpose) in favour of a more frugal and socially responsible lifestyle which is encapsulated in the slogan: 'To live more simply that all others may simply live' (Porritt, 1984: 205).

(1986: 28)

161

While seeking to avoid a crude determinism and correlation between these various factors of social, economic and political change, we can nonetheless detect some important apparent relationships. As already argued, many authors have traced the connections between the expansion of the new middle classes and the emergence and growth of *new* socio-environmental organisations (Gouldner, 1979; Offe, 1985). And in turn, it has been suggested that there is a significant, although clearly not exclusive, relationship between these phenomena and the growth of new forms of tourism.

Clearly, we are not able to present in full and unravel this complex debate. But it is interesting to begin to think about and question the dynamics of new tourism, new middle-class involvement and the campaigns and foci of new socio-environmental organisations.

ENVIRONMENTALISM AND NEW TOURISM

Environmentalism, or ecopolitics as Eckersley refers to it, is a useful way of exploring the political globalisation and global political re-ordering that Chapter 2 touched upon. Yearley (1995) identifies several reasons for assessing environmentalism as a vehicle for examining 'global political re-ordering'. First, the socio-environmental movement has repeatedly stressed the global nature, or 'worldwide-ness', of environmental problems and concerns embodied in the much-used, catch-all phrase 'think globally, act locally'. Second, environmental issues more than any others have given rise to international summits (such as the Rio Summit) and agreements and the emergence of international and supra-national bodies focused upon the environment.

Tourism adds another facet to this supposed global environmental politics and has become a focus for debates over the environmental impacts of tourism, a focus that has intensified as tourism has grown and spread. In addition, of course, much new tourism has become intimately associated with the environment, environmentalism and debates over sustainability on how to achieve less environmentally harmful forms of activity.

But it is far too crude to imply that environmentalism is one undifferentiated issue and movement. Far from it. Environmentalism embraces a broad range of views, policies and actions, and this is clearly reflected through tourism. The continuum or spectrum of environmentalism set out by Eckersley allows us to explore this relationship with tourism more systematically. The remainder of this chapter will use this framework to explore a range of positions.

Eckersley identifies five major strands of environmentalism: resource conservation, human welfare ecology, preservationism, animal liberation and ecocentrism. This spectrum is summarised in chart form in Table 6.1. For the purposes of applying this typology to the debate over new forms of tourism we have omitted the 'animal liberation' category which is of less significance to our discussion. Eckersley suggests that most environmental causes and movements fall within the two poles of resource conservation and ecocentrism, and that, in general terms, as you move from left to right in the table you are moving from *anthropocentric* views of environmentalism to *ecocentric* ones. Anthropocentrism means that these strands of environmental political theory are human-centred and seek to offer 'new opportunities for human emancipation and fulfilment in an ecologically sustainable society' (Eckersley, 1992: 26). By contrast, ecocentric environmental politics takes the emancipatory goals one stage further by acknowledging the rights of non-human life forms; in other words, saving nature for nature's sake and not for human exploitation. Eckersley concludes:

> This spectrum represents a general movement from an economistic and instrumental environmental ethic toward a comprehensive and holistic environmental ethic that is able to accommodate human survival and welfare . . . needs while at the same time respecting the integrity of other life-forms.
>
> (1992: 34)

It is not intended, however, that the continuum is strictly linear, with each position exclusive of other ideas. Some arguments within the human welfare ecology stream, for example, may be more ecocentric than some offered within the generally more ecocentric preservationist perspective. Similarly, the top row of Table 6.1 is illustrative of where organisations might be positioned, but it is not definitive. Thus, environmental organisations such as WWF may be represented in more than one stream. Nevertheless, the continuum provides a useful framework for thinking about the way in which tourism interacts with environmental politics and the politics of sustainability. It may be useful to refer to Table 6.1 as each stream is explored below a little further.

Conserving tourism resources

Resource conservation is the least controversial stream of modern environmentalism, although it is seen as anathema to the more radical ecocentric perspectives (Eckersley, 1992). It includes the national parks movement that seeks both to conserve nature and make it pay for itself, a compelling argument for environmental conservation for many cash-strapped Third World governments. For example, such a perspective is likely to encourage the costing of wildlife in terms of the potential income it can attract through safaris and game watching, an economic justification for their retention and development. Within this perspective, therefore, sustainability is conceived as 'sustainable development' and involves sustaining the environment *for* human production (the creation of national parks) and consumption (for the enrichment and enjoyment of tourists).

The notion of wilderness

The notion of wilderness is also heavily implicated in this stream of environmentalism. Wilderness is an excellent example of a socially constructed idea. To

Table 6.1 Tourism and the spectrum of environmentalism

	Resource conservation	Human welfare ecology	Preservationism	Ecocentrism
Organisations associated with this view	WWF, IUCN, UNEP, UNDP Tour companies promoting 'green' holidays Travel Adventure Society (USA) Campaign for Environmentally Responsible Tourism (CERT) Conservation Foundation	Green movement, Friends of the Earth Individual citizens, consumers and householders concerned with the state of residential areas and the state of the planet Tourism Concern	Coral Cay Conservation Elefriends. Tusk Force. Born Free Foundation Earthwatch	Greenpeace Earth First Wilderness Society
Place of origin and main ideas	USA, nineteenth century Utilitarian – 'greatest good for the greatest number' Wise use of natural resources	Industrialised Europe, twentieth century Enlightened self interest Four laws of ecology: everything is connected to everything else everything must go somewhere nature knows best there is no such thing as a free world Pursuit of environmental quality	USA, nineteenth century Reverence of nature Aesthetic and spiritual appreciation of 'wilderness'	North America, Australia and New Zealand Nature and environment are of equal importance to humans
Views on resources/ conservation aims/ sustainability	Wilderness to be managed for the greater good of people Regards non-human world in use-value terms A resource bank for industrial society to develop Develop wilderness and prevent waste	Concern with degradation of the environment Concern for health, safety and amenity Advocates 'sustainable development' – sustaining both natural resource base and biological support	Preserve nature *from* development Defence of 'wild nature' for spiritual values to humans 'Sustainability' – preservationism at any cost	Nature and wilderness is of intrinsic worth and is not a resource

Table 6.1 cont.,

	Resource conservation	Human welfare ecology	Preservationism	Ecocentrism
	Advocates 'sustainable development' Sustaining natural resource base for human production and maximum economic yield	systems for human reproduction		
Views on application to tourism	Manage wilderness as a tourist attraction Cost 'natural attractions' as economic assets National Parks (as conservation areas) can pay their own way Would encourage high-paying, well-heeled 'eco-tourists'	Concern with the environmental impact of tourism Heartland of the concerned, educated new middle classes wanting an ethically and environmentally enriching experience – alternative tourism as part of alternative lifestyles (ego-tourists)	Selective and exclusionary Volunteer holidays – (research tours) Kutay – 'Peace-Corps type travellers' Archetypal 'ego-tourists' trekking in high places, rainforests Marine preservation	Likely to ignore tourism, which is concerned as part of the problem; but is full of academics and deep green ecologists having a fine time researching and analysing environmental issues; new tourists in their own right
Policies and political programmes	Debt-for-nature swaps Conserve nature *for* development Biodiversity convention Global Environment Facility (GEF) Resource management National parks and protected areas 'Costing' flora and fauna	Critical of economic growth Policies to counter pollution and for alternative technology and lifestyles	Resistance to values of technological society Aim to create an alternative society National parks and protected areas	Concerned to protect threatened populations, species, habitats and ecosystems *wherever* situated, irrespective of their use or value to humans

Source: adapted from Eckersley, 1992 and Open University, 1995

many First World tourists it now represents ecological purity, an area free from human interference and development and generally devoid of human habitation, and it is a powerful notion that is repeatedly conjured up in tourist media such as travel brochures. Of course, for local peoples living in these supposedly wilderness areas there is a substantially different conception. As Deihl concludes of East Africa, '"wilderness" was largely a creation of Western thought since most areas they called wild were in fact used by native inhabitants' (Deihl, 1985).

A number of commentators have been heavily critical of the wilderness concept for the way in which it enshrines a division between *humans and humanity* on the one hand and *nature* on the other, and for its undeniably ethnocentric connotations. It is capable of both ignoring the ecological management undertaken by local indigenous communities and assuming that First World conceptions of management are superior. It has also resulted in the displacement and exclusion of local people from areas considered (more often than not from a First World perspective) worthy of protection. The issue of displacement is covered in more detail in Chapter 8.

Debt-for-nature swaps

As Table 6.1 indicates, there are a number of policies and programmes aimed at meeting the goals of resource conservation some of which further underline the power invested in First World interests. One of the most noted is the concept of debt-for-nature swaps. Because few commentators envisage the solution to the Third World debt problem as being repayment, First World banks are willing to sell off the debts owed to them at a discounted rate. An environmental INGO with substantial funds pays off the debt, or a small part of it, to the First World bank that is the creditor. In return, the Third World government, unable to pay off the debt in dollars, pays it off internally in the local currency to a local NGO in partnership with or deliberately set up for the purpose by the First World INGO. This payment will then be used to protect a specified natural area.

It is a relationship, as Adams argues, that greatly extends the notion of 'the one who pays the piper calls the tune':

> The principle of directing First World resources to Third World conservation has long been important in organisations such as the World Wide Fund for Nature. . . . debt-for-nature swaps . . . arise from the exposure of First World banks to Third World debt, and their willingness to sell off those debts at a discount to conservation organisations, who use them to bargain for expenditure on conservation in local currency.
>
> (Adams, 1990: 200)

While debt-for-nature swaps do not necessarily implicate tourism, they certainly do indicate the lengths environmental organisations will go to in conserving certain Third World environments. Lewis (1990), for example, documents the 'bizarre agreement' in 1987 between the Bolivian government and US-based Conservation International in which US$650,000 of Bolivian debt was bought at a US$100,000 discounted rate on the condition that the Bolivian government pay US$250,000 in local currency to create a buffer zone around the Beni Biosphere Reserve. The buffer zone, however, was not a wilderness area, but one in which cattle rearing and lumbering were major activities whose promoters were not willing partners in the deal. Additionally, the money supplied by the Bolivian government to protect the reserve drained funds away from other environmental projects – the payment of US$250,000 was equivalent to the total annual budget for the national park system.

In 1989, as Fred Pearce (1990) documents, the WWF paid US$2 million for a Costa Rican debt with a paper value four times greater. It then traded the debt for government bonds worth US$7 million to pay for the creation of the Guanacaste National Park.

Swaps, however, have covered only a tiny proportion of the Third World debt and are no longer widely considered to be a potential solution to the debt crisis. They divert the INGO's funds, energy and attentions away from tackling the structural causes of both the debt crisis and the conditions which in some circumstances force people and governments into exploiting and destroying their natural resources. They do not represent the flow of new money into a country but provide instead an illusory relief of the external debt. Moreover, they work against the local ownership and control of resources that are so critical to community tourism development by furthering the notion of private appropriation of land and resources (which in many cases in Third World countries used to be considered as common heritage), and by representing a transference of power over national resources to First World organisations.

The domination implied by debt-for-nature swaps is no less oppressive when it comes from a well-intentioned INGO than when it is imposed by the World Bank, IMF or a transnational company. But most of the criticism of debt conversion schemes appears to come from within the ranks of the INGOs themselves, which might suggest that such schemes are not promoted out of ignorance of their effects. Rather they indicate that the international socio-environmental movement is being increasingly co-opted by the financial ethic of capital accumulation and by the supposed 'realities' of the global marketplace (see the section on 'Redefining sustainability', pp. 199–223).

Biodiversity and the Biodiversity Convention

A second example of action-oriented programmes within the resource conservation stream is the commonly perceived global need to retain biodiversity. The term biodiversity has only recently come into common usage, so we have used Pearce and Moran's (1994) work (Box 6.1) to describe its meaning. It is a term which is often used implicitly and sometimes explicitly to justify the designation

---

Box 6.1 What is biodiversity?

Popular use of the term biodiversity started with a draft report by the IUCN in 1987.

Pearce and Moran (1994) describe biodiversity in terms of genes, species and ecosystems.

Genetic diversity

'is the sum of genetic information contained in the genes of individuals of plants, animals and micro-organisms. Each species is the repository of an immense amount of genetic information' (2). The significance of genetic diversity is normally related to food production, the basis for which lies in biodiversity and seed production. Control of seed production by TNCs has tended to reduce genetic diversity and make Third World farmers dependent on the TNCs.

Species diversity

'Species are regarded as populations within which gene flow occurs under natural conditions' (3). 'We do not know the true number of species on earth, *even to the nearest order of magnitude.* . . . This lack of knowledge has considerable implications for the economics of biodiversity conservation, particularly in defining priorities for cost-effective conservation interventions' (4).

Ecosystem diversity

'relates to the variety of habitats, biotic communities and ecological processes in the biosphere as well as the diversity within ecosystems. . . . No simple relationship exists between the diversity of an ecosystem and ecological processes such as productivity, hydrology, and soil generation' (5).

(extracts from Pearce and Moran, 1994)

---

of areas for some degree of protection from development and exploitation. It is important to note, however, that the notion of biodiversity and the Biodiversity Convention which came out of the Rio Summit are the subject of heated disagreement.

The main aim of the Biodiversity Convention is 'the conservation of biological diversity, sustainable use of its components, and fair and equitable sharing of benefits from the use of genetic resources' (Holmberg *et al.*, 1993: 20). Although the spirit of the treaty is more in keeping with the sustaining and accessibility of genetic resources, it has clear resonance for tourism as an economic justification for 'conservation' of the 'natural environment'.

It is implicit in the Convention that biodiversity has an economic value. In order to determine this value, however, biodiversity change must be measured and its valuation is often related to the protection of areas against destruction by humans. This is especially so where the motive of national governments in designating areas for protection arises from its hopes of developing tourism (see Chapter 9), in which case potential tourist revenue can be estimated, on the assumption that the biodiversity that the tourists are going to see is retained. Achieving this valuation in a way that is agreeable to all, however, is fraught with difficulty, for once an estimation of the monetary value has been made the problem of the distribution of costs and benefits then arises: who should pay how much to whom, for what, and what should the money paid be used for? (United Nations Environment and Development UK, 1993).

This debate illustrates the way in which environmental questions and 'agreements' are mapped onto uneven and unequal development, with the disputes falling mainly along First World/Third World lines. Writing in *Third World Resurgence* on the Biodiversity Convention, Shiva argues that it started principally as an 'initiative of the North to "globalise" the control, management and ownership of biological diversity . . . so as to ensure free access to the biological resources which are needed as raw material for the biotechnology industry' (quoted in Holmberg *et al.*, 1993: 22). In this sense, holiday environments can also be viewed as the raw materials of which Shiva writes.

The Global Environment Facility (GEF)

The Rio Summit offered the First World the ideal mechanism to achieve this globalisation of the control and management that Shiva refers to: the Global Environment Facility or GEF. The GEF was set up in November 1990 by the World Bank, the United Nations Development Programme (UNDP) and the United Nations Environment Programme (UNEP) to assist the so-called developing world in funding projects which either protect biodiversity against destructive development or promote development which does not destroy biodiversity. The GEF, however, is not a development agency. As Pearce and Moran explain:

> It operates via many development projects, but it modifies them so that the technologies used are cleaner than they otherwise would have been. Its purpose is not development as such, but the capture of global environmental value – the value that comes from reducing the 'global bads' of climate change, biodiversity loss and ozone layer depletion.
>
> (1994: 132)

The Rio Summit allocated to the GEF the role of financial administrator for the Biodiversity and Climate Change Conventions which arose out of the conference. Again, Shiva is highly critical of this role and of the World Bank's part in it:

> The erosion of biodiversity is another area in which control has been shifted from the South to the North through its identification as a global

problem . . . by treating biodiversity as a global resource, the World Bank emerges as its protector through GEF . . . and the North demands free access to the South's biodiversity through the proposed Biodiversity Convention. But biodiversity is a resource over which local communities and nations have sovereign rights. Globalization becomes a political means to erode these sovereign rights, and means to shift control over and access to biological resources from the gene-rich South to the gene-poor North. The 'global environment' thus emerges as the principal weapon to facilitate the North's worldwide access to natural resources and raw materials on the one hand, and on the other, to enforce a worldwide sharing of the environmental costs it has generated, while retaining a monopoly of benefits reaped from the destruction it has wreaked on biological resources. The North's slogan at UNCED and the other global negotiation fora seems to be: 'What's yours is mine. What's mine is mine'.

(1993: 152)

In 1992 Oliver Tickell articulated the suspicion of the socio-environmental movement, that control of the funds by the World Bank could only lead to the imposition of a First World agenda on the allocation of those funds. He also pointed out the contradiction in the GEF's approach:

The main qualification for receiving a GEF grant for preserving biodiversity is apparently to be running a World Bank project that threatens or destroys that biodiversity, like a giant dam, or a plan to develop logging in untouched forests. Marcus Colchester of the World Rainforest Movement estimates that 70 per cent of GEF funds are tied to mainstream Bank projects, and are actually subsidising social and environmental destruction.

(1992: 3)

Despite the misgivings of the socio-environmental movement and others, Fernandes (1994) claims that 'Already the leading northern NGOs appear to have developed strong coordinating GEF links with the World Bank, UNDP and UNEP' (24); and Chatterjee and Finger claim that the WWF is now the most consulted NGO on GEF projects (1994: 155).

Fernandes' analysis of the GEF suggests that a large number of integrated sustainable tourism projects received funding under phase I of the GEF. He quotes Soares of the Brazilian Institute for Economic and Social Analysis, who concludes that many GEF programmes, which are often linked to significant ecotourism and sustainable tourism components, represent 'mainstays for a (development) model that unceasingly reproduces conditions for the planet's deforestation, even while preaching its conservation' (Soares, 1992: 48).

Fernandes also cites other fundamental flaws in the GEF mechanisms such as the stimulation of destructive competition between organisations working in the field of biodiversity, a failure to involve local communities, an over-dependence on international consultancies (see 'Redefining sustainability', pp. 199–223), and the creation of a number of 'paper parks' – a reference to the designation of parks by national governments which have few resources to provide management and protection systems on the ground (see Box 9.2, p. 285). Moreover, the link between the World Bank and INGOs such as WWF, IUCN and Conservation International leads to these organisations adopting an approach more characteristic of the World Bank and advocating corporate schemes for a range of environmental programmes and projects (including tourism and ecotourism developments) which give 'total management control' to the 'private or NGO sector'. As a result they fail to even 'recognise the existence of village conservation movements opposing development projects . . . much less to acknowledge their effectiveness in pressurising governments and corporations' (Lohmann, 1991: 98–9, cited in Fernandes, 1994: 26).

Conservation with a people dimension?

The human welfare ecology stream has the same general perspective of 'sustainable development' but goes further in its concern for environmental quality. It is most clearly reflected in the 'green movement' and the individual concern with the state of the environment. It is this perspective that most readily captures the much-expressed concern over the environmental impacts of tourism and how diverse new tourism destinations such as the Himalayas, Zanzibar or Bali (as we discussed in the previous chapter) are being overrun and corrupted both culturally and ecologically.

It is also a stream that is reflective of the debates over the most appropriate way to travel and the most virtuous form of alternative tourism. As Morrison of the traveller's magazine, *Wanderlust*, suggests, rhetorically:

> to make matters worse we now have to suffer that greatest hypocrisy of all
> – the lamentable cries of anguish from a latter-day explorer bemoaning
> the commercialisation of a part of the world that they were instrumental
> in opening up. It's a sadly familiar cycle. After the explorers come the
> travellers. After the travellers come the tourists. And after the tourists come
> those blasted explorers once again, this time heading up some conservation
> group to wail about the exploitation of a once virgin paradise.
>
> (1995: 68)

In a similar vein, Eckersley suggests that the human welfare ecology stream appears to be a 'peculiarly late twentieth century phenomenon' that can be attributed in large part to the 'emergence of "post material" values borne by the . . . New Middle Class' (1992: 36). It is an especially strong current of environmentalism in green politics in the First World, and tends to appeal to 'enlightened self interest', an individualistic concern centred upon '*our* survival, for *our* children, for *our* future generations, for *our* health and amenity' (Eckersley, 1992: 38). This is strongly reminiscent of the arguments put forward on the need to protect tourism resources from further development, from tourists discovering it, and so on. In other words, it appears to offer a thoroughly First World perspective on *our* needs and aspirations and why we should encourage sustaining natural environments for, among other purposes, *our* holidays. As Adams (1990) concludes, conservation is 'dominated by global concentrations of wealth and power, and centralized decision-making. Despite the well-meaning rhetoric of environmentalists advocating sustainable development, it is not the Third World that stands to gain most' (200).

Preserving Third World tourism products?

The third stream of environmentalism, preservationism, can be contrasted to resource conservation which, Eckersley argues, aims for nature conservation to enable development. Preservationism, by contrast, seeks to preserve nature *from* development. Like the two previous streams, however, it tends towards anthropocentrism in that it recognises 'wild nature' and 'wilderness' areas as being of spiritual and aesthetic value to humans, on whose behalf it should be preserved.

There is an increasing number of Third World environments that are subject to preservation and to which 'discerning tourists' are attracted. The significance of authenticity in the discussion of tourism has already been emphasised in Chapter 3, where the importance of aestheticisation (the ways different facets of the Third World become an aesthetic experience) in forming a critique of new tourism was argued. The drive to preserve particularly natural environments, but also indigenous cultures, finds resonance in these concepts.

Beneath the push towards preservation is the perception that wilderness areas are rapidly being lost and the need to protect those areas remaining for the consumption, in part, of informed First World tourists. In this way preservationism tends towards exclusionism for not only does it rely on the containment of visitor numbers to ensure the retention of a stable environment (the ecological carrying capacity) but it also requires fewer tourists to maintain the 'reverence' and 'spirituality' of 'wild' areas (as Edwards' sentiments suggest, see p. 142). Furthermore, in Chapter 5 ('The scramble for Third World destinations', pp. 147–54) it was suggested that it is exclusionary in another important respect in that it requires the ability to interpret and understand the environment (and the notion of sustainability); hence Brooks' (1990) reference to the prerequisite understanding of 'cultural values and ecological balance' in 'remote and unspoilt regions', and Brockelman and Dearden's (1990) view that only 'serious naturalists and Nature enthusiasts' should be considered for trekking in areas of Thailand.

Underlying these sentiments, it appears that preservationism tends towards a high degree of ecological and social selectivity. As Eckersley argues, 'wilderness appreciation has developed into a cult in search of sublime settings for "peak experiences", aesthetic delight, "tonics" for jaded Western souls' (1992: 40).

And in so doing it has singled out those places 'that are aesthetically appealing according to Western cultural mores' (1992: 40), including mountain ranges, ecologically rich rainforest areas and diverse marine environments. In other words, First World geographical imaginations are filled out with environments that are regarded as somehow sacred and worthy of our attention. Again, in parallel to the two environmental streams discussed earlier, it is First World citizens who are expressing their perceptions, their choices and exercising their will. It is a further illustration of the way in which uneven development is played out, as areas become targeted and coveted for their ecological content while other areas fall from our view and attention.

From his thoroughgoing review of the social dimensions of policies and initiatives for environmental protection, Colchester (1994) summarises four fundamental problems with what he terms the 'classical conservationist' approach (an approach that appears more akin to preservationism than resource conservation). First, humans come second in the conservationist's view of nature preservation and this has some serious implications for Third World communities affected by the push towards nature preservation (as Chapter 8 discusses). Second, and reflecting the discussion on the social construction and representation of wilderness, the generally holistic (or cosmovision) view of nature held by indigenous peoples is radically different from the cultural notion of wilderness held by 'western conservationists'. Third, the view of indigenous peoples held by conservationists has been 'tinged with the same prejudices that confront indigenous peoples everywhere' (1994: 56), and this is highlighted in Box 6.2 which focuses on these views and associated attitudes. Again, there are overtones here of the ethnocentrism of the environmental debate, also known in other contexts as environmental racism. And finally, conservationist interests have sought state legitimation for enforcing their view of conservation on local peoples.

Machlis and Tichnell (1985) argue that preservationist approaches to protected areas often result in what they refer to as 'militaristic defence' strategies or what Pleumarom (1994) terms 'militarization', arguing:

> Tourism, particularly in the context of 'sustainable development', often only serves to exacerbate the entrenched asymmetry of unequal relations in the world. Militarization is the most visible tendency of this process. For example, 'legitimate' violence in the name of resource control, such as 'shoot-to-kill' actions in Kenya's game parks . . . a tendency reinforced by the provision of para-military training for park rangers and anti-poaching equipment by international conservation groups.
>
> (1994: 146)

As Pleumarom suggests, environmental INGO's have been implicated in this militarisation process. Box 6.3 illustrates this process with evidence of the WWF's alleged involvement in the hiring of mercenaries to protect the black rhinoceros population in Namibia.

Box 6.2   He's destroying his own rainforest . . .

He's destroying his own rainforest. To stop him, do you send in the army or an anthropologist?

This hurtful, insensitive slogan began a fundraising campaign by the Worldwide Fund for Nature in 1994. It first appeared in the *Financial Times*, presumably for the benefit of WWF's growing band of corporate supporters, and set the tone for the advertisement that followed: glibly 'pro-nature' and implicitly anti-people . . .

Yet the WWF advertisement provides a useful warning, as it offers a perfect illustration of the dominant conservationist mindset. Unfortunately for indigenous peoples, most Western conservationists still cling to a romantic, Eurocentric conception of nature as an empty, unspoiled wilderness, separate from and uncontaminated by humanity . . . in which the indigenous inhabitants of these 'wildernesses' are at best an inconvenient disruption of the great romantic myth, at worst a menace to be repelled by barbed wire and guns . . .

Not surprisingly, the world's indigenous peoples are bemused and outraged by the behaviour of Western conservationists who think they can manage their lands better than they can. They see the wilderness view of conservation as a new form of colonialism, with the same devastating results as military conquest or slavery. They also profoundly resent the idea that areas of land they have inhabited for generations – often for thousands of years – should be regarded as empty . . .

Like logging and mining, conservation has meant driving indigenous people from their lands by a mix of repression and fraud, but whilst miners and loggers are usually explicit about their intention to exploit, conservationists pacify public opinion with the language of environmental concern . . .

Conservationist rhetoric also makes politicians sound virtuous: call anything 'sustainable' these days and it is bound to win plaudits. Currently the European Commission is considering awarding a grant of 15 million Ecus (£12 million) for the 'sustainable development' of Paraguay's Chaco region. The fact that the Enxet Indians who live there have not been consulted does not seem to concern them . . .

According to a 1985 survey by the IUCN, 70 per cent of such 'wilderness' regions are inhabited. In Latin America the figure rises to 86 per cent. Sometimes, as a supposedly benign alternative to resettlement, people have been 'allowed to remain' on their land under certain strictly imposed conditions. Usually the intention of such schemes has been to encourage tourism by pandering to Western ideas of 'noble savages'. The people concerned are thus dehumanised and presented as exotic species to be 'conserved' . . . as long as they play by the rules . . .

In South Africa the last group of 'Bushmen' (also known as San or Khoi-Khoi) were told that they could stay in the Kalahari Gensbok National Park

Box 6.2
continued

provided they maintained a 'traditional' lifestyle and used 'traditional' methods of farming – in other words, as long as they continued to look like 'Bushmen'. Needless to say, this was not a success. . . . The 'Bushmen' decided they wanted to have new housing, new clothes and hunting dogs. . . . The wardens of the National Park concluded that 'their desirability as a tourist attraction is under serious doubt, as is the desirability of letting them stay for an indefinite period in the park. They have disqualified themselves . . .'

Lately it has become fashionable for conservationists to talk about 'consulting' local people, and to acknowledge the 'role' of local peoples in 'managing protected areas'. For many, such concern has come too late. . . . In practice the conservation movement has subjugated indigenous peoples to state or corporate control. It has violated their rights and, for the most part, failed in its own objective of environmental protection.

(Survival International, 1995)

National parks and area protection

One of the principal mechanisms for advancing preservationist policies has been through national parks and other categories of protected areas. National parks were first established in the USA in the nineteenth century and began a growth that is charted in Figure 6.3. Internationally, the categories most often recognised are those defined by the International Union for the Conservation of Nature and Natural Resources (IUCN) in 1985 and listed in Table 6.2. Categories I–III are referred to as 'strictly protected areas' and were formed to maintain biodiversity and natural formations. Categories IV–VIII allow some degree of human use and controlled exploitation (IUCN, 1985). To the IUCN's categories should be added the biosphere reserves designated by UNESCO's Man and Biosphere (MAB) programme. Biosphere reserves are protected areas of environments internationally recognised for their value for the preservation of genetic diversity. Additionally, national governments may devise their own categories and add other standards of protection to those internationally recognised.

These categories of protection have a degree of legal standing, but, in Third World countries at least, this rarely translates consistently into appropriate security of protection on the ground, and many protected areas are suffering 'serious and increasing degradation as a result of large scale development projects, expanding agricultural frontiers, illegal hunting and logging, fuelwood collection and uncontrolled burning' (Wells and Brandon, 1992: 1).

In a manner not dissimilar to the Biodiversity Convention's promotion of debate and disagreement, the creation and protection of national parks, biosphere reserves and other areas have often been at the core of conflicts over land in Third World countries. Thrupp (1990), for instance, views tourism development (particularly its emphasis on wilderness protection through the creation

Box 6.3 WWF bankrolled rhino mercenaries

The World Wide Fund for Nature (WWF) has admitted that it provided funds to armed anti-poaching units in Namibia set up by a clandestine and proscribed operation run by a team of British mercenaries.

Documents obtained . . . reveal details of three payments that the WWF has previously denied making and which an independent inquiry failed to reveal.

The WWF's admission came a week after Scotland Yard's international and organised crime branch submitted a file on the secret programme, code-named Operation Lock, to the Director of Public Prosecutions . . .

The WWF was anxious to dissociate itself from Operation Lock when it was uncovered last year. It had been set up in 1987 by Prince Bernhard [of the Netherlands, a former WWF president] . . . and Dr John Hanks, then the WWF's Africa programme director, to infiltrate the illegal trade in rhinoceros horns and identify individuals and countries involved.

They were concerned that little was being done to stem the decline in numbers of black rhino from 100,000 in 1960 to just 4,000 in 1987.

Dr Hanks had appointed KAS Enterprises Ltd. to run the operation. KAS's chairman was the late Sir David Stirling, founder of the Special Air Service. Many of its staff were former SAS members. It is thought they had little experience in rhino conservation.

In order to distance itself from the project, the WWF appointed Lord Benson . . . to conduct an independent inquiry. His report to the board of WWF International in Geneva is believed to have concluded that no WWF personnel other than Dr Hanks knew of the scheme and no WWF money was involved.

However, . . . a request for 109,400 Swiss francs (£43,760) was made to the WWF on 13 October 1989 by the Namibia Nature Foundation for 'urgent short-term' funding for anti-poaching units in Etosha and Damaraland. Dr Hanks said last week that those units were probably set up by Operation Lock and would have been armed for self-defence . . .

Robert SanGeorge, director of communications, . . . confirming the information received, . . . said a total of Sfr. 157,160 (£62,864) was paid between Nov. 1989 and Feb. 1991. The WWF says it learned of Operation Lock in summer 1990.

'It would appear that we innocently and unwittingly funded anti-poaching units that were set up as part of Operation Lock,' said Mr SanGeorge.

Asked how his investigation failed to identify payments made to anti-poaching units established by Operation Lock, . . . details of which were contained in a file labelled 'Anti-Poaching Units' at the WWF headquarters in Geneva, Lord Benson refused to comment.

(Boggan and Williams, 1991)

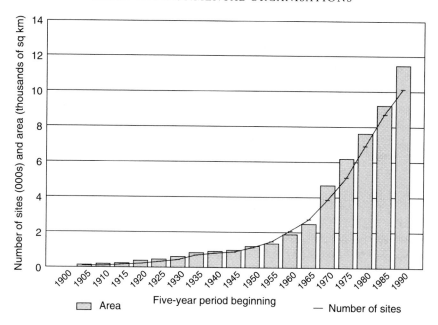

Figure 6.3 The global growth of protected areas
Source: World Conservation Monitoring Centre, Cambridge
Note: Covers growth of protected areas for IUCN management categories I–IV for
sites larger than 1,000 hectares

of parks and reserves) as largely serving the interests of the privileged upper-middle classes, mainly First World tourists and scientists.

Box 6.4, which recounts the observations of Chris McIvor from a debate about conservation in Zimbabwe attended by both international environmental agencies and local representatives, illustrates a few of the important points concerning such conflicts. First, there is a clear association of conservation with colonialism: reserved and protected areas were part of the system of securing raw materials (such as timber) for the imperial infrastructure (Colchester, 1994). Second, there are sharply opposed views of the meaning and value of nature. As Colchester (1994) observes, national parks and other protected areas have 'imposed élite visions of land use which have resulted in the alienation of common lands to the state' (5). This has also resulted in the displacement and resettlement of people to make way for park conservation – the case of the Maasai in East Africa stands out here, but other examples, as well as that of the Maasai, are illustrated and discussed in Chapter 8. Third, there is the question of who exactly gains from the construction of parks. It rarely seems to be the local people and, indeed, part of the answer appears to be found in the removal of local rights and a loss or denial of ownership. Instead it is the rich consumer in the industrialised North with leisure and wealth to be a tourist in the Third World who gains from the designation of national parks. Survival International

Table 6.2 Protected area categories

Category	Type	Management objective
I	Scientific reserve/ strict nature reserve	Protect nature and maintain natural processes in undisturbed state. Emphasise scientific study, environmental monitoring and education, and maintenance of genetic resources in a dynamic and evolutionary state
II	National park	Protect relatively large natural and scenic areas of national or international significance for scientific, educational and recreational use
III	Natural monument/ natural landmark	Preserve nationally significant natural features and maintain their unique characteristics
IV	Managed nature reserve/ wildlife sanctuary	Protect nationally significant species, groups of species, biotic communities or physical features of the environment when these require specific human manipulation for their perpetuation
V	Protected landscapes	Maintain nationally significant natural landscapes characteristic of the harmonious interaction of people and land while providing opportunities for public recreation and tourism within the normal lifestyle and economic activity of these areas
VI	Resource reserve	Protect natural resources for future use and prevent or contain development that could affect resources pending establishment of management objectives based on appropriate knowledge and planning
VII	Natural biotic area/ anthropological reserve	Allow societies to live in harmony with the environment, undisturbed by modern technology
VIII	Multiple use management area/ managed resource area	Sustain production of water, timber, wildlife, pasture and outdoor recreation. Conservation of nature oriented to supporting economic activities (although specific zones can also be designed within these areas to achieve specific conservation objectives)

Source: IUCN, 1985

also round on conservation and its effects on tribal peoples: 'When national parks are set up to protect wildlife, tribal peoples are often the first casualties. They are thrown off their land. Denied the right to graze their herds or hunt for food they sink into poverty and despair' (undated: 4).

Box 6.4 *Conservation and imagination: clash on the environment*

. . . despite the fact that both sets of contributors were to address the issue of 'conservation' one was left with the impression that they were speaking a very different language.

The Western environmentalists largely concentrated on the destructive impact of human activity. They lamented the disappearance of natural habitat, the loss of areas of wilderness . . .

Their prescriptions echoed this standpoint. Land should be set aside in national parks where human interference was kept to a minimum. More resources should be spent on policing these areas, on preserving the wilderness that Africa can offer the rest of the world.

One ecologist . . . warned Zimbabweans to beware the evils of development. Nature, he claimed, needs to be protected from economic exploitation so that society can enjoy the aesthetic and recreational benefits of an unspoiled countryside.

By contrast, Zimbabwean participants seemed to see no inherent contradiction between conservation and development . . .

There was very little talk by local environmentalists of the recreational and aesthetic or the fact that the vast majority of visitors who frequent protected areas come from outside the continent. 'Conservation for us,' claimed one Zimbabwean, 'means the wise management of natural resources for economic use. It does not mean the absence of use at all costs.'

There are several reasons why arguments for a preservationist approach to the environment and the exclusion of human activities from protected areas are unlikely to find much support amongst African populations. The history of early conservation in Africa is indistinguishable from the history of colonialism and in particular the eviction of indigenous communities from land and resources that they once enjoyed.

In Europe where the majority of people are urbanised and where industry is the main engine of economic growth, the concept of nature as a resource to be enjoyed by urban consumers is understandable.

In Africa, however, where the majority of people still depend for their subsistence on agricultural production . . . the concept of nature is very different. The land and its wild animals are not a source of aesthetic enjoyment but a resource to be managed so that people can survive.

As one Zimbabwean environmentalist concluded: 'It is impossible to talk of wildlife preservation to a farmer whose fields have been raided by elephant and buffalo. Unless there is some tangible economic return to support his family, wildlife is a threat not an asset. When Western environmentalists talk to us about the aesthetic and recreational appeal of our landscape and the need to preserve our wild animals in their natural habitat, we wonder if they would continue to have such ideas if they shared our poverty. For most Africans the land is not an arena for leisure pursuits but a means of livelihood and survival.

(McIvor, 1995)

Ecocentrism?

The final stream of environmentalism is ecocentrism.[2] Unlike the other environmental streams, ecocentrism does not place humans above other life forms and instead regards nature and environment of intrinsic value and of equal importance. As Table 6.1 documents, this is a position adopted by the major environmental organisations such as Greenpeace and Friends of the Earth who take an holistic approach to nature. The similarities to preservationism in terms of programmes adopted are, however, detectable. Ecocentric environmentalists are especially concerned to protect threatened species, habitats and ecosystems and 'strongly support the preservation of large tracts of wilderness as the best means of enabling the flourishing of a diverse non-human world' (Eckersley, 1992: 46). Although organisations such as Greenpeace and Friends of the Earth have not really entered the debate over tourism, and indeed as Table 6.1 suggests, may wish to ignore tourism as an environmental problem, the relationship to new forms of tourism of the activities undertaken by organisations such as Earthwatch, the Ecotourism Society, Programme for Belize and Coral Cay Conservation (whose goals are similar to those of Friends of the Earth and Greenpeace) is clearly identifiable (see, for instance, Box 5.5, p. 139).

There are three overarching conclusions that can be drawn from this review of the different streams of environmentalism. First, when we talk of environmentalism, we are talking ostensibly about First World perceptions of environment, ecology and nature, and in particular the views formulated and advanced by organisations based in the First World and disproportionately represented in the new middle classes. As Table 6.1 suggests, these environmental streams have also permeated debates over contemporary tourism and the formulation of supposedly sustainable forms of travel.

Second, the idea of the geographical imagination is a useful means of understanding how and why elements such as the 'environment' and 'conservation' are seen in radically different ways by different groups or interests and how certain interpretations are more powerful than others, and have the ability to impose themselves on others. This is especially important in the context of tourism, as the perceptions of local people and communities of their environments can be dramatically different from those of the environmental groups that seek to protect them and the tourists that visit them. (See Box 6.4 for an illustration of these differences.)

Third, in a geographical sense, environmentalism more than any other political theory has drawn in both global and local dimensions. We are asked to think not only of our local environments but to consider the ways in which these environments are part of a global ecosystem. It is to the consequences of this that we turn next.

ENVIRONMENTALISM AND POWER

In Chapter 3 it was argued that environmentalism and ecology have been identified as one of the ways in which power is transmitted, and neo-colonialism was introduced as a way of representing this and indicating the coexistence of a number of different forms of colonialism. It was suggested that thinking about environmentalism and sustainability, at least in part, as one such form of neo-colonialism has produced new ways of formulating the contemporary debates on environmental issues, and the views of Wolfgang Sachs (Box 3.2, p. 51) were presented in doing this. The ways in which this has been applied to the development of tourism were also suggested.

Box 6.5, concerning the impact of what Daltabuit and Pi-Sunyer (1990) call 'new, "soft path"' tourism (and which they also refer to as a 'more intensified ecological tourism' (10)) upon the indigenous Maya communities in the Yucatán region of southern Mexico, helps to illustrate these points. The authors assess the fortunes of local people in supplementing the now well-established mass tourism industry based upon Cancún with the emergence of these new forms of tourism. They refer to 'new correlations of power' and the effects of 'hegemonic' metropolitan cultures. Their work is also of importance as they refer to the 'environmental movement as a political force and powerful cultural construct' (1990: 10). One case study is of Coba, a village situated close to the remains of a major Classic Maya metropolis in Quintana Roo, Mexico, outside the general orbit of tourism until recently and centre of what Daltabuit and Pi-Sunyer call the phenomenon of 'archaeotourism'. The story of Coba is a familiar one: the declaration of a national park, the expropriation of local lands, the building of a large and comfortable foreign-owned hotel complex (Villas Arqueológicas) and unilateral solutions imposed upon local people who are excluded from any processes of discussion and consultation.

Doubts must also be raised over the nature of environmental concerns and how these are translated into practice. A global concern for the environment and the call to 'think globally, act locally', while lofty and harmless in practice, have a tendency to become a crusade (to 'think globally, impose locally') that is devoid of notions of social justice and a concern for local peoples' perceptions. Take, for example, the invitation to purchase a piece of rainforest which appeared in the newsletter of the Programme for Belize and which is shown in Figure 6.4. As was pointed out in Chapter 3, this example would appear to display the antithesis of local control. A brief description of the Programme for Belize is given with the advertisement in Figure 6.4.

In this context, Shiva sees the global debates around environmental concerns, such as those at the Rio Summit, in a restrictive sense and as more akin to an application of eco-colonialism:

> Unlike the term suggests, the global . . . was not about . . . [the] life of all
> people, including the poor of the Third World, or the life of the planet.
> . . . The 'global' of today reflects a modern version of the global reach of

a handful of British merchant adventurers who, as the East Company, later, the British Empire raided and looted large areas of the world.

(1993: 150)

Box 6.5 Tourism development in Mexico

Communities are experiencing intensified pressures to participate in the tourism industry as it undergoes a shift from highly localized resort development to a mode that increasingly stresses the marketing of the physical and human environment, including the archaeological remains of pre-Hispanic Maya civilization . . .

First, will the further penetration of tourism result in improved material and social conditions for the local population? Second, will these communities be given an opportunity to play a significant role in determining their own future? And, related to the latter question, can the Maya generate sufficient political power to protect their property rights and minimize the social and cultural costs of tourism?

From the beginning, the expansion of tourism in this part of Mexico . . . has been very much the work of national and international agencies and institutions, both public and private. . . . Fundamentally, local people and local resources have been approached as elements to be 'managed' by planners engaged in macroeconomic strategies . . .

. . . in places like Quintana Roo – if history and experience are any guide – a 'managed' environment is bound to translate into an environment managed by outsiders chiefly to satisfy the needs of outsiders. It is a short step from external direction to what can be termed a process of *appropriation*, by which physical environment, and within it human societies and historical remains, become subtly redefined as global patrimony – universal property . . .

Environmental tourism, much like other forms of 'exotic' travel, appeals 'to the deepest recesses of the Western imagination' (Bruner 1989: 440) . . . [a] particular fantasy of communion with nature . . .

These issues and questions – fundamentally matters of power and its representation – have been very much on our minds as we consider environmental and economic priorities as defined by outsiders and compare such plans and visions to the realities of life experienced at the local level in Coba, a Maya village and major archaeological site.

. . . much that now makes Quintana Roo so appealing as an ecological treasure house and location of well-preserved Maya antiquities is attributable to environmentally benign patterns of subsistence and the . . . inaccessibility.

Affluent, educated, and dressed in fashionable expedition gear, the visitors who lodge at the villas clearly represent the up-market end of the business. Altogether, Coba is as close to a soft-path archaeological tourist location as one is likely to find . . .

(Daltabuit and Pi-Sunyer, 1990)

Figure 6.4 Rainforest for sale
Source: Programme for Belize advertisement
Note: This poster appeared on billboards across London and in the newsletter of the
Programme for Belize (PFB). The PFB's overall goal is to promote 'conservation and
economic development of the natural resources of Belize, with emphasis on the long-
term sustainability of the resource base' (PFB, 1989: 1). The organisation raises funds
for the purchase and management of areas of the Belizean rainforest. It now owns and
protects over 90,000 hectares of tropical forest and wetlands, and all their wildlife.

And drawing conclusions from their work on tourism developments in Mexico (Box 6.5), Daltabuit and Pi-Sunyer describe the contemporary version of this global reach:

> Environmental concerns clearly strike a deep chord in the educated Western public. Environmentalism often utilizes a powerful rhetoric that is virtually identical to that of nationalism. With it, however, comes an element of privileged discourse that harks back to the formulations that gave such authority to development theory, especially as it was applied to the Third World. In both cases, ideology legitimises interventionist policies. Problems are defined and solutions formulated not within the societies in question but by outside experts who are accorded extensive power and prestige.
>
> (1990: 10)

Colchester (1994) reaches a similar conclusion, arguing that the fashion for linking the 'global' and the 'environment' has encouraged international and state interventions – and this quite clearly carries connotations of the exercise of international power.

Environmentalism and discourse

One way of representing the power reflected in environmentalism is through the notion of discourse (already discussed in Chapters 2 and 3). Adams argues (1990) that in adopting the terminology, or discourse, of sustainability, environmentalists have 'attempted to capture some of the vision and rhetoric of development debates' (3). But decontextualised from power and a more critical understanding offered through a political economy perspective, he argues that such usage has taken on a disturbing naivety and that sustainable development (as defined by Brundtland), although superficially attractive, remains a 'better slogan than it is a basis for theory' (1990: 3).

It is an emerging aspect of this discourse that the activity of tourism can be conveniently linked with terms such as 'environment', 'environmental management', 'biodiversity management' and 'ecosystem protection' to imply a notional sustainability of development. IUCN and WWF reports and pamphlets regularly talk of the protection of areas for and by the development of nature tourism. For example, the IUCN Central America Office's publication *Recursos* refers to the objectives for the IUCN's work in the Talamanca region of Costa Rica: 'The community will use the land to pursue and create conservation activities, scientific tourism, research and demonstration plots, with the aim of stimulating new development projects in the area' (1989: 58). And the WWF-UK's annual review for 1991 refers to 'great potential for tourism' exhibited by the Gashaka-Gumpti reserve in Nigeria which after years of decline still has bright prospects, especially for tourism. 'The Nigerian government has now declared Gashaka-Gumpti as a National Park and WWF has asked to assist in this work' (1991: 15).

184

An added problem has been the intellectual weight afforded to the discourse of First World environmentalists. In Chapter 3 (Box 3.2, p. 51) we introduced Sachs' arguments that the concept of the ecosystem as discourse has provided scientific credibility to the environmental debate and movement. In a similar vein, Yearley argues that such problems are posed by the manner in which expert knowledge and scientific diagnosis and reasoning are employed in environmental management, a problem that can be applied to the issue of biodiversity.

> Given the centrality of science, it is understandable that the discourse of science will affect the way that environmental problems are thought of. ... Unless there are powerful reasons to the contrary, scientists assume that natural processes are consistent throughout the natural world ... the point is that science aspires to universally valid truths ... a universalizing discourse.... This orientation has left its stamp on the overall international discourse of environmental management.
>
> (1995: 226)

It is interesting to reflect on Yearley's use of 'discourse' here. Clearly, he is signalling that 'science' and its use in addressing 'environmental problems' and 'environmental management' is a further way in which power is expressed. As he goes on to argue, doubts have been raised about the way in which the First World has analysed and interpreted 'global environmental questions'. A similar line of reasoning is equally applicable to the debates over both the environmental impact of tourism and tourism's role in addressing environmental degradation.

Cloning

There is a final route through which a First World conditioning of Third World environmental issues manifests itself – especially along the development and environmentalist channel, one which could be labelled 'cloning'. Reservations have been raised about the manner in which environmental INGOs reproduce themselves at the local level. It is, in one sense, a well-crafted technique of propagating self-like organisations and claiming that the concerns of some organisations are also reflected, as by magic, in the Third World. The Third World is teeming with the 'sons and daughters' of development and environmental INGOs, many of which are created in a self-like image. This cloning, or 'parrot syndrome', as Jon Tinker, President of the Panos Institute, describes it, is illustrated further in Box 6.6. As Tinker implies, part of the answer must lie in acknowledging that First World agencies, organisations and institutions do not have all the right solutions, and it is necessary to move beyond the somewhat patronising viewpoint that we can 'help others from developing countries to help themselves', as Coral Cay Conservation put it (*Expeditions* brochure, 1996). This cloning has its reflection in the field of tourism where there is a strong sense that the First World has the blueprints for successful environment-friendly tourism, as illustrated in Boxes 5.5 on the Ecotourism Society and 7.16 on the Adventure Travel Society (pp. 139 and 235).

Box 6.6   The parrot syndrome

Environmental clones

A genuine global shift towards sustainable development in 1992? One barrier is the parrot syndrome: Northern environmentalists recruiting Southern NGOs to endorse Northern analyses and policies. Whales, the fur trade, global warming, Antarctica: the fact-studded, rhetoric-laden, morally impregnable and full-colour-lithographed NGO case often seems irresistible. Southern NGOs are too easily seduced into supporting it uncritically.

They mirror Northern thoughts back to the North, and we call it a global network. But back home, the Southern NGOs begin to talk with strong Northern accents and rapidly become discredited.

You can often detect the parrot syndrome when Latin American wildlife NGOs talk about rainforests. Some of them have lost their domestic credibility, because they now sound like North American clones.

The remedy? More self-restraint among environmentalists in the UK and elsewhere in the North: less eco-evangelism, less certainty that we are right, more listening. And greater efforts to help Southern NGOs prepare and disseminate their own analysis of development issues.

(Jon Tinker, President, Panos Institute, 1992)

We return to the local dimensions of new tourism in Chapter 8; suffice it to note at this point the way in which the notion of a benevolent 'local community' has been so universally subsumed by First World organisations and commentators. So much so that it could even be argued that this is a further reflection and extension of commodification; Third World local communities have emerged as commodities utilised by the myriad INGOs that have emerged in the late twentieth century, to secure grants and aid, and on their behalf widespread fundraising is undertaken. As Korten (1996) concludes, while NGOs can clearly have an important role in building local economies and in advocacy for policies that strengthen local control, not all NGOs are created equal.

CONCLUSION

Environmentalism and environmental issues are central to the debates over contemporary tourism and the emergence of new, alternative types of activity that seek to escape the environmental harmfulness associated with mass tourism. Rather than cataloguing types of environmental degradation and the responses to it, this chapter has tried to demonstrate the scope of environmentalism and how this is reflected in the debates and types of action related to tourism and notions of sustainability and sustainable development.

Our discussion started from a recognition of broader contemporary change, and charted the linkages between environmentalism, the new middle classes and new forms of politics, especially as expressed through the socio-environmental movement. This is an important starting-point for comprehending the nature and content of the debate and the advocacy of new forms of tourism.

Although not strictly linear and exclusive, the environmental types identified by Eckersley help capture the continuum of environmentalism and environmental politics. Above all, this is a means of demonstrating that environmentalism itself is a highly contested arena, with socio-environmental organisations differing markedly over the most appropriate interpretations, policies and programmes, and especially how these relate to 'sustainable tourism'.

Central to the discussion is the manner in which these debates are transmitted through INGOs. This is important, for as later chapters discuss, while it is the most powerful countries together with the World Bank, IMF, other supranational agencies and the transnational tourism industry, that represent the centres of power and are often presented as the protagonists in discussions of Third World tourism, much less critical attention has been applied to the activities of the multinational INGOs and smaller NGOs. These organisations, often mobilised under the cloak of benevolence and respectability around issues of environment, have tended to escape criticism. However, the formulation of environmental politics in the First World and the drawing of other, Third World, places into a global environmental order are significant elements in the power jigsaw (see Figure 2.5, p. 39) and the way in which power is reflected through new forms of tourism.

7

THE INDUSTRY
Lies, damned lies and sustainability

Of all the players in the activity of tourism, those in the tourism industry itself have traditionally faced more blame than all others. They have been the target for some much-deserved criticism; and they have also been the easy scapegoat for many negative impacts of tourism developments.

This chapter considers the tourism industry's adaptations to the new forms of tourism and the different means by which it has made and is making these adaptations. In order to set the scene and to describe the structure of the industry which serves as a platform for these adaptations, the next section covers some relevant features of size and structure of both the conventional mass, and the new or alternative, forms of the industry.

We recognise that in themselves the two descriptive labels, 'mass' and 'new', are inadequate and that there is considerable blurring at the divide between them. There is great uncertainty about the relevant definitions, as Chapter 4 illustrated – what is considered as a 'new' form of tourism and as 'sustainable' varies from person to person, value to value, and within the industry. This blurring of the distinction between the two types is increasing as two processes unfold: first, the large companies offering conventional mass tours are increasingly diversifying their activities to offer exclusive tours with a difference; and second, as the growth in new and alternative tours continues, a number of specialist companies have started to emerge as dominant within this sector and, we would argue, to take on certain characteristics more commonly associated with the large-scale operators.

Notwithstanding these difficulties of definition, this chapter develops the critique of new tourism outlined earlier in the book as it applies to the industry. The following section looks especially at the differences between mass and new forms of tourism in terms of their size and structure. The section 'Redefining sustainability' (pp. 199–223) examines some of the ways in which both forms of tourism manipulate the notion of sustainability to their own ends, and the last section of the chapter (pp. 223–34) briefly describes some of the new personnel involved in new forms of the industry and some recently emerged features of the industry.

SIZE AND STRUCTURE

The effect of scale of operation

An outline of the size and structure of the tourism industry is an important starting-point because, as well as being the world's largest single industry, it is also highly fragmented and diverse. It is composed of several different branches: tour operators; travel agents; accommodation providers; carriers; tourism associations (both NGOs and market-oriented associations); destination organisations (such as tourism chambers of commerce); and consultancies. Each of these can be further subdivided. Figure 7.1 attempts to capture the significance of this scale and diversity.

This fragmentation is one of the factors which has led to minimal environmental regulation, while at the same time ensuring that plenty of examples of environmental 'good practice' can be found. These can be and are employed as marketing tools for the public image of the industry.

Given its size and structure, it is in fact tempting to ask, as Tourism Concern does,

> what exactly is 'the tourism industry'? Is there any such monolithic thing as the title implies? In fact, what we call tourism really embraces a vast and diverse range of activities, from large-scale mass or package tours to small-scale, individually-tailored holidays; from internal domestic visits to family or friends, to international or intercontinental journeys, to business trips and 'sun sand and sea' recreational breaks; from activity, sports, nature, health, 'green' or alternative holidays, to culture or adventure
>
> (Tourism Concern, 1992: 1)

We accept the fact of this diversity and range of activities within the industry and believe that the power held by different branches of the industry varies with size and scale. At its simplest level this means that transnational companies wield great power and influence over Third World governments while small-scale hoteliers have little clout, although at the local level this is not always true; transnationals and international industry organisations are able to influence the practices of the industry and the legislation of national governments, while local catering establishments cannot bring about changes in the industry even though they may adapt their own practices in response to external pressures; the northern tour operator has ready access to the clientele, while the Third World tour operator expends much energy attempting to capture the passing market of First World tourists. As stated here, this is greatly simplified, and of course there are small-scale companies and establishments which are highly significant within their local communities. But in general the scale of operation reflects the level of power and the ability to influence other relevant organisations in tourism such as government.

These differing levels of power are reflected in different ways of adapting to change. Companies vary considerably in the way they absorb the practices of

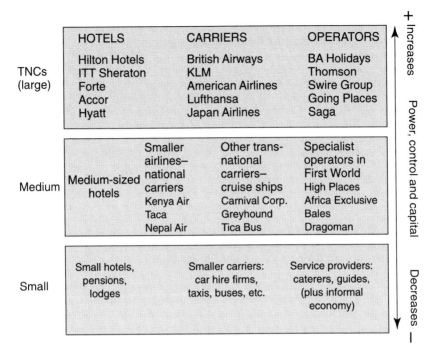

Figure 7.1 Elements of the tourism industry
Note: The companies named are merely examples in the category: the list is not
intended to be accurately representative, still less inclusive

new tourism or redefine sustainable practices so that they merge into existing
practices with relative ease.

It can be argued that the greater the size and power of the company the
stronger the profit maximisation motive – although this latter term can cover a
number of business management strategies – and the more likely it is that the
industry will use consultation, codes of conduct, internal management restructuring and other ploys as public relations exercises to 'green' their image and as
alternatives to introducing significant changes in practice. Such changes would
reduce the impact of their development and move towards sustainability, but
would also run the danger of reducing profits. The imperative and methods of
sustaining profits have already been discussed in Chapter 2 ('Sustainability and
global change', pp. 22–37). Conversely, the weaker the profit motive, the more
willing the operators of the company will be to change practices. This can be
argued as a general rule, but it is clearly one that has many exceptions.

It is important to add that ethical issues are not necessarily uppermost in the
minds of those who run small-scale, specialist tourism businesses. Any consideration of the effects of size, however, must obviously take in the power of the
transnational corporations involved in tourism and their ability to determine the
nature and direction of developments in the industry.

Transnational companies (TNCs)

Lea (1988) pointed out that 'the three main branches of the industry – hotels, airlines and tour companies – have become increasingly transnational in their operations in the 1970s and 1980s, to the point where these large enterprises dominate all others' (1988: 12). Hong (1985), Lea (1988), and Madeley (1996) have all claimed that the operations of TNCs dominate and control the development of tourism.

TNCs achieve this domination in a number of ways. For the hotel industry in Third World countries, for instance, Hong identifies five forms of TNC participation (see Box 7.1). Essentially, these forms of participation allow the TNC to exercise overall or substantial control of tourism activities through some form of contractual relationship while investing the minimum amount of capital in the development. This control of all or many aspects and activities of the industry is referred to as vertical integration within the industry. As Britton outlines, for instance, airlines illustrate this process with their

> resources to buy into hotels, tour operators and other transport operators to reap the benefits of vertical integration: and by forging alliances, set up computerised reservation systems which create captured international integrated (interindustry, for example, airlines, rental car, and accommodation) networks.
>
> (1991: 457)

When this relationship is extended across all three sectors of the industry, encompassing the carriers (most frequently airlines, in the case of tourism to Third World countries), hotels and tour operators, it becomes evident that most of the finance paid into the industry by the tourist is controlled and retained by the TNC. A tourist travelling on holiday from Europe to a Third World country, for instance, spends the bulk of the cost of the holiday on the flight and accommodation. In general, the amount spent on consumables, and even durables, purchased locally will be small in comparison with the costs of flight and accommodation.

In many cases, the hotels, airlines and tour operators used are owned and operated by TNCs whose headquarters are in the industrialised world. The way in which these linkages affect tourism to Third World destinations is illustrated by Paul Gonsalves:

> it is possible for a tourist to leave Japan on JAL [the Japanese airline], be transferred from a Third World airport by a Honda car to a Japanese-owned hotel, to be accompanied throughout his or her tour by a Japanese guide (from Japan, not just a local who speaks Japanese), eat at Japanese-owned restaurants, shop at a Japanese supermarket, and return by JAL to Tokyo, in order to tell his or her friends what a wonderful place the Third World is! . . . Given the complexity of the global economy,

it is likely that the above story is a simplification of reality: for example, the car could have been a Mercedes, the hotel a Club Med, and the restaurant a McDonalds. What is inescapable, however, is that *ownership, control and therefore benefits, from Third World tourism, accrue mainly to the rich industrialised nations from where the tourists originate.*

(1995: 35–6; emphasis in original)

Box 7.1   Five major forms of TNC participation in the Third World hotel industry

There are five forms of participation by TNCs in hotels in developing countries:

1 *Ownership or equity investment*: . . . where the TNC owns a part or the whole of the share equity of the hotel. This is the most direct form of TNC control . . .
2 *Management contracts*: 74% of TNC involvement in the Third World is through management contracts. According to the UN Commission on Transnational Corporations (UNCTC) [now defunct] report on *Transnational Corporations in International Tourism*, 'The management contract is the most far-reaching form of involvement of TNCs in developing countries . . . '
3 *Hotel-leasing agreements*: In these agreements, the TNC pays the owner a proportion of the hotel's profits after deducting the expenses incurred in the operation of the hotel.
4 *Franchise agreements*: These agreements allow the owner of a hotel to use the name, trademarks and services of the transnational hotel chain in return for a fee. . . . The hotel is promoted as a member of the hotel chain and has access to the chain's worldwide sales, communications and reservations systems.
5 *Technical service agreements*: These may relate to particular aspects of the establishment and management of a hotel. This may include market surveys, feasibility studies to determine the size, type and facilities required by the hotel, . . . personnel and staff training and financial or operational management planning and control systems and security.

This interlocking of interests and ownership within the various components of the tourist industry results in the domination and control of the industry by TNCs. . . . Thus 'the real beneficiaries are the rich industrialised tourist generating countries which control the entire industry – hotel cartels, airlines, tour operators and agencies' (UNCTC).

(Hong, 1985: 18–21)

Gonsalves' examples are very generalised, but they highlight the First World ownership and control of tourism activities. Clearly, then, only a relatively small proportion of the money spent by the tourist to a Third World country is spent in that country. This feature of the industry is known as 'leakage', which refers not only to the purchase of imported goods and services by tourists but also to the imports of goods and services by hotels and other tourism establishments or organisations and to the repatriation of profits by foreign owners of hotels and other services. Leakage of course is not a feature of the tourism industry exclusive to tourism to Third World countries.

The level of leakage is significant because it reflects the economic power held by TNCs in comparison with local communities and local governments. But estimates of the precise extent of leakage vary according to source, location and situation. A few such estimates are given in Box 7.2, but it should be noted that the calculation of leakage can be very complex, subject to problems of data collection and availability and based on arguable assumptions and definitions. In general, most estimates suggest that greater than 50 per cent of all tourist money paid either never reaches or leaks out of the Third World destination country. Ways of hiding this negative feature therefore have to be found, and some of these are illustrated below (see 'Redefining sustainability', pp. 199–223).

We shall not dwell on this leakage here. It has been well documented over the last two decades and is an accepted feature of tourism, especially of the operations of TNCs in Third World countries. So we move on to a consideration of TNC operations as they relate to the new forms of tourism and their attempts to achieve sustainability.

In some cases, the First World tourist who flies to a Third World destination (and more than 80 per cent of international tourists travel by air) will stay in a hotel owned by a TNC based in the First World, possibly even the same one which flew them out there in the first place. The quote above from Gonsalves illustrates this for the case of mass tourism.

Such tourists may not necessarily form part of the growing mass of new or alternative tourists who seek adventure, nature, wildlife, culture, authenticity and otherness. Increasingly, however, they are enticed on trips away from the safety of First World standards in the hotel enclave by the prospects of exotic flora and fauna in nearby national parks or by a riot of colour and culture in a local market. This is part of the process of diversification by the large tour operators offering conventional holidays which was mentioned above. Increasingly such companies are using local contacts, organisations and companies or their own couriers or staff to offer variations on the standard, conventional theme of hotel accommodation, hotel food, hotel bar and hotel swimming pool. Through this process, even the most unashamedly hedonistic tourist (who wants nothing more than the swimming pool and the bar) can find themselves lured, even if only for a short time, into the category of new tourists. In this small way, even the conventional mass forms of tourism are being affected by the new, alternative and supposedly sustainable forms of tourism.

An important aspect of TNCs of relevance to the tourism industry and its attempts to address the issue of sustainability is their lack of accountability. Carothers quotes a former staff member of the United Nations Centre on Transnational Corporations (UNCTC) who explains that the issues at stake were:

that multinationals fear even the semblance of public scrutiny. They shun any serious discussion of critical issues: their global market dominance, price fixing practices in small countries, wage cuts and job losses in the Third World, huge commercial debt repayments, and other 'negative' matters. A peak of absurdity was reached at the final preparations for the Earth Summit when there was heavy lobbying to remove the term 'transnational corporations' from the draft text of Agenda 21. . . . Third World debt, environmental destruction and the need for millions of new jobs cannot be addressed without understanding how multinationals operate. Governments and citizens need to know what transnationals are doing, within a legal framework that holds them accountable.

(1993: 15)

Box 7.2 Estimates of leakage from tourist revenues in Third World countries

- The World Bank estimates that 55 per cent of gross tourism revenues to the developing world actually leak back to developed countries (Frueh, 1988).
(Boo, 1990, vol. I: xv)

- The study on Antigua found high leakage of the tourist dollar, especially through the repatriation of profits, leakages from hotels in the form of commissions paid to travel agents, imports of food and beverages, interest payments to foreign banks, and repatriated earnings of foreign employees. Other money leaks were also through payments to foreign auditors, insurance companies, and other service companies.
(Hong, 1985: 22–3)

- In St Lucia, out of the $66.9 million of gross tourist receipts captured in 1978, 45 per cent flowed out of the country as first round leakages.
(Spinrad, 1982: 85)

- In most of the Caribbean, the level of what are known as 'leakages' is very high, averaging at around 70 per cent, which means that for every dollar earned in foreign exchange 70 cents is lost in imports. In the Bahamas, a senior tourism official suggested in 1994 that the leakages for that country might be as high as 90 per cent. More diversified economies such as Jamaica's have been more successful in blocking the leakages. The Organisation of American States assessed Jamaica's leakage at 37 per cent in 1994, a far more respectable figure than is usual in the region.
(Pattullo, 1996: 38–9)

- A leakage of 77 per cent has been estimated for 'charter operations' to the Gambia. A study published in 1978 by the Economic and Social Commission for Asia and the Pacific estimated the leakage was between 75 to 78 per cent when both the airline and hotel were owned by foreign companies, and between 55 to 60 per cent in the case of a foreign airline, but locally-owned hotel.
(Madeley, 1996: 18)

With this in mind, the UNCTC aimed to increase the TNCs' contribution to development, to provide technical assistance to Third World countries on foreign investment issues and to produce an international code of conduct for TNCs. The world of industry, commerce and finance, however, assisted by organisations such as the ultra-conservative Heritage Foundation and by US President Ronald Reagan, continually attacked the UNCTC. Despite the fact that the Centre had achieved very little beyond establishing a few non-binding principles for foreign investment, the United Nations, under its then new Secretary General, Boutros Boutros-Ghali, relegated the importance of its work in 1992 and finally disbanded it in 1993.

With the disbanding of the UNCTC and with the recent push by the tourism industry for self-regulation (see 'Redefining sustainability', pp. 199–223), we now appear to be further than ever from the legal framework mentioned above.

Independent travel and independent tour operators

To tap into and cater for the growth in independent travel (see Chapter 4) specialist tour operators have grown in number and significance. Since 1976 in the United Kingdom some of these operators have been represented by the Association of Independent Tour Operators (AITO), which now has over 150 smaller specialist companies as its membership.

AITO's aims and some relevant background information are given in Box 7.3. In its own words,

> The significant difference between AITO members and the largest tour operators is that, while AITO members cannot expect to compete on equal financial terms with the giants of the industry, they are in a class of their own when it comes to the innovative flair of the specialist. Invariably, the specialists are the pioneers of every new holiday concept – followed, at a later stage, by the large companies.
>
> (1996a: 2)

The AITO stresses this distinction in their list of membership requirements:

> They should be able to demonstrate independence from mainstream tour operating companies. . . . They should match the profile of AITO members, i.e., should be specialist companies at the smaller end of the spectrum, with hands-on management as far as the day-to-day running of the company is concerned.
>
> (1996a: 10)

This distinction is important to the AITO because it allows its members to distance themselves from the widely perceived unsustainability associated with the practices of the large-scale mainstream tour operators. In turn, this distinction and distance give them not only a status of specialists offering personal service but also allows them to claim, almost by default, environmental and ethical principles for their own.

Box 7.3 The Association of Independent Tour Operators (AITO)

AITO's aims are to:

- ensure public confidence in holidays booked with its members;
- inform and improve standards among its members;
- encourage environmental awareness among its members and to promote environmentally sustainable tourism;
- help their members market their holidays; and
- protect and represent the views and problems of the smaller, specialist tour operators.

'The majority of AITO members are small and owner-managed. They therefore score when compared with their mass market competitors in the areas of product knowledge and personal service. (AITO's largest member carries slightly less than 200,000 passengers p.a. but still fulfils the specialist/hands-on criterion.)'

Among other criteria for AITO membership, companies need to conduct a minimum of 50 per cent overseas business and be able to demonstrate independence from mainstream tour operating companies.

In recent years AITO has come to be recognised as the official voice of the smaller or specialist tour operator whose views had so seldom been represented . . . by those who regulate the travel industry. . . . AITO has become increasingly important on the 'political' front . . . as a respected body of opinion representing companies jointly carrying around 1.9 million passengers.'

For its members, AITO serves as a social forum, lobbying group, press contact, training stimulus, source of information and promotional tool. AITO now also promotes the Campaign for Real Travel Agents (CARTA), a scheme to link independent tour operators with independent travel agents.

Green tourism

'AITO was involved, in 1990, in the establishment of Green Flag International [see Box 7.5], a non-profit-making company set up to encourage tour operators to understand the importance of environmentally sustainable tourism. AITO is convinced that the future of the travel industry lies in taking environmental issues seriously. Green tourism is promoted in the AITO Directory and AITO intends to continue to back initiatives in connection with the environment.'

All AITO members must abide by a Code of Business Practice.

(Association of Independent Tour Operators, 1996a)

There can be no doubt that the AITO values environmental sustainability, however it may be defined, very highly. It lays considerable stress on its general environmental aim, its promotion of green tourism, its membership of Green Globe, and its involvement in related debates. It promotes a code of conduct for tourists and it emphasises the importance of ethical considerations in the business practices of its members.

It is, however, an organisation designed to help its members market themselves and improve their performance (maximise their profits). In itself, this is fine, for it is clear and explicit and profits are not something to be avoided. But what we need to ask is whether this conflicts with its promotion of sustainability. One of its guidelines for tourists, for example, is 'Get your holiday off to a green start – if possible, travel to and from your airport by public transport' (1996b: 42). This is resonant of Michael Heseltine's view of the worst and most irritating environmental polluter as the car driver who empties the contents of his ashtray into the road (BBC, 1991). Heseltine's inability to see the half ton of metal already on the road which belches out toxic fumes and threatens safety is so short-sighted as to be risible. In the AITO example above, the most polluting element of the holiday is the air journey. The code of conduct ignores this leg of the journey because it is essential to the holidays it is selling, and the AITO chooses instead to focus attention on a relatively trivial leg of the journey. Long-haul operations overwhelmingly involve air travel, an environmentally damaging form of transport. As Hall and Kinnaird have noted:

> global travel to ecotourism destinations undertaken in fuel-hungry aeroplanes is in itself incompatible with ecological sentiments. As the very support upon which all life depends is under threat as a consequence of our Western lifestyles, it is acknowledged that patterns of consumption must shift away from fossil fuel burning and the use of non-renewable resources. The extolling of ecotourism development in faraway lands . . . may be thus viewed as paradoxical.
>
> (1994: 111)

This paradox calls into question the claims of sustainability made by operators of new forms of tourism. Of course, the AITO is unlikely to suggest to potential customers that they should stay at home. But the example illustrates both where its real interests lie and the limits placed on its 'greenness' by the nature of its business, purpose and ethos.

We would suggest that the 'Independent' in the AITO is a relative term which cannot be defined without reference to the political and ideological contexts in which it is set. Likewise, the environmental sustainability it pursues is subject to constraints, not least of which is the fact that its definition varies with many factors, and should be interpreted only with reference to its political, economic and social contexts.

The new tourist, the independent traveller, the backpacker or the trekker, is one whose image, fashion and consciousness necessitate a clear, and sometimes explicit, acknowledgement that they seek sustainability in their holiday pursuits, with a minimum of negative impact. They may even be aware of the likely incidence of financial leakage from the destination country, and may well attempt to spend their cash in local shops rather than in the TNC-owned hotel parlours and kiosks. But for them too, the bulk of their holiday spending goes on the long-haul leg from home to Third World country. Even if they attempt to stay in and eat in small local establishments, travel by public transport and buy local crafted rather than mass-produced goods, they know they have the economic clout to spend their way out of difficulty or hardship. In terms of leakage, then, the new tourist is only marginally less associated with TNC control of the industry than the conventional mass tourist, and their money is only marginally less likely to leak out of the destination than that of the hedonist mass tourist enclave.

Consider the following two types of tour:

1 a Club 18–30 tour;
2 a trekking expedition to Nepal.

Assume that two average tourists – one seeking sun, sand, sea and sex, the other seeking something alternative, authentic and sustainable – live in the same town in England and earn similar amounts of money. Table 7.1 makes a qualitative assessment of the sustainability of each type of tour. The assessment is simple, intuitive and unscientific, and many of the categories of impacts depend upon definition. It uses simple qualitative scales for each type of impact in order to provoke debate and precisely because we do not wish to imply that there exists an indisputable definition of sustainability.

The conventional but superficial wisdom might tell us that the trekking expedition makes less of a negative impact on the environment than the

Table 7.1 A qualitative assessment of some differences between a conventional mass tourist package and a typical trekking package

Impacts often linked with sustainability	Club 18–30	Trek
Distance travelled to destination	1,000–2,000 km	7,000–8,000 km
Level of pollution associated with mode of travel (air)	high	high
Length of visit	1–2 weeks	3–4 weeks?
Cost of tour paid to operator in UK (£)	200–500	1,500–3,000
Daily expenditure at destination (£)	medium/high	low/medium
Contact with local population	limited	limited
Number of jobs created in destination community	medium/high	low
Quality of jobs created in destination community	low	low
Secondary production and services created in destination community	medium	medium
Social dislocation caused within destination community – dependency on tourism	possibly high – very dependent	limited – very dependent
Cultural impacts	limited	possibly high
Direct ecological damage at area of contact	high	high
Indirect ecological damage in surrounding areas (e.g. deforestation, changes in farming practices)	low/medium/ high?	high

conventional hedonistic tour. Table 7.1 gives cause to question this conventional wisdom and the general perception that trekking, and possibly other new forms of tourism, are sustainable. Over-simplified this comparison may be, but it suggests the need for further research to give a sounder footing for these highly generalised impact levels.

REDEFINING SUSTAINABILITY

We would not argue that the profit motive of private companies in capitalist countries necessarily negates or dominates other motives. There are many examples from around the world of good environmental practice allied with profitability; there are examples of unquestionable altruism on the part of profit-maximising companies. Moreover, the very idea of publicising and promoting examples of good practice is eminently sensible. But the profit maximisation motive does have a tendency to subvert and subjugate other considerations, ethical and environmental. It is essential to keep this in mind in any analysis of the tourism industry.

The industry has responded to the growing importance of the notion and use of the term 'sustainability' in a range of ways which are examined in this section. The framework for the analysis here includes, where appropriate, companies which cater for conventional mass types of tourism to Third World destinations as well as those offering new types of holidays and tours for the new middle classes. This is because the focus in this section is on the term sustainability, and it is not just new forms of tourism to Third World countries which claim to be sustainable. The mainstream industry is also attempting to green its image, and this attempt is linked with the process of diversification which some large tour companies are undertaking. This section therefore looks at the techniques employed by both these forms of tourism, if indeed the two are so easily distinguishable and so mutually exclusive.

Many of the techniques of relevance here have been briefly discussed in Chapter 4. Broadly, those listed in Box 4.11 which are under some degree of control by the industry rather than other bodies fall into the following categories: advertising; industry regulation; environmental auditing; consultation techniques; and codes of conduct. To these can be added internal management strategies.

Advertising

For members of the tourism industry advertising is a means of claiming sustainability. Here we do not analyse the language of tourism advertising – this has already been covered elsewhere, especially in Chapters 3 and 4. We shall, however, look at some of the claims, at the media used for the purpose of advertising, and its effect on control over the industry.

Figures 2.1, 3.2, 3.3 and 5.1 illustrate some of the language of new tourism, as do Boxes 4.6 and 5.6. That language may be different in vocabulary from the language of mass tourism, but it is almost as narrow in its range. Instead of words such as *pleasure, relaxation, carefree, resort,* and so on, the new tourism plays heavily on words such as *conservation, ecology, responsible, environmental,* and so on. Pratap Rughani's description of tourist brochures as 'pleasure propaganda, selling escapism in a tone of juvenile orgasm, where "the natives smile welcomingly", everything is commodified, available and *tremendous*' (1993: 12) seems to apply to new tourism as much as it does to conventional mass tourism.

The medium for advertising most commonly used by operators and exponents of the new tourism (apart from their own brochures) is that of magazines and journals. In particular, those magazines which will lend credibility to operators' claims of sustainability, environmental friendliness and cultural sensitivity are targeted as suitable vehicles for publicity: *New Internationalist, Resurgence, Green Magazine, New World* (newsletter of the United Nations Association, UK), the *Geographical Journal, Geographical Magazine, NACLA Report on the Americas, Wanderlust,* and the publications of the Sierra Club and the National Audubon Society. Various solidarity organisations such as the Cuba Solidarity Campaign, Nicaragua Solidarity Campaign and the Central America Human Rights Committee also occasionally include with their newsletters advertisements or publicity fliers for tour operators which proclaim a solidarity association with their aims and objectives.

Association with socio-environmental organisations allows operators or agents the chance to tap the conscience of the new middle-class tourists from the First World. Their advertisements offer the temptation of guilt-free, low-impact travel at the same time as providing a specific purpose to their tour, such as learning the language, researching the ecology, visiting specific development projects, promoting a solidarity twinning arrangement, joining a delegation on a specific study tour, or working in a brigade. In this way, the operators and agents are simply functioning in much the same way as the large-scale mass tourism operators who have identified their market population sector and then target their advertising at them through appropriate media channels. The market may be more specialised (niche) and fragmented and the operators may be smaller in scale, but the appeal to escapism (from the daily work routine) is very similar, even if implicit rather than explicit, and the commodification of the tourism product (the wildlife, the national park, the scenery) is just as strong an appeal as mass tourism's appeal to the sun, sand, sea and sex. As Tom Selwyn writes, under the subheading of 'The omnipotence of the instantaneous', 'the invitation is to make [the same kind of] instant choices: to suck, fuck and then quickly move on to the Dominican Republic' (1993: 127).

Many of the new operators deliberately align themselves with campaigning organisations which focus on environmental, social and ethical issues. Others produce their own newsletter or magazine, as distinct from their annual brochure, which feature articles on general tourism issues as well as news about their tours. It is an important indicator of their credibility to be able to show that they support some pertinent international organisation such as WWF or a local conservation organisation at the destinations they visit. Table 7.2 lists various features associated with a number of British-based new tour operators – new in the sense that they cater for the new middle-class desire for trekking, travelling or trucking to Third World destinations.

One important characteristic of their tours which nearly all the new tour operators report on is the size of their groups. Small group sizes indicate an awareness of the impact of tourist groups and thereby add to the environmental credentials of the operator. The practical considerations of managing a touring group may be a significant factor in group size, but it makes good public relations to show sensitivity by stipulating a maximum group size regardless of this consideration.

Some of the campaigning organisations themselves solicit the advertisements and attentions of the new tour operators. The National Audubon Society, for example, has a Marketing and Licensing Department which invites service providers and producers to use their name in a pamphlet entitled *Looking for a New Niche . . .* , in which it expounds upon 'the Selling Power of the Audubon Name' (Audubon, undated). As it states in the pamphlet, 'The NATIONAL AUDUBON SOCIETY has been licensing its name for fifteen years. Now more than ever before, our name is a powerful marketing tool for increasing sales, strengthening brand loyalty, and enhancing corporate image.' It would be difficult to find a clearer, more explicit confirmation that the new nature and conservation-associated forms of tourism are firmly rooted in capitalist accumulation than this. To question the sincerity of the ethical basis of this type of marketing and operation would not seem to be too surprising. As Wight says,

> There is no question that 'green' sells. Almost all terms prefixed with 'eco' will increase interest and sales. Thus, in the last few years there has been a proliferation of advertisements in the travel field with references such as ecotour, ecotravel, ecovacation, ecologically sensitive adventures, eco(ad)-ventures, ecocruise, ecosafari, ecoexpedition and, of course, ecotourism.
>
> (1994: 41–2)

In their view of how (eco)tourist companies should set about 'marketing their product', Richard Ryel and Tom Grasse (1991) give an even clearer prescription of how the new, environmentally friendly companies should fit into the prevailing mode of capitalist accumulation. In an article which at times reads rather more like an instruction manual to the new ecotourism companies than a supposedly objective analysis of the marketing of ecotourism, they state:

> Nature travel companies should therefore invest in repeated advertising with their most productive advertising media. . . . Advertising that complements editorial content also enhances the effectiveness of advertising. . . . Ecotourism companies should keep abreast of upcoming editorial coverage in order to take advantage of special features that focus on their destinations.
>
> (1991: 173–4)

Table 7.2 Selected characteristics of 'new' and specialised tour operators

Tour operator	Links with/supports	Own magazine	Specified group size
Abercrombie and Kent	Donates £5 per booking to various wildlife charities; Friends of Conservation; sponsored Royal Geographical Society (RGS) conference on 'Fragile Environments: People and Tourism'	*Sundowner*	Maximum of 15 on accompanied tours
Africa Exclusive	Projects associated with the RGS; water pumps in Hwange Game Reserve (Zimbabwe)	*Horizons*	Individual, tailor-made tours
Bales	CERT (Campaign for Environmentally Responsible Tourism); The Egyptian House; (£1 per person)	Client-based Loyalty Club newsletter	Specifies a minimum number of 15; no maximum specified; takes older and disabled members
Andrew Brock Travel	—	—	Individual tours + groups of average size 10–12 and maximum 18
Dragoman	Survival International; CERT	*Dragodirect*	23 or 24 depending on truck
Encounter Overland	WWF; Save the Gorillas; Tusk Force; Born Free Foundation – Elefriends project; Project Tiger; rainforest preservation in South America	Client-based newsletter	Depends on tour, but always specified
Exodus Biking Walking Discovery	Intermediate Technology – Fuel for Food programme; groups supported changes annually	Exodus newsletter	Dependent on itinerary and activity; maximum 16 often specified
Explore	Green Globe; WWF; Friends of Conservation (Africa)	Explore newsletter	Depends on tour; average 15–16; a few tours specify a maximum of 24
Guerba Expeditions Africa	WWF-UK; CERT; Project Life Lion; African Foundation for Endangered Wildlife; Rhino Ark Charitable Trust; Diane Fossey Gorilla Fund	Guerba newsletter	Maximum is dependent on activity; truck size 22; trekking tours 16; coach tours 26
High Places	Río Mazán Project (Ecuador)	High Places newsletter	Trekking maximum 12; expeditions smaller

Table 7.2 cont.

Himalayan Kingdoms	Helped Tourism Concern with code of conduct; Kathmandu Environmental Education Project; annual charity support	Two client-based newsletters per year	Maximum of 12; average of around 8
J. & C. Voyageurs	Occasional sponsorship of NGOs such as WWF and local organisations; has no charity budget	—	Specialist groups in Africa up to 8
Naturetrek	World Environment Partner 1996	Occasional	Minimum 4; maximum usually 16
STA Travel	—	—	Individual travel
Trips	Member of Tourism Concern; Programme for Belize; Rainforest Adventure for City Kids; Plan International; Rainforest Action Costa Rica	World Comic (six monthly)	Individual, tailor-made, but have organised tours of up to 8
Truck Africa	CERT; African Medical and Research Foundation; Chipangali Wildlife Trust	—	22 or 26, dependent on truck size
Worldwide Journeys and Expeditions	Kasanka Trust; World Pheasant Association; Galapagos Trust; Friends of Conservation; KIDAI Rhino Trust	New newsletter, 1996	Maximum group size 8–10; but 85% are tailor-made for groups of 2–4

In-house brochures, newspaper supplements, magazines and journals are not the only media available to the new tour companies. Increasingly, advertisements for specific tours are appearing on travel, tourism and environment-related bulletin boards on the internet (as noted in Boxes 2.5 and 2.6 and Appendix 1). It is a particular feature of such advertisements that they appeal to the researcher or enquirer in the new middle-class tourist. Examples are given in Boxes 7.13 and 7.14 later in this chapter where 'new academic tourism' is discussed. Major advantages of this medium are its cheapness and its audience, a population which contains a large body of potential and current ecotourists and ego-tourists (see Chapter 5).

Industry regulation

It was argued in 'The tools of sustainability in tourism' (pp. 115–22) that the issue of regulation of the industry can be represented as a struggle for control of the industry between different interest groups. These may be many and varied – as has already been noted, the industry itself is highly fragmented with many associations of hoteliers, travel agents, tour operators, caterers, transport companies, and so on, at local, regional, national and international levels. Some of these groups undoubtedly have a role to play in promoting the attainment of ethical standards of practice, for the fragmentation is such that it would be impossible for all but the most bureaucratic of governments to regulate for all related practices and to enforce the legislation as well. Moreover, there is a clear and undisputed place for national and international legislation on a number of matters, such as airline safety and safety matters relating to other aspects of tourism. In most other areas, regulation for sustainability is a concept as contested as sustainability itself. And this contest leads to the current debate around the issue of self-regulation. It is essentially between two camps: those who believe that the industry should pursue voluntary self-regulation on issues relating to sustainability; and those who believe that regulation should take the form of government imposed and enforced statutory legislation. It is not as simple as the descriptions of these two opposite camps would suggest – there are many combinations of company self-regulation, articles of association and government legislation which can be promoted – but the argument is often presented as a simple dichotomy.

Rebecca Hawkins and Victor Middleton of the World Travel and Tourism Environment Research Centre (WTTERC), which was set up by the World Travel and Tourism Council (WTTC) in 1991, have pointed out that 'Despite concern about the environmental impacts of tourism . . . the industry overall has scarcely been affected by international regulation (1994: 104). In 1993, they identified six major categories of international environmental regulation and control which affect the tourism industry. These are listed in Box 7.4. Their identification of these categories and their general conclusion that the effect of international regulation on the tourism industry is very limited lead, almost inevitably, to a defence, or even promotion, of industry self-regulation. In other words, their supposedly objective outline of the advantages of self-regulation (no disadvantages are listed) becomes part of the contest itself.

The areas of the industry affected most by this struggle are those which can be externalised in the industry's accounting procedures, where the industry has an impact upon the environment, society, culture and other factors which are not costed financially.

It is self-evident that regulation of the activities of different branches of the industry constrains these activities. In the eyes of the industry and under the doctrine of the 'free market' (see Chapters 2 and 9), constraints on these activities are for the worse. They inhibit competition and consequent price reductions, they create 'unnecessary' bureaucracy, they cause delays, they may alienate those who work in the industry, and they stifle its performance and effectiveness.

But without these constraints, the industry is free to pursue profits with no

---

Box 7.4 *Categories of international regulation and control of relevance to the travel and tourism industry*

1 Agreements which deal with the right to free time and a safe environment as an aspect of human rights (e.g. the Stockholm Declaration, 1972).

2 Agreements documenting the environmental impacts of travel and tourism (e.g. Manila Declaration, 1980) – these are now being addressed more directly in international agreements such as the Alpine Convention and the Antarctic Treaty.

3 General environmental policy dealing with specific emission, pollution and ozone layer issues which have implications for the environmental practices of some travel and tourism companies – e.g. Protocol for the Prevention of Marine Pollution of the Mediterranean Sea by Dumping from Ships and Aircraft (1978); International Civil Aviation Organisation's Rules (1986) limiting aircraft emissions of smoke, unburnt hydrocarbons, carbon dioxide and oxides of nitrogen.

4 Policies developed for specific areas, such as the Mediterranean Basin, the Great Lakes and Antarctica as a reaction to general environmental damage and in which tourism is or could be a major issue – e.g. the Barcelona Convention (1978), the Convention for the Protection and Development of the Marine Environment of the Wider Caribbean Region (1986).

5 Regulatory and self-regulatory global policies developing as a result of the Rio Summit, intended to respond to the new ethic for sustainable living and to ensure that growth remains within Earth's capacity – e.g. Agenda 21; the Climate Convention.

6 Emerging self-regulation within travel and tourism corporations – usually implemented on a self-interest, voluntary basis, although it may have implications for membership of some organisations.

(Hawkins and Middleton 1993: 165)

---

regard to the external costs, the negative impacts on the environment, the culture, or the society. It is also free to use its voluntary attempts at self-regulation as a public relations exercise or marketing ploy. And without official, non-industry based, monitoring and inspection, it can deceive its consumers into believing that its operations are environmentally friendly or ethically sound.

Of course, it is not as simple a problem as stated by these two distinct and polarised camps. The issue may be more fairly and faithfully represented by a continuum of views between the two. But, as McKercher points out, 'In a free market system, such a diverse and highly unregulated industry as tourism will likely continue to defy most efforts to limit its expansion' (1993: 11). The diversity which McKercher talks of refers to the highly fragmented nature of the industry already noted. 'Effective control measures can only occur through integrated programmes that incorporate federal, state and local legislation' (1993: 11). Any attempt to regulate such an industry would have to be clearly targeted unless it is to attract justified criticism from industry members.

At the same time, this fragmentation of the industry is one of the factors which permits operators within it a certain inconsistency. When government tries to regulate the operations of private companies, there are few industries in which it is allowed to do so without vociferous opposition from companies which vigorously uphold the benefits of voluntary self-regulation. When tourism companies are asked about the industry's responsibilities, on the other hand, their answers are somewhat at variance with their attitudes to regulation. As a Tourism Concern/WWF study of a number of companies showed: 'All operators stated that national governments had some responsibility, and nearly 60 per cent of operators said that governments had total responsibility. This view was echoed by travel agents, carriers and hotels' (Forsyth, 1996: 31). This clearly points to the need for some form of authorised regulation rather than voluntary self-regulation.

The establishment of Green Globe by the WTTC (see Box 7.5) was seen by the proponents of self-regulation as helpful because

> it enabled tourism companies to seek advice on environmental matters from experts who also understood the travel trade. . . . Furthermore, such progress had been achieved, by and large, without the intervention of outside bodies. This demonstrated the commitment of the tourism industry to sustainable tourism.
>
> (Forsyth, 1996: 6)

As Brian Wheeller points out: 'what they [the WTTC] do, they do very well, indeed excellently. The quality of their high profile publicity is what one would expect from such a professional body' (1996: 15). But in the same paragraph, he also opines that the WTTC represents business interests which advocate the message 'no outside regulation, we can regulate ourselves', and who are acting, not altruistically, but only in the immediate interests of their members.

Noel Josephides, Managing Director of Sunvil Holidays and chairman of the AITO Trust, supports this view of Green Globe:

> the underlying reason for its launch is to prevent, by having in place a self-regulatory system, any government interference in the workings of the industry. There is no doubt that the large global players recognise the increasingly harmful impact the industry is having on the environment, which is now exciting considerable interest and anxiety among the media and inevitably the regulators. They also know that this unwelcome interest will interfere with the current freedom and market dominance they enjoy. If they have the Green Globe scheme in place before too many questions are asked, they will be able to hide behind the façade of self-regulation.
>
> (1994: 10)

It is perhaps too easy to interpret the WTTC's wholesale encouragement of self-regulation as promoted cynically, purely in self-interest and in pursuit of short-term profit-making. It could also be interpreted as a genuine attempt to help

Box 7.5   The Green Globe scheme

Background and aims

The Green Globe scheme is an international environmental management and awareness programme developed for the travel and tourism industry by the World Travel and Tourism Council (WTTC). It arose from the Rio Earth Summit in 1992 (see Box 2.3) and was launched in 1994.

Since its launch, it has absorbed Green Flag International (GFI), an organisation which described itself as 'acting as a bridge between the travel industry and conservationists' in its efforts to mould 'a more environmentally sustainable tourism industry' (GFI, 1990: 1). Indeed, the former director of GFI, Dick Sisman, is now Green Globe's director.

Green Globe has been designed to encourage all travel and tourism companies, whatever their size, sector, location or level of environmental activity, to make a commitment to continuous environmental improvement.

Programmes

Green Globe promotes a number of programmes in pursuit of its aims:
Green Globe Industry Partners;
Green Globe and the Consumer;
Green Globe Destinations;
Green Globe Projects; and
Green Globe and Communications Technology.

It offers achievement awards, training, networking, publications, advisory and information services, branding and marketing assistance.

Membership

'Membership is granted on an annual basis to companies which declare their commitment to environmental improvement and show progress in achieving their stated objectives. . . . As the international consumer awareness programmes associated with Green Globe gather pace, it is expected that . . . Green Globe will be one of the principal ways by which the discerning tourist will choose a holiday company or destination in future' (United Nations Environment and Development UK Committee, 1994: 7).

'Greenwash'

Under the headline of 'Greenwash Contradictions', Tourism Concern reported on the launch of Green Globe (1994a: 13). It pointed to the contradiction between Green Globe's advocacy for companies of sound environmental principles through self-regulation and the WTTC's lobbying and promotion of the needs to expand travel infrastructure, liberalise policies to encourage growth in the industry, and remove physical, bureaucratic and fiscal barriers to travel.

the industry adapt to what may become environmentally essential regulation. The former interpretation suggests that business is simply trying to avoid the inevitable; the second that it is prescient in trying to adapt to it.

Whether government legislation would really help to reduce the uneven and unequal nature of tourism development may be debatable (see Chapter 9). But self-regulation led by bodies such as the WTTC and the World Tourism Organisation (WTO), whose stated aims are the promotion of the tourism industry rather than its restraint, is likely to lead to policies which further the pursuit of profits in a business world where profit maximisation and capital accumulation is the dominant form of operation.

Environmental auditing

An important technique available to the industry for the purposes of adjusting or amending either its practices or its image is that of environmental auditing. Goodall describes an environmental audit as:

> a management tool providing a systematic, regular and objective evaluation of the environmental performance of the organisation, its plant, buildings, processes and products. In essence, environmental audit and EIA [Environmental Impact Assessments] have the same goals and are complementary tools in the struggle to achieve sustainable tourism.
>
> (1992: 62)

While we consider the principle of environmental auditing to be laudable and the practice to be worth further work and refinement, we would argue that Goodall's description of the technique as 'objective' is rather naive. To be considered as objective, it would need to account for who is carrying out the audit, for whom and for what purpose. We would further argue that the broad context of the tourism operation (regional, national, international) needs to be considered. It is normally the case, however, that environmental audits are conducted on hotels, airlines, tour operators or other branches of the industry with the major point of reference being the efficiency, economic or otherwise, of the operation. Reference to the wider distributive effects of the operations are rarely made. Instead, they address issues such as the recycling of cans, bottles and other materials, the reduction of water use by, for instance, not changing towels every day, the installation of energy-efficient lighting systems, and reduction in the use of toxic chemicals. As Geoffrey Lipman, WTTC President, has said:

> Hotels are plunging in to reduce, re-use and recycle and the adoption of sound management. Car rental companies in remote parts of the world are switching to lead-free petrol, and airlines are committing billions of dollars to quieter, less polluting and more fuel-efficient aircraft.
>
> (1992)

These examples illustrate the narrow way in which the industry perceives efficiency: there is an underlying implication that such changes in practice should be made for the purpose of cost reduction; and where the environment is explicit as a consideration, it is so only in relation to the natural environment. Social, cultural, distributional effects may be more difficult to measure, but they could not be much further from consideration than they are at present. We stress that such initiatives are laudable and are worthwhile in themselves; they are to be encouraged. But their ability to solve most, if not all, of the problems caused by the tourism industry is imaginary.

It is the purpose of the technique of environmental auditing to improve the day-to-day environmental practices of the industry. Indeed the International Hotels Environment Initiative (IHEI) developed its Charter for Environmental Action in the International Hotel and Catering Industry in 1991 with this in mind. The Charter is shown in Box 7.6.

Again, it is important to stress the need for industry to adapt itself along the lines of the Charter. But, given the list of signatories to the Charter, it might be worth questioning whether there is a hidden agenda behind such initiatives. It is difficult to believe that the companies listed are willing to adapt their procedures and practices to an extent which would bring about a change in operating motive and which would allow for the explicit costing of socio-environmental factors. The power structures within the industry and their motives for operating are unlikely to be altered by techniques such as environmental auditing and little if any redistribution of benefits will take place. Even for a hotel like the Intercontinental (whose group is a signatory of the Charter) in Managua, the capital city of Nicaragua, its environmental policy on water use appears rather diluted, to the point of meaninglessness, when one contrasts its regular draining of the clean water in its luxury swimming pool with the broken supply of poor quality drinking water to many of the *barrios* surrounding it.

It will be noted, however, that the examples used here so far relate to the large-scale, luxury market, that end of the market which caters to the wealthy tourist and the expenses-paid business traveller. The new, still relatively wealthy middle-class traveller to the Third World, and the specialised operators who cater for them, are less likely to stay in such luxury accommodation. Their accommodation is less likely to be subject to environmental audits, despite the increase in recent years of (even Third World) government inspection of such facilities, largely because the small to medium-scale accommodation companies do not have the capital resources or the internal management structure to conduct them.

The technique of environmental auditing, then, may more appropriately be seen as fine-tuning of the system, or at least one end of the system. And it is a fine-tuning exercise which allows the companies concerned to claim sustainability as theirs. To give these claims more substance, well-known environmentalists or environmental consultancies are brought into the development of such initiatives and programmes as an exercise in consultation (see 'Consultation and consultancies', below).

Box 7.6 *Charter for Environmental Action in the International Hotel and Catering Industry*

Recognising the urgent need to support moral and ethical conviction with practical action, we in the hotel industry have established the International Hotel Environment Initiative to foster the continual upgrading of environmental performance in the industry worldwide.

The Initiative will endeavour to:

- provide practical guidance for the industry on how to improve environmental performance and how this contributes to successful business operations;
- develop practical environmental manuals and guidelines;
- recommend systems for monitoring improvements in environmental performance and for environmental audits;
- encourage the observance of the highest possible standards of environmental management, not only directly within the industry but also with suppliers and local authorities;
- promote integration of training in environmental management among hotel and catering schools;
- collaborate with appropriate national and international organisations to ensure the widest possible awareness and observance of the initiative and the practice it promotes;
- exchange information widely and highlight examples of good practice in the industry.

The Charter signatories are Presidents, Chairmen and Chief Executive Officers of the following transnational hotel companies:

Accor
Forte PLC
Hilton International
Holiday Inn Worldwide
Conrad International Hotels
Inter-Continental Hotels Group
Marriott Lodging Group
Société des Hotels Méridien
Wharf Hotel Investments Limited
Ramada International Hotels and Resorts
ITT Sheraton Corporation

The credibility of claims to sustainability made by companies which are the subject of environmental audits or similar techniques is somewhat strengthened when the audit is conducted by an 'independent' organisation. In 1992, for example, the San José Audubon Society (SJAS), a Costa Rican chapter of the US National Audubon Society, began to monitor the activities of tour operators in Costa Rica:

with the objective of clarifying what constitutes an ecotour. The organisation hopes to satisfy the tourist's desire to know where to go for an ecotour, and be assured satisfaction with the product. Also Audubon seeks to help the operator to effectively market a particular ecotour or lodge practising ecotourism.

(SJAS, 1992)

Together with Beatrice Blake, author of the guidebook *The New Key to Costa Rica*, the Rainforest Alliance and the Institute for Central American Studies, the SJAS researched and put together a list entitled *Recommended Ecotourism Companies* that is updated and published monthly. To be included, companies must comply with the SJAS's Code of Environmental Ethics for Ecotourism. Monitoring their compliance is achieved by observation, reports and question-naires, and is carried out by representatives of Audubon, volunteers, community organisations and tourists.

As a conservation organisation, the SJAS may not be truly independent, but its independence from at least the influences of the industry lends weight to the value of its audits of the companies it examines. And despite the organisation's own internecine political difficulties, this example serves to illustrate that such an exercise can be fruitfully conducted and produce a meaningful and useful result.

With similar purpose, Green Horizons Travel is a travel agency set up in the UK to sell holidays with companies which meet criteria based on environmentally sound and sustainable principles. Box 7.7 uses the company's own words to set out its purpose and *modus operandi*. The credibility of Green Horizons Travel is strengthened by its inclusion in the process of environmental auditing of a panel comprising the Environmental Advisor to the AITO, the Director of the Campaign for Environmentally Responsible Travel and the founder member of Tourism Concern; and this obviously adds weight to the recommendations made by the company. (The issue of the co-opting of environmentalists and their consultancies for the purposes of public relations is examined in the next sub-section.) It should be borne in mind, however, that most companies catering for the new middle-class traveller offer essentially the same service as the large-scale, vertically integrated operators in terms of flights, accommodation and tours.

Consultation and consultancies

Consultation is another technique which can be used by the industry either to assist in efforts to achieve sustainability or as a public relations exercise designed to demonstrate 'genuine' motives while at the same time making few if any significant changes to established practices and procedures. A number of techniques of consultation were listed and briefly discussed above in the section 'The tools of sustainability in tourism' (pp. 115–22). One of these, the Delphi technique, will be illustrated and discussed in Chapter 8 and sources of information

---

Box 7.7 Green Horizons Travel

'Green Horizons Travel has been set up to offer a range of holidays with operators who are taking action to increase the benefits for local communities in the destinations they visit and minimise the adverse effects of their operations on the environment . . . we are the first retail travel agency to offer such a service . . .

Our recommended operators have been selected following an examination of the companies' brochures and research based mainly on the response to a questionnaire, designed by ourselves, to assess each one's environmental responsibility . . .

Assessment has been based on the following criteria:

- Operators' environmental policy and monitoring of their impact;
- Support of local businesses and community;
- Information given to clients before/during holidays;
- Support and active involvement with environment, community and conservation organisations;
- Operators' overall commitment to environmental concerns including environmental housekeeping.'

Green Horizons Travel information pack offers its readers:

- a background statement about the company;
- a list of 'Tourist DOs and DON'Ts';
- a list of organisations acting to reduce the negative impacts of tourism;
- a list of around 30 tour operators that meet their criteria for inclusion in their books;
- a questionnaire for tour operators.

(Green Horizons Travel, 1995)

---

regarding the others are given in Appendix 2. Here, however, the term 'consultation' is taken as having a number of distinct directions which are shown in Figure 7.2 and Table 7.3. The three major types of consultation identified here are not exclusive, and the illustration in Figure 7.2 is partial and much simplified. The governmental level in the consultation procedure has been deliberately omitted partly in order to maintain its simplicity and partly to sharpen the focus on the industry – environmental consultancy companies operating on behalf of the tourism industry can be seen as part of the industry itself.

The exercise of consultation is closely intertwined with the issue of participation, which is examined in some detail in Chapter 8. In particular, the types of consultation shown in Figure 7.2 and Table 7.3 and discussed here should be referred to Jules Pretty's typology of forms of participation shown in Table 8.1 (p. 241).

Linking the words 'consultation' and 'consultancy' is significant here, for environmental consultancies are sometimes used by the industry as a vehicle of

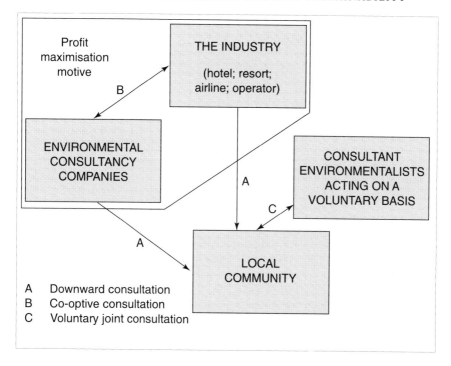

Figure 7.2 Consultancies and consultation

Table 7.3 Three forms of consultation

A Downward local consultation	From the company to local people who will be affected by the tourist development. Can come from either the private consultancy firm operating on behalf of other bodies or directly from the tourism company
B Co-optive consultation	Where a tourism company enlists the advice or association of a person or organisation with ethical/environmental credentials to assist the company
C Voluntary joint consultation	Where a community enlists the help of an independent (i.e. unpaid) environmental consultant or organisation

consultation with local people who may be affected by the industry's tourism proposals. This relationship between the industry and its environmental consultants is important in setting the type and quality of consultation which takes place, and therefore whether it can be seen as consultation which moves towards a presumed definition of sustainability or as a consultation exercise designed to avoid a move towards sustainability.

The consultation which often takes place cuts out the local community. The industry prefers to consult with environmental consultancies, and in some cases those companies large enough to do so may establish environmental departments within their own management structure (see 'Internal management strategies', p. 218).

The kind of consultancies relevant to this field include not just those dealing directly with travel and tourism (such as Green Globe – see Box 7.5) and technical and engineering firms which carry out EIAs on tourism development projects, but also those whose remit is more general. These latter include Forum for the Future, an outline of whose aims and activities, including a Green Futures Consultancy, is given in Box 7.8. It was established in 1996 by three well-known British environmentalists, Paul Ekins, Sara Parkin and Jonathon Porritt. The Forum's aim of promoting 'best practice in sustainability' gives it a reason to approach industry itself without waiting to be asked.

Forum for the Future 'does not see itself as a mass membership organisation', choosing to work instead with a small core of supporters, all of whom are well-known and who can give the organisation kudos and access to different arms of the world of business and finance, the captains of industry. Its 'corporate partners', some of which are listed in Box 7.8 and which include two major international companies involved in the tourism industry, Centre Parcs and Taj International Hotels, support the Forum, financially, and in return gain access to leading environmentalists, such as the three directors. In this way, they can green their corporate image.

The Way Ahead (a radical green publication by a faction of the British Green Party) cynically describes one of the Forum's early initiatives, a conference on 'Business as Partner in Social Development' in India in March 1996: 'The venue was a five star establishment in New Delhi. The accommodation rates at the Taj Palace Hotel are astronomical, but do include free "airport transfer and daily cocktails". Clearly those "partners" to be "socially developed" were not expected to be involved' (1996: 4).

This form of consultation is obviously severely restricted in its reach. It is designed for the upper echelons of business decision-makers and seems to accept that, at least at this stage in the development of the notion of sustainability, other voices can be ignored. It firmly maintains the decision-making at its present level. It is a two-way co-optive consultation, mutually beneficial to both partners. In return for implementing some of the environmental best practices promoted and publicised by Forum for the Future as well as supporting it financially, business is able to improve its environmental credentials and to promote the belief that it is addressing the problems of sustainable development in what *The Way Ahead* calls an 'inherently unsustainable' system.

Of course it is not one of the Forum's stated goals to give green credentials to industry by association, and to be fair to the Forum it does not aim to set itself up as a consultation body between the developers and the developed. It aims to inform, influence and inspire these 'decision-makers and . . . key agents of

---

Box 7.8   Forum for the Future

Early in 1996, Forum for the Future was set up with a vision 'to inspire and motivate people by promoting the positive solutions to today's environmental problems, and to accelerate the move to a safer, fairer and genuinely sustainable way of life.'

The organisation has initiated six major schemes:

- a *Green Futures* publication 'to showcase latest examples of best practice from around the world' which is circulated to 'a very large number of UK decision-makers';
- Forum Scholarships aimed at young people 'aspiring to "green" leadership in any sector';
- the development of a Best Practice Database on Sustainability;
- a Green Futures Consultancy to advise businesses, local government, the voluntary sector and professional associations on sustainable development priorities;
- The Sustainable Economy Unit, a policy think-tank to show how sustainability and prosperity can be jointly achieved; and
- Forum Associates and Coalitions: partnerships, associateships and corporate supporters who wish to pursue the organisation's goals.

Forum for the Future has a formidable array of famous names as its trustees and council members and its 'corporate partners', who have 'a proven commitment to environmental issues', include B & Q, the National Westminster Bank, the Body Shop, Tesco, the Post Office, Wessex Water, Taj International Hotels and Centre Parcs.

(information and extracts from Forum for the Future leaflets, 1996)

---

change to meet the challenge of achieving genuine sustainability'. This is a daunting aim, and no doubt they will accumulate many examples of the effectiveness of their influence and no doubt these will all have merit in their own right. But to achieve a degree of sustainability (or commonly agreed good practice) in a few cases, to have a catalogue of best practice examples, is precisely what the industry would want in order to justify self-regulation, to work against government legislation, and to continue doing business as usual without the systemic, structural change that is required to prevent continued concentration of wealth and power in the hands of those who already take the decisions. As a Green Party cynic remarked, 'Perhaps industry sees these consultancies as the chance to wear a clean conscience as well as a good suit?' (personal communication, 1996).

Tim Forsyth's work for Tourism Concern and WWF surveyed eight consultancies whose total or partial role was to advise tour operators on the establishment of tourism abroad.

Half of consultants questioned believed there was a commercial advantage to be gained by being environmentally aware before industry standards and market expectations make it compulsory. . . . Two also wrote 'sympathy leaflets' or brochures for companies. These were partly to educate tourists about destinations, but also to add value to services offered.

(1996: 25)

Again, the association between environmental credentials and the profit motive, using the vehicle of consultancies, could not be much clearer.

Other forms of consultation can and do take place, as Figure 7.2 shows. Moreover, there are examples of healthy and generally beneficial consultation exercises which are more widely participatory and which can genuinely claim to move development practice towards a greater degree of sustainability. We believe that in any analysis of consultation exercises in tourism developments, it is important first to consider the different forms that consultation may take and second to examine the exercise's inclusivity/exclusivity, its direction of consultation (that is, who initiates it) and its beneficiaries.

Codes of conduct

It was remarked in Chapter 4 that codes of conduct publicised by tour companies, or others, can be misused as exercises in public relations. They can be seen as attempts to jump on the 'green' bandwagon for the purposes of establishing environmental credentials. This ploy can be used to show an environmentally responsible attitude to the supposedly discerning consumer or to head off potential antagonism and bad publicity from critical environmentalists.

The WTTERC has established a database of codes of conduct relevant to the travel and tourism industry, such has been the recent growth in their number especially in the international sphere. A report by Mason and Mowforth (1995) suggests that most codes offer no measurable criteria and conform to no widely accepted set of standards. A good many of these have been devised by tour operators merely to give some palliative to the pressure exerted by the environmental lobby and in anticipation of potential criticism. They may also serve as attractors to potentially discerning and critical customers who are seeking a relatively ethical holiday, allowing them still to travel to faraway and exotic places, have a holiday and return environmentally enriched. In these senses, the code of conduct becomes more a marketing ploy than a set of standards by which to guide a company's behaviour and practices.

A good example of the ineffectiveness of a code of conduct is that produced by Thomson Tour Operations, the Thomson Holiday Code. This appears on the inside back cover of all their brochures, most of which are large, and replete with colourful photographic images that suggest unlimited hedonism. The blandness of the advice given in the code of conduct and its positioning within the brochures seem to suggest that it is included more as a public relations exercise than as a committed stance on environmental protection. Thomson's commitment to the environment is examined further below.

WWF's *Beyond the Green Horizon* (1992) (prepared by Tourism Concern) contains a chapter on 'Marketing tourism responsibly', which includes a list of recommendations for the tourism industry to follow. This list serves effectively as a code of conduct for the industry's marketing practices. The same chapter also points out that: 'A European Commission directive[1] on package tours stipulates that brochures "may not make misleading claims, but must provide clear and precise information . . . " on such practical aspects as price, transport, accommodation and visa requirements' (Tourism Concern, 1992). Although it should be recognised that some detail may have to be sacrificed for communicability, it is also necessary to acknowledge that codes of conduct can clearly be used to make the kind of misleading claims referred to by this directive.

A suitable system of reference for codes of conduct, made easily accessible to tourists and others who wish to verify the claims made by any member of the industry would help to improve the credibility that can be placed in codes of conduct. But any such system would need to take account of for whom they are intended, who initiates them, and who monitors their use if indeed any assessment is made of their effectiveness. To this end some system of environmental auditing of codes of conduct would be helpful. The WTO or WTTC are best placed and best resourced to achieve this, but as has already been established at several points in this book, these industry-based organisations represent the interests of their members and companies which provide their funding.

Where joint efforts are made to devise codes of conduct, the balance of power on the coordinating body often favours representatives of the industry. 'Interested parties' tend to move the aims of such efforts towards the externalisation of as many costs as possible and away from genuine environmental, social and ethical considerations. As an editorial in *The Ethical Consumer* magazine stated about an 'ecolabelling' scheme: 'It seems obvious to us that for any such scheme to appear, let alone be, impartial, representatives from industry are the last people you want on the board' (Ethical Consumer, 1994).

We believe that, for the tourism industry and its codes of conduct, this may over-state the case. There is a clear argument for industry representation in order to move away from simple exhortation towards incorporation of business practicalities (but see below, '"Reality" and "practicality" in achieving sustainability', pp. 221–3). Moreover, the active involvement of the industry is important, for without it, no codes are likely to be effective. Despite which, we believe that the problem of vested interests, as described above, also applies to the design, operation and monitoring of codes of conduct in the tourism industry. As with other techniques of sustainability, they enable the industry to redefine the concept for its own purposes so that changes to operating practices need only be minimal.

Internal management strategies

The internal management systems and strategies of tourism and other companies may be manipulated in order to allow for the environmental and social costs of operations and projects. These costs are those which have customarily been externalised by industry, that is, costs which do not have to be borne by the company but which are borne by the wider community. Many companies large enough to afford the creation of a new department to deal with these externalities have restructured themselves in recent years to include departments, divisions or branches which, supposedly, account explicitly for environmental matters arising from their operations.

One such company is British Airways (BA) whose environmental strategy is characterised in Box 7.9. In BA's own words:

> Environmental management is the means by which an organisation such as British Airways addresses environmental issues. All the environmental activities and initiatives are brought together into systematic programmes. This includes the formulation of policy, the organisation of responsibilities, reviews and audits, procedures, advice and support, training and communication and the measurement of performance.
>
> (1995: 20)

In 1992 BA received an award for its annual *Environment Report* from the Chartered Association of Certified Accountants (ACCA), and there is no doubt that the company exposed itself in making public so much information from its environmental audit. To gain such awards and to produce impressive reports

---

Box 7.9 British Airways' environmental strategy

Since 1989, British Airways has:

- appointed a Director of Safety, Security and Environment and a Head of Environment;
- initiated a corporate 'Good Neighbour' goal;
- undertaken a series of reviews and audits of its main airport locations and their impacts on the local environment in terms of noise, gaseous emissions, congestion, waste and tourism;
- sponsored annual Tourism for Tomorrow Awards;
- sponsored the WTTERC;
- conducted an environmental audit of its related tour operator company, British Airways Holidays;
- initiated a Light Green programme which allows employees to suggest ways of reducing the environmental impacts of the company; and
- produced an annual *Environment Report* which makes public a considerable amount of data about its operations.

---

and large amounts of data, BA created an Environment Branch which meets regularly with directors of other branches, offers awards for environmental achievement (to those it calls its 'Environmental Champions', of whom there are 300 around the world), produces relevant publications, and requests external consultants to undertake research when appropriate.

This is its internal management strategy for the environment, and in 1995 the direct costs of its Environment Branch (including consultancy work) amounted to £425,000. This represents just under 0.06 per cent of the company's operating profit of £728 million in 1995/6, although it should be added that these are only the direct costs of the branch created specifically to cover environmental issues. Doubtless other departments undertake work which can be categorised as 'environmental'; doubtless also, environmental factors have influence in many technical decisions concerning such things as engine noise or emission controls.

But the question which has to be asked is whether this internal environmental management strategy is genuine and effective or whether it is a ploy to display a corporate conscience for the sake of its public image. As one of British Airways Holidays' (BAH) Environmental Coordinators has said, 'it seems that the tourism industry is castigated for what it doesn't do, and what it does do is viewed critically as PR' (personal communication, 1995).

Despite sympathy with this problem of interpretation, despite all the fine words about integrating environmental considerations into the work of all branches of the company and despite all the impressive reports, it is still important to ask whether the creation of BA's Environment Branch has had or is likely to have a significant effect on the company's operations. Is it possible that changes that have been attributed to the influence of its Environment Branch would have occurred in any case as a result of changing regulations and changing public awareness? And has the growing importance of the public image of airlines had a significant role in the company's perception of the need to create such a branch? Beyond these questions, the real test is whether the creation of such a department really affects the ethos and operations of the company. In the case of BA, the answer to this will depend heavily upon one's viewpoint and employment, especially if the latter is with BA or a BA subsidiary such as BAH. A definitive answer is not offered here; but it is stressed that it is important to ask the questions in any analysis of the tourism industry. It is also important to recognise that such management restructuring exercises offer the potential for deception as well as genuine change.

There is no clearer illustration of this type of use of internal management restructuring as a ploy to claim false environmental credentials than the World Bank's actions through 1987 and 1988. The World Bank in itself is not a private enterprise or a transnational corporation of course, but in essence it operates on behalf of such companies and actually leads rather than follows the objectives which guide them. It is therefore used here, briefly, to illustrate a point that is relevant to the rest of the industry.

At the start of 1987 the Bank employed 2,700 staff above secretarial level, of whom only three were trained ecologists. Their task was to monitor the environmental and social implications and impacts of over 250 new projects each year. In response to mounting awareness and criticism from many quarters throughout the 1980s, the Bank created a new Environment Department which employed some forty people. By the end of 1987 it began to assess environmental threats in the thirty most vulnerable developing nations, and to design initiatives against desertification and deforestation. Particularly through its then-President, Barber Conable, it also began to express a commitment to environmental and social criteria. At the time, the Bank's Acting Director of its Environment Department, J. Warford, spelled out this new commitment:

> World Bank staff have clear instructions to carefully consider the environmental consequences of projects, and, if necessary, amend or reject them on environmental grounds. The Bank has recently announced plans to substantially increase its staff devoted to this work, and will recruit additional ecologists, biologists, anthropologists and other expertise as necessary to ensure that our projects and policies are consistent with environmental objectives.
>
> (personal communication, 1987)

Despite its admission of past mistakes and failure and despite its newly stated commitment to reform, the Bank remained wedded to large-scale projects and major investments in energy, infrastructure and industrialisation. Its commitment to local residents and indigenous people has taken the form largely of resettlement programmes and cash compensation. Its commitment to more conspicuous environmental issues is also highly dubious – in 1995 a billion-dollar World Bank package to help prop up private Mexican banks was partly funded by cutting previously approved Bank loans for the environment and other 'not such high' priority projects (Chatterjee, 1995). As Ian Linden, General Secretary of the Catholic Institute for International Relations, makes clear:

> Unfortunately this intention of amendment did not extend . . . from the diakonic to the evangelistic. Poverty programmes were merely clamped on to the main body of doctrine. The World Bank would try to help the strangers, widows and orphans, but the content of their crusade, the Great Doctrine of structural adjustment, remained the same.
>
> (1993: 3)

A further example of environmental management revamping is given on pp. 225–6, where the credibility of the creation, use and effectiveness of Thomson Tours' post of Environment Manager is briefly examined.

It is often pointed out that awareness of an issue is the necessary stage before action. In this context this means that company directors and presidents will talk about the environment before changing their practices. It may take longer than the save-the-world groups would like, but the change will occur if given time.

This argument might be based more on wishful thinking than on evidence, for the industry, more than any other branch of tourism, will change rapidly once it perceives some factor or phenomenon that will bring advantage to it.

A reorientation of the aims, ethos and operations of the tour industry implies the need for structural rather than cosmetic change in its management operations. Unless such structural change is evident, then there must be serious doubt about the intentions behind the creation of new management divisions or departments, especially where they are given severely limited budgets. Such moves as the creation of a new department and the design of a new mission statement may be necessary steps in the process of change but are not in themselves indicative of a significant shift in approach. Indeed, they may simply reflect the industry's wish to redefine sustainability in order to allow its operations to continue as usual.

With this in mind, the obvious suggestion is the centralisation of some form of rating system, independently designed and applied, to which tourists and other users of the industry (such as suppliers) could refer. The problem with such a system is that it suggests some form of regulation, and, as already noted earlier in this section, most of the industry is fixed determinedly against regulation imposed by others. As Jonathan Croall says, 'Sadly, such a scheme is likely to be firmly resisted by the industry' (1996: 5). This only serves to further throw into question the motives of industry-based moves to alter its management structures. The industry will gain credibility for its internal changes only if it subjects them to external criteria and assessment. And even then, if it is done under the guise of a consultancy, as seen earlier, there is a chance that the consultancy will act as an integral part of the industry with its currently prevailing ethos of profit maximisation.

'Reality' and 'practicality' in achieving sustainability

The tourism business community is much the same as other sectors of business in its invocation of 'business realities' in order to justify or excuse its resistance to change and to external influences. 'Commercial practicalities', 'the real world', 'the need to keep the competitive edge', and similar phrases are used to argue against regulation from government and interference from environmentalists and conservation and human rights organisations.

Recently, the arrival on the scene of many companies which claim to operate a form of sustainable, environmentally friendly and ethical tourism has begun to dent this excuse. The alliances made by these companies with environmental and conservation organisations, the stipulation of maximum group size for tours, the promotion of codes of conduct, attempts at weak forms of consultation, and other ploys to portray environmental and ethical credentials have shown to companies which typified the no-change attitude that a degree of change is possible. But what needs to be asked is whether such change as has occurred has really affected the prevailing mode of profit maximisation as a motive for and a means of operating? Or has it simply served to preserve the existing social, economic and political structures in which the problems of tourism identified by many authors over the last twenty years are inherent?

We have tried here to question whether the techniques used by the industry to convey an atmosphere of change and to claim sustainability are cosmetic and superficial regardless of whether they originate from the conventional mass tour companies or the new adventure tour companies. It is important to question techniques such as consultation, advertising, codes of conduct, environmental auditing, and the restructuring of internal management systems. Are these techniques used effectively to bolster respect for the prevailing mode of production and service and all the problems that this creates? Or do they genuinely address the problems in such a way that the system no longer creates them?

Have consultancies such as Forum for the Future, formed by environmentalists who are often considered to be, or at least are portrayed as, radical, simply lent their good name to a variety of companies in return for minor, marginal changes in practice and a healthy consultancy fee? How valid is their justification that the real world is run by businesses whose mode of operation dominates, and that therefore we must work with and inside them in order to bring about change? Is this the only practical strategy of action to pursue? And if so, how effective is it likely to be in altering the basis of the system of capital accumulation, which we have suggested throughout is the origin of the problems caused by both mass and new forms of tourism development?

The excuse of 'the real world' or 'business practicalities' can be used as a justification for doing nothing or making the least change possible. These 'realities', however they are defined, can be used to persuade those in search of change to work with the industry in order, jointly, to examine and implement the ways in which change can be made, sustainability sought, and impacts reduced. It is precisely this 'request that environmental groups change campaigning "from exhortation to a discussion of practicalities and tools"' (Forsyth, 1996: 22) which is used as a ploy to rein in the demands of the environmental lobby, to persuade it that cosmetic change is adequate, and then to subvert it by using it (the environmental lobby) as a public relations ploy.

It can be contended that both new tour operators, in their attempt to present a distinct environmental image, and conventional tour operators, in their attempt to take on environmental credentials, are in fact simply carrying on business as usual but with selected 'add-on' features and marginal changes to established practices. As one of BAH's area managers described his company's policy on social and environmental issues, 'It is basically a PR exercise in many respects because obviously there is a growing awareness of green issues' (personal communication, 19 September 1996). In an examination of environmental valuation, Colin Price summarises both the way in which the industry usurps the notion of sustainability and the logic it uses to justify this:

> The mood of the 1990s has made it mandatory for public, corporate and private bodies to embrace the sustainable development idea. The let-out

clauses of weak and metaphorical sustainability ensure that this need not be financially burdensome. Sustainability objectives will be adopted by the politically astute, while continued application of discounting underwrites a 'business as usual' practice.

(1993: 142)

What Price calls discounting refers to the ignoring of externalities such as environmental and social costs. This captures our own analysis of what is happening in the industry. But regardless of the contentious nature of this view, it is crucial that the questions above are asked before the veracity and sincerity of the industry's stated policies are accepted or rejected.

NEW PERSONNEL AND NEW FEATURES OF THE NEW TOURISM INDUSTRY

The following pages give thumbnail sketches of a variety of personnel and features involved in the provision of new forms of tourism. Although they are based on real characters, are sketchy and stereotypical, they are illustrative of many of the features and phenomena of new tourists, new tour companies and new tourism that have been discussed in these pages.

The new tour operators and their teams

Because the scale of operation and the size of the new tour operators is generally smaller than those of conventional mass tourism (there are exceptions) it is possible to identify individual exponents and employees of the new tourism business, whereas the latter are characterised by large departments. Indeed, many of the new operators which offer independent or small group travelling, trucking and trekking tours are closely associated with one or two individual characters or with a family. High Places, for instance, is directed by Mary and Bob Lancaster, who lead a team of about twelve tour leaders, all of whom are identified with a photograph and potted biography in their brochures. Similar potted biographies are given by many of the trekking and trucking companies such as Explore, Truck Africa, Naturetrek and Guerba Expeditions. A number of the potted biographies for the Explore team were given in Box 5.4 (p. 134).

David Sayers Travel, specialists in botanical and garden travel, is a very small company which plays strongly on its personal approach, inviting potential clients to visit its Director, David Sayers, and his partner at their Lincolnshire home. Bales, an older and well established tour company catering for the less adventurous, older and more wealthy clients, also exploits the family basis of the company to convey its personable and approachable style.

Leadership characteristics played upon by some companies which offer expedition adventures, including overland trucking, mountain climbing, biking and rafting tours, are the youthfulness, vigour, success and machismo of their guides or leaders. 'Dave has a Guinness World Record for his 10,000 km trans-

Africa ride' (Exodus Biking Adventures). '[H]e also descended the infamous Rhondu gorges on the Indus River in northern Pakistan, featured in the television documentary "The Taming of the Lion"' (Ultimate Descents). 'He has made a number of films of Chris Bonington's expeditions and has been on eleven trips to the Himalayas' (Himalayan Kingdoms). One 'lives close to the Cairngorms in Scotland where he and his team run the Avalanche Patrols each winter' (High Places). 'Chris has worked as a mechanic in many areas of the world, including a stint in the Yukon as a gold mine engineer!' (Truck Africa). 'A veteran of two Everest expeditions and many other Himalayan climbs' (Guerba Himalaya). 'A skilled photographer, Chris has captured an enormous variety of African wildlife on film' (Naturetrek).

Clearly, lifestyle is an important feature of the directors, leaders and partners of the new tour operators, and this is as much so in the USA as in the UK. One company which uses the term 'lifestyle' in its name, Lifestyle Explorations, is featured in Box 7.10. A few insights into the thoughts of its Director, Sara Laing, complement the information taken from its brochure and appear, in part, to reflect the characteristics of neo-colonialism discussed in Chapter 3. Clearly, it is interesting to contrast the notion of arranging tours with the purpose of selling land and property in Honduras and emphasising its cheapness to potential US purchasers or investors, with the fact that this country is characterised by chronic poverty, malnutrition and landlessness. Such juxtapositions are symptomatic of the paradox that underlines so much tourism and further emphasise the relative power and powerlessness that is involved.

Box 7.10 Lifestyle Explorations – Third World for sale

HONDURAS – A GREAT PLACE TO VISIT; AN EVEN BETTER PLACE TO LIVE

Ever wish you had made your move to Tahiti or Hawaii years ago, before development, tourism and inflation? However, there are still places in the world where you'll find lush, unspoiled nature, low prices and an opportunity to create the life you've always wanted.

Honduras is one of those places. . . . Honduras is a naturalist's dream with mountains, pine forests, bass-filled lakes, pristine beaches and islands. . . . The friendliness, honesty and generosity of the people are legendary. A growing number of them speak English as well as Spanish.

Honduras today is stable and democratic. What's more, the government has recently established a 'pensionado' program with many advantages for North Americans relocating here.

Experts are calling Honduras 'the number one "sleeper" investment country' in the western hemisphere. To start a small business, you only need a small investment. And the law makes it easy for foreigners to own businesses here.

Your dollars go a lot further here. Imagine buying a prime lot for $3,000, where you could build a two-bedroom, two-bath home for $20,000 to $30,000.

Box 7.10
continued

Or renting a comfortable apartment for $150 to $175 per month. One person can live well on $500 a month . . .

Medical care is excellent with US trained physicians and with state-of-the-art equipment. A visit to a private doctor costs under $6, and the total for an operation, including hospital stay, might be about $1,000.

There's a small but growing expatriate community. You'll find Rotary Clubs, International Women's Clubs, church services in English.

People who travel with us aren't ordinary travellers. Ours is a tour for people who don't like tours! . . .

In Honduras . . . you'll also visit North Americans, talk to them and hear their stories. At our on-site seminars you'll meet highly placed officials. You'll get bottom-line facts and figures difficult to get elsewhere. Find out about taxes, residency, estate planning, real estate, health care, banking and insurance . . .

(from Lifestyle Explorations brochure, 1991)

Sara Laing, tour director of US-based Lifestyle Explorations is also a self-styled environmentalist and compiles its *Earthpilgrim's Journal*. In this, she writes 'Fragile Paradise of Honduras - As in all the other scenic and quaint areas of the world, these places are threatened to be overrun by tourists and developers, but may be saved from this by intelligent and insightful legislative action.'

The new environmental managers

Reference has already been made to the way in which large tourism companies are restructuring their management systems to include environment departments. Along with the creation of such new departments come new posts. Thomson Tour Operations, for instance, whose new Environment Department was noted in the last section, employ an Environment Manager, a post held in 1995 by Paul Thornton, after earlier employment with the Disney Corporation and ten years with Thomson Travel, predominantly in marketing management, but also in 'a variety of other projects for the company including strategic and marketing issues as well as having . . . launched Thomsons on to the Internet' (Commonwealth Institute symposium, 'Managing Tourism: Education and Regulation for Sustainability', 16 November 1995: biographical notes of speakers, 1995).

Reference has already been made to Thomson's code of conduct for the tourist, printed at the back of all its brochures, and of the code's palliative nature and lack of significance. A few of the statements made by the company's Environment Manager during a conference at the Commonwealth Institute in 1995 served to underline the company's lack of commitment to environmental sustainability and its use of the terms 'environment' and 'sustainability' to demonstrate publicly a supposedly ethical lining to their policies. In response to questions regarding the responsibility to educate about the impacts of tourism, Thornton explained that they were there to sell holidays and that the question of 'commercial realities' meant that there was no culture within the tour operating business to consider the impact of tourist developments on the environment. Despite the WTTC President's statement that hotel companies were plunging in to reduce, re-use and recycle, Thornton suggested that tourists should have as many baths as they want and because they have paid for it they should be able to 'abuse [the environment] if they wish'. He also suggested that Thomson already does a great deal for their customers by producing their leaflets in a number of different languages, but at the same time he considered this effort to be more than what was required.

There is of course an argument expounded by the free marketeers that sustainability can only be achieved if market forces are given the freedom to operate. But even the free marketeers acknowledge that environmental, social, cultural and other ethical factors are externalised or ignored by the market. The logic of the argument then assumes that environmentally unsustainable practices can only be rectified after the fact, that is, after they have produced the particular form of pollution associated with that factor. But sustainable development, as far as it can be defined, is something that happens before or during the fact rather than after it.

It seems odd, then, that such a large and renowned tourism company, which is ostensibly trying to pursue the goal of sustainable development through its day-to-day practices, should appoint an Environment Manager who espouses the cause of the free market. It has to be added, however, that other (types of) environmental managers exist.

The new service providers

Not only are the tour operators, the airlines, the hotel chains and travel agencies attempting to adapt their practices, publicity, language and clientele to the new tourism, so too are the smaller-scale service providers, the hoteliers, lodge owners and managers, restaurateurs, minibus companies, guides, and even craft salesmen and saleswomen. In fact, it is not just a case of practices being adapted by those who already offer the service, the service providers themselves are changing.

Take the example of the Illusig family from Italy, for instance. Jacopo Illusig describes and explains the reasons for his family's recent move to set up a pizzeria in Puerto Viejo de Talamanca on the Caribbean coast of Costa Rica (see Box 7.11). The family are illustrative of a new wave of First Worlders seeking to try out their luck and their entrepreneurial skills on a new 'frontier'.

For the Illusig family, the move was opportunistic; they sought to escape their disenchantment with life back in Italy. In other cases, the move is linked with altruistic motives, such as a desire to save the tropical rainforest. Also in Costa Rica, for instance, Amos Bien, an American ecologist, 'decided that the best way to convince people not to cut down rainforests is to demonstrate that conservation through tourism and ecologically sound management is the best use of land

---

Box 7.11   *The Illusig family in Costa Rica*

'In Italy, we worked so that we could earn, eat, dress ourselves and pay taxes. We had to dream about going on holiday. Then he [my father] went to Spain to open a pizzeria. We were in Spain for one year. It was a different kind of tourism, one of drunks, fiestas, wild holiday living.

'Then some people told us that Costa Rica was very pretty and very cheap and that it was possible to buy a small restaurant easily. He returned to Italy and sold the house before leaving for Costa Rica. He was here for a couple of months and then he called us and told us to come. I arrived first, in April 1994, and then the rest of the family.

'I find it enchanting here; we live differently; it's calmer and quieter; there are no fiestas, or if there are, they are calm fiestas. We earn less than we did in Italy, but there's more time here - to go to the beach, to rest, to read a book, to live. There's less money, but there are fewer costs. It's a different life – you work four or five hours a day and say that you have worked hard; in Italy, you work twelve hours a day and say that you haven't worked very hard.'

Have you observed a growth in the number of tourists since your arrival?

'From what I have seen during my year here, there has been a big increase. Every month, there seems to be a new place opening, a new lodge, a new restaurant - always more services, followed by more tourism, and so it goes on. I believe that more services, more infrastructure will bring more tourists – it's going to grow. It's not always good competition.'

(author's transcripts, 1995)

---

from an economic standpoint' (Blake and Becher, 1991: 171). The result is Rara Avis, a 1,500-acre private forest reserve, established in 1983, close to the Braulio Carrillo National Park.

But Bien has since been joined by many of his fellow Americans in Costa Rica, not all with the same sincerity of motive as his. Indeed, the North American influence is in strong evidence in most of the tourist locations within Costa Rica's remaining natural ecosystems, especially in the business of providing the services that the tourists require.

This is not to say that all the service providers for the new tourism are from the First World, still less from North America. A majority come from within the national boundaries, but even in many such cases the seeping influence of the First World and its business values and practices have an effect through the mass media, conferences, and general contact with more and more First World tourists and operators. Gradually, as more contact at conferences and trade fairs takes place and as more linkages between the companies in the First World and the service providers in the Third World are made, the balance between what the supplier can offer and what the visitor wants tips towards the latter. The service providers purchase more and different items and expand their range of services. And slowly the prevailing and dominant value system shifts from that of the caterer to that of the catered for.

The new financiers

OPIC is the Overseas Private Investment Corporation, a US-based organisation that seeks to link US investors with investment opportunities overseas. Annually it transfers slightly more money from the US private sector to the Third World private sector than the United States Agency for International Development (USAID) manages with public money to public organisations in the Third World. Box 7.12 gives more details of the mission and means of OPIC, particularly about its Ecotourism Award, which has now been discontinued as a special branch of investment.

In order to qualify for an OPIC investment scheme, which would help those requiring investment to find it, any potential Third World developer must submit at least 25 per cent of the shareholding stock to the US investor. Effectively, this condition surrenders considerable control and power over the development to the investor. 'There is no requirement that the foreign enterprise be wholly owned or controlled by US investors. However, in the case of a project with foreign ownership, only the portion of the investment made by the US investor is insurable by OPIC' (1995: 3).

Referring to environmental problems and the lack of US investment in Puerto Rico, Howard Hills, then Vice President of OPIC, addressing the First World Congress on Tourism and the Environment in Belize City, stated that 'ignorance of how we want to operate is no excuse' (author's transcripts, Hills, 1992). Such a display of cultural arrogance betrays the ideological and cultural dominance which often underpins relationships between First World and Third World developers.

OPIC conducts environmental impact analyses for any schemes which apply for its assistance. The standards of these analyses are described as 'living up to those of the World Bank' (author's transcripts, Hills, 1992), the irony of which will be clear to all who have followed the constant, at times bitter, and damming criticisms of World Bank environmental impact assessments and its environmental and social policies by a range of environmental and campaigning organisations.

It is also of great assistance to potential overseas investors to have the services of a local investment organisation such as the Foundation for Investment and Development of Exports (FIDE) in Honduras. FIDE has focused on the creation of conditions that would attract investment under the Export Processing Zone (EPZ) concept. The Honduran EPZ Law of 1987 allowed privately owned and operated industrial parks and free trade zones to be established anywhere in the country, and the free trade zone concept was subsequently extended to include tourism developments. FIDE offers its overseas clients who wish to invest

---

Box 7.12 OPIC's Ecotourism Award

'OPIC's Ecotourism Award recognises business ventures in the tourism sector that are planned and developed with sound environmental values . . . in carrying out its mission of promoting development through investment, OPIC seeks to encourage US investors to make environmental sustainability and prevention of environmental degradation a priority. Increasingly, US investors and tourism sector developers are recognising the commercial as well as the social value of sound environmental management. Sustainable development also promotes understanding in cross-cultural matters including land and resource use, employment of local population, and protection of the integrity of tourist destinations.'

OPIC operates an Environmental Investment Fund which is designed to promote investment in five sectors: ecotourism, sustainable agriculture, forest management, renewable and alternative energy, and pollution prevention.

In 1992 OPIC presented its annual Ecotourism Award to Eric and Maggie Schwartz who own and operate La Selva, a jungle lodge in the Oriente region of Ecuador, part of the Amazon Basin. OPIC's description of their work is as follows:

'They have recognised the importance of safeguarding the environment for the people of Ecuador and the pleasure of their guests, and their environmental ethics are an integral part of a successful business strategy in a developing country. La Selva has demonstrated the viability of private sector initiative in protecting the environment, and contributed to the preservation of a precious natural resource in Ecuador.'

(information and extracts from *1992 Ecotourism Award*, leaflet prepared by the Overseas Private Investment Corporation, 1992)

---

in the country a service which includes individually tailored itineraries, accompanied site visits, meetings with government, banking and international officials, contacts with possible partners and suppliers, appointments with lawyers and relevant consultants, the identification of buildings and sites for rental or purchase and assistance with immigration, housing, health and schools.

Overseas investment is clearly a two-way process. It can be attracted as well as pushed. FIDE has offices both in Tegucigalpa, the capital of Honduras, and in Florida.

'New' academic tours

Private companies have been involved in providing travel services to groups of school and college students for as long as school journeys and fieldwork trips have been operating. But recently, three types of change in this sector of the market have been witnessed. First, Third World destinations are increasingly considered suitable for this type of tour; second, the industry is beginning to exploit the possibilities for studies of the natural world and issues of environment, society and development in Third World settings; and third, operators now instigate the tours themselves, without waiting for schools and colleges to organise them first.

Not only are companies actively seeking school and college groups (see Box 7.13) but they also initiate 'research tours', attracting First World professionals and tapping into the last decade's rise in environmental awareness (see Box 7.14). Nature, Third World culture, difference and otherness are clearly being sold in the examples given in the two boxes. As Richard Cahill formerly of Eco-Tours de Panamá (see Box 7.13) says, 'there is a great market for student travel' (personal communication, 1995).

Beyond the advertisements featured in Boxes 7.13 and 7.14, new organisations, which cannot be classified either as educational institutes or as tour operators, have begun to exploit this sector of the market. Ecopaz, for instance, is the Centre for Ecology, Peace and Justice Studies and is based in Teresópolis, Brazil. They advertise a two-week travel and study seminar to Brazil at a cost of $1,195 (see Box 7.15). Fred Morris of Ecopaz, however, followed up an earlier and similar notice on the internet with the admission that 'we at Ecopaz took some heavy hits . . . for having posted information about our Ecopaz Travel/Study Seminars, as we were accused of being "hot-shot operators" trying to exploit a new hot item' (personal communication, 1995). Excerpts from his justification for the posted information or advertisement (depending on your viewpoint) are given below, and are presented here simply as an illustration of a feature of the new forms of tourism to Third World destinations.

> [A]n experience in travel can create a larger awareness of the nature of the problems facing the worldwide environment and begin to stimulate a sense of solidarity with all peoples and the planet. . . . For nine years, during the 80s, I did 'political tourism' in Central America, taking North Americans to Costa Rica, Nicaragua and Honduras to enable persons to see for themselves the realities of that part of the hemisphere, realities quite different from what was being presented in the US media. . . . More than 600 persons participated in [these] and . . . returned home with a new sense of solidarity with the people of that region, as well as a new understanding of the political realities . . . while efforts at persuading tour operators to be more sensitive toward the environment and the peoples they are visiting are all important, it is more important that increasing numbers of tourists have a chance, as part of their travel experience, to come to a real understanding of the social, economic, political, environmental realities of the places they are visiting so that they can become part of the solution to the environmental problems.
>
> (email communication, 1995)

The word 'realities' echoes the discussion on reality and practicality earlier in this chapter. Its use implies that there is only one reality and that the user of the

230

Box 7.13   Academic tours

Eco-Tours de Panamá

Richard Cahill, former Sales Manager of Eco-Tours de Panamá, explains:

We have been working on new products and one of them is conducting a 'Rainforest workshop' for universities. The idea is to conduct field studies for students interested in learning about tropical forests and their inhabitants. We have run other programs for universities at a smaller scale, but we feel there is a great market for student travel that receives credit [for academic qualifications]. The program would be led by a professor and we would give the logistical help and guide expertise.

[The students] can come from the United States . . . , and the idea is to bring them down here where they would make a specific study. . . . At the end of the two weeks they would receive . . . a credit or certification, depending on what the authorities in the United States require.

(personal communication, 1995)

Operation Crossroads Africa

'AFRICA: Internships Work Projects Travel Study Camps'
Crossroads participants do not go to Africa to impose their own Western values, but to seek comprehension of African values; and they are challenged to adjust to local ways of doing many things.

Students generally arrange with their schools to receive credit (typically 5 to 10 units) for their summer experience.

An important aspect of the Crossroads program is for participants to 'represent' their campus or community and to share their experiences on return.

You will ' . . . experience Africa from the inside out', as one participant described her experience. 'This is not an African tour.'

(email communication, 1996)

University of Costa Rica – eco-farm

We have been entrusted by American friends with the task of planning, creation and administration of an eco-farm on a 50–70 hectare site on the northern Pacific coast of Costa Rica (dry tropical forest). The farm will also host a center for eco-agricultural education and research.

[After 1998] The project will then offer ecological education to university students and school children from the United States and Costa Rica during short stays at the farm. . . . During their stays they will be able to do organic farming and other eco-touristic activities.

(email communication, 1996)

---

word is the one who has it. Throughout all of this book, this idea is disputed, especially with regard to intangible, highly politicised notions such as sustainability, environment, development and even tourism.

Box 7.14   Research tourism

AMAZON, EDUCATION, RESEARCH, ECOTOURISM,
ECOLOGY, PRESERVATION
Research, Education and Ecotourism Trip
14 Day Excursion in the Brazilian Rainforest

PURPOSE: To experience, understand and explore conservation, research,
education and ecotourism opportunities in the Brazilian Amazon.

The trip will be guided by Hilton P. da Silva, MS, MD, a native from the
Amazon Basin currently finishing his PhD in Medical Anthropology/ Public
Health at the Ohio State University, and an expert in Amazon conservation and
biodiversity. Other scholars and scientists from Brazil will join us along the trip.

PLACES TO BE VISITED INCLUDE:
– The Ferreira Penna Research Station
– A Native Caboclo settlement
– Belem
– The Museu Paraense Emilio Goeldi
– Mosqueiro, a fresh water beach
– Scenic airplane trip over Marajo Island
– and Brasilia

Since this is a not-for-profit activity the trip's price is only $2,890.00 (DBO) per
person for the entire fourteen day period.

(email communication, 10 June 1996)

New tourism conferences

Conferences are a profit-oriented business in themselves and should rightly
be considered in a chapter examining the tourist industry. There is no better
illustration of the way the 'conference business' has seized upon issues associated
with new tourism than that provided by Belize in the early 1990s.

In July 1991 the Caribbean Eco-Tourism Conference was held in a hotel on
the outskirts of Belize City (this is more fully reported in Box 9.9). It was
followed later that year by the First World Eco-Tourism Congress, then, from
27 April to 2 May 1992, also in Belize, by the First World Congress on Tourism
and the Environment. As one conference participant wryly remarked, if you
change the title slightly each time you can claim that they are all the 'First'.

The 1992 conference registration fee ensured that only the professionals, the
tour operators and those with their fares and fees paid for them were able to
attend. Delegates heard two days of presentations and plenaries which allowed
little or no opportunity for in-depth analysis. Questions were unrelated to each
other and debate never got off the ground. The two days of fieldwork gave
delegates the chance to trample through the pristine environments that they
value so highly and to wrap themselves in sustainability and appropriateness. It

---

Box 7.15 Ecopaz travel/study seminar

[Ecopaz has] an effective program for helping persons of the Northern Hemisphere come to a fuller understanding of the problems facing a developing country like Brazil, with regard to the environment. We believe that a firsthand look at the realities of Brazil can motivate persons to become more seriously involved in the efforts to avoid the destruction of the earth's ecosystem. It is for this reason that we are sending you the following information about our next Ecopaz Travel/Study Seminar.

. . . a group of 16 persons will meet in Rio de Janeiro to start an adventure of learning and sharing in Brazil. They will spend two weeks informing themselves of the realities in Brazil with an eye toward being more effective on returning to their homes in favourably influencing corporate and national behaviour toward the developing nations of the South . . . they will have the opportunity to see some of the most impressive places in Brazil.

The program includes the following:

- Orientation to Brazil, touring of Rio, including Corcovado, Sugar Loaf and Copacabana Beach;
- Meetings with Brazilian environmentalists;
- Time for shopping. Evening visit to folklore show in Rio;
- Travel to Recife, economic capital of the Brazilian northeast;
- Meetings with leaders to hear about social/economic/environmental problems facing the Brazilian people. Beach time and shopping;
- Travel to Manaus. Walk through the rainforest in National Park;
- Touring of colonial Manaus;
- Meet with indigenous peoples' representatives and government officials;
- Fly to Iguaçu Falls on frontier with Argentina and Paraguay.

Cost: US$1,195 per person. Costs include all lodging, all meals, local transportation, program and translations. Air travel to and from Brazil is not included.

The maximum group will be 16 persons.

(email communication, 15 September 1996)

---

also helped to relieve them of some more of their money in favour of Belizean hotels. Indeed, some felt that it was worth asking whether filling hotel beds was in fact the whole point of the exercise. As one local business delegate put it, 'I wish we could put on ten of these a year. We fill all the beds. We all do well out of it' (author's transcripts, 1992).

The major organisers of the conference were Bob Harvey, Diane Kelsay Harvey and the Belize Tourist Board. The former two, US citizens, went on to organise the Second World Congress on Tourism for the Environment in Venezuela (1993) and the Third World Congress in Puerto Rico (1994). All these conferences were run as profit-making enterprises and all were dominated by US participants, US interests and US groups. The spin-off for the Belize Tourist Board was the major publicity exercise that this represented for the country.

The conference reflected many of the different facets of new forms of tourism to the Third World. Like many other Third World countries, Belize is extremely keen to attract international tourists; the international conservationist movement is eager to mould it into sustainable and exclusive shape; and the business community wants to milk it of every cent it can generate. In among all this enthusiasm, Belize seems to have dropped its trousers to reveal its natural resources and its still-beautiful environment to the up-and-coming ecotourism industry, with the hope that well-meaning conservationists can serve as the prophylactic to prevent potential environmental diseases.

The Adventure Travel Society Inc. (ATS) is another company whose major role involves the organising of seminars and an annual conference on adventure travel and ecotourism. A profile of the ATS is given in Box 7.16 which uses the company's own publicity material to illustrate the aims and role of such companies and their association with new forms of tourism.

Conferences are of course an eminently suitable medium for the sharing and transfer of ideas, debates and developments in the field of tourism. We are not arguing that they should not be held, nor that they should not be run as a business. Rather, we wish to point out a new trend in conference content linked to new forms of tourism and the notion of sustainability. It may be argued that such regular gatherings of interested parties to discuss sustainability may serve as mass exercises in self-deception and self-assurances that 'we' are getting there.

Furthermore, it is noticeable that in many cases the fact that the conferences are run as profit-making businesses is not fully acknowledged by the conference organisers. As the Belizean hotelier quoted above pointed out, the motive for conferences may be more to do with the attraction of extra tourists than with the sharing of ideas. In such cases, the conferences themselves are merely an extension of the conventional tourism business and should be recognised as such.

CONCLUSION

This chapter has examined ways in which different branches of the tourism industry have adapted their operations to absorb the notion of sustainability. The techniques used in claiming sustainability of operations vary partly according to the size of operation, ranging from the vertically integrated TNCs to the small-scale lodges and service providers. Clearly, sustainability is no longer the exclusive claim of new forms of tourism. Techniques such as advertising, regulation, environmental auditing, consultation, codes of conduct and management strategies are employed to 'green' the image of large- and small-scale operators and to sell the products of both mass and new forms of tourism.

Box 7.16 *The Adventure Travel Society and its annual congress*

Since our 1989 beginnings and through five Congresses, the Adventure Travel Society (ATS) has been one of tourism's strongest advocates of adventure travel and ecotourism. From the outset the co-founders, Jerry Mallett and James Pearson, both of Denver, Colorado, brought their lifelong love of the outdoors and respect for conservation to ATS.

ATS has greatly assisted the tourism industry by continually focusing on the importance of sustainable economic policies for the future of adventure travel and ecotourism. In addition to the industry, vacationers worldwide have benefitted and will continue to benefit from ATS' attention to sustainability issues.

Our mission

ATS is dedicated to promoting natural resource sustainability, economic viability and cultural integrity through the development of tourism.

Our philosophy

We believe that nations and regions can develop a tourism-based infrastructure without compromising natural resources. Furthermore, we contend that adopting a policy of protecting a region or nation's environment and indigenous cultures through responsible management will result in a sustainable tourism effort.

What we do

We are a professional corporation that provides tourism consulting, marketing seminars, and produces the Annual World Congress on Adventure Travel and Ecotourism.

ATS has three distinct roles:

- Creator and producer of the annual World Congress and Expo on Adventure Travel and Ecotourism, an international conference and trade show . . . ;
- An international trade association. . . . Among other benefits, ATS publishes a newsletter three times per year to keep its members informed about the latest developments in adventure travel and ecotourism;
- The ATS staff also presents its highly acclaimed seminar titled 'Selling Adventure Travel' to travel agents, tour operators, travel writers, associations, retailers and government offices.

(extracts from Adventure Travel Society, 1996a, 1996b)

A crucial question to be considered with all these techniques concerns the extent to which they promote genuine change in practices or cosmetic change which serves as good publicity but which makes little effective difference. It has been suggested here that this will depend, at least in part, on the motive behind the operations. Where the profit maximisation motive externalises all other possible motives and factors, sustainability will most likely be redefined to fit in with a business-as-usual approach. At the same time, the automatic assumption made by new forms of tourism that their operations are environmentally friendly and sustainable has also been brought into question.

The chapter has also briefly described and questioned the roles of a number of recent features of and employment possibilities associated with the tourism industry, especially those linked to the drive for sustainability and new forms of tourism to the Third World.

8

'HOSTS' AND DESTINATIONS

For what we are about to receive . . .

In 1963, Katherine Whitehorn wrote in the *Guardian*: 'The only unspoilt village is the one no outsider has ever visited, not even you.' While this is extreme in its denial of the dynamic element and benefits of social integration and acculturation, it makes the point about the effect of visitors and tourists on local communities. More than thirty years on, there is a vast body of work that demonstrates that local communities in Third World countries reap few benefits from tourism because they have little control over the ways in which the industry is developed, they cannot match the financial resources available to external investors and their views are rarely heard. This chapter focuses on these local communities which receive tourists and looks at their levels of power, control and ownership of tourism. Again, the analysis is made through the key themes of the book, uneven and unequal development, relationships of power and globalisation, and the key words, sustainability, new forms of tourism and the Third World.

In the chapter title the word 'hosts' is in inverted commas. This draws attention to the implication that there is a willingness on the part of those who receive guests and possibly even an assumption that they have a degree of control over tourist developments in their community. As has already been discussed and is well-documented elsewhere, it is not often the case that local people derive benefit sufficient to outweigh the disbenefits of their community receiving tourists. Chapter 3, particularly, illustrated the uneven and unequal relationships of power within local communities. The terms 'destination community' and 'visited population' are also used interchangeably with the word 'host', but 'hosts' is used in the title because, as will be seen in this chapter, there are examples of communities managing to take a degree of control of, and to exercise power over, the developments of tourism in their localities.

The body of this chapter examines the different levels at which local communities participate in tourism and the levels of ownership and control that they, and others, hold over the resources of the tourism industry. The relationships of power between local populations and the tourists, the governments, the industry, the NGOs and the supranational institutions produce effects which reflect and promote the unequal development of visited populations and these other players in the activities of tourism. The differences in the approaches taken in pursuit of community control and government control are also outlined.

The word 'destinations' is used in the chapter title because all too often the local communities visited by tourists are viewed precisely as that – places, to be collected, as if the people who live there are either irrelevant or at best incidental to the place. Alternatively, where 'experiencing the local culture' is considered to be important as part of the tourist experience, then the local people may be considered as objects or commodities. In no instance is this more so than in the case of organised tours to visit tribal peoples discussed below. Pratt's notion of transculturation and the ways in which cultural domination is transmitted from one group (the tourists) to another (the visited population) are then discussed. Throughout the chapter, we examine the demands made upon the hosts or visited populations that they be 'authentic'.

LOCAL PARTICIPATION IN DECISION-MAKING

The two words, 'local' and 'participation', are regularly used together to emphasise the need to include and involve local people; and it is this juxtaposition of the two words which implies, paradoxically, that it is local people who have so often been left out of the planning, decision-making and operation of tourist schemes.

As discussed in Chapter 4, one of the criteria often agreed as essential to the condition of sustainability in any 'new' tourist scheme is the participation of local people. From around the world, however, many examples of the relative lack of power held by local people in tourism developments in their locality have been documented – Brandon (1993) cites over fifty schemes, 'many of [which] had initiated nature tourism activities, but few of the benefits went to local people' (135). (See also Johnston (1990), Wells and Brandon (1992) and West and Brechin (1991).) This exclusion of local people from involvement and decision-making in, and from the operation and benefits of tourism can be seen in some of the examples cited in this chapter (see Box 8.1, for example) and elsewhere in this book.

Most of the general agreement regarding the need for local participation, however, appears to come from positions of power, especially from those who have a remit regarding sustainable development: the environmentalists, other NGOs, government officials, politicians, and World Bank officials. As Jules Pretty points out:

> In recent years, there have been an increasing number of comparative studies of development projects showing that 'participation' is one of the critical components of success. . . . As a result, the terms 'people partici-pation' and 'popular participation' are now part of the normal language of many development agencies, including non-governmental organisa-tions, government departments and banks. It is such a fashion that almost everyone says that participation is part of their work.
>
> (1995: 4)

Box 8.1   The curse of the tourism industry

Marketing men put curse of tourism industry on Mayas

Order a prawn cocktail in a hotel in Chetumal, south-east Mexico, and it will probably come smothered in 'Mayan sauce'. A trivial example, but one that shows how the tourist industry, helped by Latin American governments, is turning a great pre-Columbian civilisation and its present-day descendants into a marketing concept.

But critics, including Mayan organisations, claim that archaeological sites and indian villages face being turned into a giant theme park, and that the millions of indigenous inhabitants have no part in decision-making. 'The bottom line is that they are just exploiting the resources of our people,' says Greg Cho'c of the Kekchi council of Belize. 'Mayan people are not involved and cannot influence the project.' . . .

The aims of the Mayan World scheme . . . include improving the quality of life of local residents, protecting the environment, and safeguarding historical and cultural heritage. But the Mexican government's own archaeological and cultural institute, the INAH, is sceptical. 'They have no awareness of what ecology is,' says the director of the local INAH office, Adriana Velásquez. 'If they put up a palm-thatched hut they think it's "ecological".'

She cites the once-unspoilt Xcaret ruins, which have been turned into a park for day-trippers from the up-market resort of Cancún. The entrance fee is about £13, out of the reach of local people . . .

Rolando Pérez, a Quiché Maya, is one of about 30,000 Guatemalan refugees living in south-east Mexico. . . . Mr Pérez . . . believes white and mixed-race people want to eliminate the indians. 'They see us as an obstacle to development,' he says. 'They just want to build big hotels for the tourists. They're the ones that benefit, not us.'

Local initiatives, such as a village guesthouse scheme started by Mayan villagers in Belize, have been ignored, says Stewart Krohn, managing director of Channel 5 television in Belize. . . . 'If you go to a meeting of the Mundo Maya you won't find a Maya there, except maybe serving dinner,' Mr Krohn says. 'The Mayan people are just being used as low-cost labour. If I was a Maya, I'd put sugar in their gas tank.'

(Gunson, 1996)

And Survival International has noted that 'it has become fashionable for conservationists to talk about "consulting" local people. . . . This looks good on paper, but [is] hardly an adequate substitute for land ownership rights and self-determination' (1996).

Thus, phrases such as 'targeting local people' and 'eliciting community-based participation' (Brandon, 1993: 136), and sentiments such as 'environmentally sustainable development . . . rests on gaining local support for the project' (Drake, 1991: 132) and 'projects must provide direct benefits to local peoples'

(Epler Wood, 1991: 204) come from the perspective of the project planner, who is usually from the First World (as are all the above examples). The planners are often associated with a major INGO (such as WWF, Conservation International and the Ecotourism Society in these cases) or a supranational institution such as the World Bank (as in two of these cases) and all seek their own form of sustainability through their appropriate projects.

Not that the sentiments behind their academic papers and project plans are not laudable, but the push for local participation comes from a position of power, the First World. The principle of local participation, however, is easier to promote on paper from a distance than it is to put into practice at the local level. A range of difficulties, such as conflicting interests and the existence of local power bases and élites (discussed later in this chapter) serve to complicate and confound the good intentions of First World planners.

By contrast, the only forms of local participation that are likely to break the existing patterns of power and unequal development are those which originate from within the local communities themselves. This chapter provides a few such examples, but even these illustrate the fact that local circumstances always manage to complicate the best of intentions.

Local circumstances, the unequal distribution of power between local and other interest groups, and differing interpretations of the term 'participation' are reflected in Pretty's typology of participation, discussed below. This is especially helpful in developing an understanding of the factors which affect the development of tourism schemes in local communities, and the case studies illustrated in this chapter are referred to the typology.

Pretty's typology of participation

Jules Pretty's typology (given in Table 8.1) describes the type of involvement of each of his seven levels of participation and offers a critique of each level. The types of participation range from manipulative participation, in which virtually all the power and control over the development or proposal lie with people or groups outside the local community, to self-mobilisation, in which the power and control over all aspects of the development rest squarely with the local community. The latter type does not rule out the involvement of external bodies or assistants or consultants, but they are present only as enablers rather than as directors and controllers of the development. The range of types allows for differing degrees of external involvement and local control, and reflects the power relationships between them. For local people involvement in the decision-making process is a feature of only the *interactive participation* and *self-mobilisation* types, while in the *functional participation* type most of the major decisions have been made before they are taken to the local community.

It would be easy here to make the prescriptive assumption that the greater the degree of local participation, the better (by whatever definition) the project. There are those, however, who might disagree with this assumption, especially,

Table 8.1 Pretty's typology of participation

Typology	Characteristic of each type
1 Manipulative participation	Participation is simply a pretence: 'peoples' ' representatives on official boards, but they are unelected and have no power
2 Passive participation	People participate by being told what has been decided or has already happened: involves unilateral announcements by project management without any listening to people's responses; information shared belongs only to external professionals
3 Participation by consultation	People participate by being consulted or by answering questions: external agents define problems and information-gathering processes, and so control analysis; process does not concede any share in decision-making; professionals under no obligation to account for people's views
4 Participation for material incentives	People participate by contributing resources (e.g. labour) in return for food, cash or other material incentive: farmers may provide fields and labour but are not involved in testing or the process of learning; this is commonly called participation, yet people have no stake in prolonging technologies or practices when the incentives end
5 Functional participation	Participation seen by external agencies as a means to achieve project goals, especially reduced costs: people may participate by forming groups to meet project objectives; involvement may be interactive and involve shared decision-making, but tends to arise only after major decisions have already been made by external agents; at worst, local people may still only be co-opted to serve external goals
6 Interactive participation	People participate in joint analysis, development of action plans and strengthening of local institutions: participation is seen as a right, not just the means to achieve project goals; the process involves interdisciplinary methodologies that seek multiple perspectives and use systemic and structured learning processes. As groups take control of local decisions and determine how available resources are used, so they have a stake in maintaining structures and practices
7 Self-mobilisation	People participate by taking initiatives independently of external institutions to change systems: they develop contacts with external institutions for resources and technical advice they need, but retain control over resource use; self-mobilisation can spread if governments and NGOs provide an enabling framework of support. Self-mobilisation may or may not challenge existing distributions of wealth and power

Source: Pretty, 1995

but not exclusively, those who represent a vested interest in a particular development project – the development agencies, governments, supranational institutions, or operators for instance. In these cases, some of the lesser types of participation might be considered preferable. It is precisely this point which emphasises the importance of the power relationships involved in any (tourist) development project, and the fact that Pretty's typology reflects this underlines its value.

At this point, it is worth contrasting a number of examples of local participation in tourism developments in order to illustrate the manifestations and effects of different levels of involvement. In the following paragraphs we have attempted simply to describe the situations of each case study in the appropriate box and in the text to relate it to Pretty's typology, which allows us to make a consideration of the power vested in each interest group and their relation to the local community.

Box 8.1 includes excerpts from an article by Phil Gunson on the large-scale Mundo Maya (Maya World) project which covers five southern Mexican states plus Guatemala, Belize, El Salvador and Honduras. It is described in Mundo Maya publicity material as 'a ground-breaking tourism and regional development initiative ... [which] seeks to improve the lot of area inhabitants with low-impact projects which give visitors the opportunity to explore the area' ('Mundo Maya Travel Guide', undated: 4). In 1991 the project initially received $1 million from the European Commission to promote three kinds of tourism in each country: cultural tourism, coastal tourism and eco/adventure tourism. The project promotes infrastructural improvements, new hotel construction, archaeological projects and extensive international marketing through very glossy brochures, in-flight magazines, and travel trade shows.

As Box 8.1 makes clear, there appears to be little or no attempt to involve local communities in decision-making. As the editor of *Tourism Link* (a journal of the Belize Tourism Industry Association) explained, 'full decision making powers for all Mundo Maya affairs lie in the hands of only five persons – basically the top public sector tourism officials of each country' (Arnold 1992: 4). The fact that this statement came as part of an article of complaint by private sector representatives about public sector control of the project rather underlines the irrelevance of local communities in this squabble for power. According to Pretty's typology, this example might be classified as *manipulative participation* or *passive participation.*

Box 8.2 provides an account of a small-scale tourism scheme in a community of mostly Salvadoran exiles in Costa Rica. Although there has been a degree of external assistance in this case, the idea for the scheme arose from within the community itself and all the tourists' activities are under the direct control of the community. Moreover, it is one of the advantages of this type of tourist scheme that money for services rendered goes direct to those who render them without being 'creamed off' or cut down to a minimum by middlemen and agents. Although the community received considerable assistance in its early

Box 8.2   Finca Sonador, Costa Rica

The Longo Maï movement, based in France, aims to give war refugees a positive and productive home and work environment rather than a temporary and transient camp. In 1978 the movement helped a group of Nicaraguan refugees fleeing Somoza's terror to form a small community in southern Costa Rica. After Somoza's overthrow, they returned to Nicaragua, but they were soon replaced by Salvadoran refugees fleeing the same type of state terror.

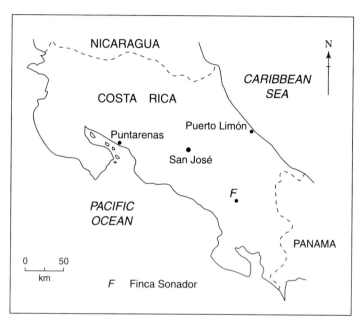

The community is now called Finca Sonador and has a population of around 300, mostly Salvadorans with a few Costa Rican campesino families, refugees from poverty and landlessness in their own country. The Finca is a relatively self-sufficient agricultural village which produces coffee and a few other products for sale beyond the village. In low income months their traditional survival agriculture is based on corn, beans, rice and yucca.

Since 1992, the Finca has attempted to diversify its economic activities by attracting a few ecotourists. A small number of families are willing to accommodate visitors and have the space to do so, although in some cases visitors may find themselves sharing a room with children.

Prices for meals and accommodation are negotiated with the family, although there is a non-binding guidelist of tariffs. By the standard of the northern professionals who form the majority of visitors, prices are ridiculously low. Other activities offered in the village include horseriding, guided tours,

Box 8.2
continued

involvement in farming and general inclusion in fiestas. These ensure that the tourist money is distributed further afield within the community. Advertising for the scheme is largely by word-of-mouth, although a publicity sheet is posted in the Quaker lodging house in San José.

The small scale of the scheme would seem to be an essential feature both for the tourist and the host. For the ecotourist, it is important that the experience has its air of exclusivity in the sense that this is not the usual tourist experience – it is a feather in the cap of the tourist and is therefore good for their image. For the host, it earns a little extra income with little if any extra cost and does not disrupt the community's or the family's way of life.

(Mowforth, 1996)

years, its tourism venture has been largely unassisted, which classifies this scheme as *self-mobilisation* in terms of Pretty's typology.

These two examples come from Central America, but later examples in this chapter are from Africa and the Himalayas, as well as Latin America. Readers might also find it useful to refer other case studies presented elsewhere in this book to this typology.

Participatory appraisal and inquiry techniques

In their efforts to involve local populations in the planning, decision-making and operating of tourist schemes, planners and academics have developed a range of techniques. Some of these were listed in Box 4.10 and briefly mentioned in 'The tools of sustainability in tourism' (pp. 115–22). As was pointed out, techniques which allow for consultation and participation are still young in their development and suffer various shortcomings. It is debatable that none of the relatively sophisticated techniques that have become available recently are able to improve on the traditional and well-used technique of the meeting. Local communities the world over traditionally use both formal and informal meetings to debate the courses of development and issues which may affect them. Of course meetings are not always all-inclusive; for example, women and children are excluded from many but by no means all such meetings.

More sophisticated survey techniques, public attitude surveys, stated preference techniques and contingent valuation methods all suffer the disadvantage of being conducted, administered, promoted and publicised by persons outside the local community affected by a tourism development. They are tools used by professionals administering the surveys on local communities, who by definition do not therefore enjoy control over it. It should not come as a great surprise that both inputs and results are often open to dispute. In terms of Pretty's typology such techniques may help to improve the level of participation, but they are unlikely to attain a high level unless they focus on the degree of decision-making devolved to the local community as well as its active involvement in the operation of the scheme. There is little doubt, however, that with their 'systemic and structured learning processes' (Table 8.1) they can increase the likelihood of sustained success of schemes. A number of the examples given in this chapter illustrate this.

These techniques represent recent attempts to involve local people in research, policy appraisal and implementation themselves. But in reviewing the origins of participatory appraisal, Chambers (1994a, 1994b and 1994c) notes that rural appraisal techniques can be traced to the late 1970s and early 1980s as a reaction to the 'biased perceptions derived from rural development tourism (the brief rural visit by the urban based professional) and the many defects and high costs of large-scale questionnaire surveys' (1994b: 1253).

Accompanying the general fashion for 'local participation' discussed above, recent years have seen the development of a trend for participatory approaches to inquiry and research. Thus, sustainability indicators (as mentioned in 'The tools of sustainability in tourism') are geared to the local level. But sustainability indicators are only one in an armoury of new techniques of local participatory analysis and policy tools. Participatory Action Research (PAR), Participatory Research Methodology (PRM), Participatory Rural Appraisal (PRA), Rapid Appraisal (RA), Rapid Assessment Procedures (RAP), Rapid Assessment Techniques (RAT), Rapid Ethnographic Assessment (REA), Rapid Rural Appraisal (RRA), and a bewildering array of other acronyms and initials have entered into use. Although they are often formally stated to involve many steps in the process, essentially they all follow the three-step procedure of participatory inquiry, collective analysis and action in the locality. (For a more detailed outline of the general procedure, see the work of the International Institute for Environment and Development (IIED).)

The principle of local participation underlies all these techniques, but they involve differing degrees of participation often in different stages of the appraisal, and the same technique may be interpreted and implemented in different ways by different people. A specific example is given in Box 8.3, which includes extracts from a 1996 advertisement (on the internet) for members of a team to conduct a PRA in Ecuador. While the request for support or assistance in this case came from the community, through the Ecuadorean NGO EcoCiencia, to Jon Kohl at Yale, there is an implication that it is the 'outside tecnicos' and professionals who, out of largesse, will solve the problems of the local community. The leader has a clear preference for the establishment of an environmental interpretation centre, and it is uncertain whether the idea for this arose from within the community and/or is what the local people had in mind. Nevertheless, it is likely that this example would be classified as functional or interactive participation by Pretty's typology. A little more precision in the classification may be possible with more information regarding the number and kind of major decisions which were made before the process began to involve the local community.

Box 8.3   Recruiting a PRA team for Ecuador

This summer I will be working with an Ecuadorean NGO called EcoCiencia and will be leading a Participatory Rural Appraisal (PRA) in a small peasant community located in Pululahua Geobotanical Reserve, a 30 minute bus ride north of Quito.

EcoCiencia has held some successful campamentos with kids and parents but now sees potential and has interest in establishing a more permanent environmental education program with complete local participation. They suggested possibly a small andean zoo.

I'm taking the zoo suggestion one step further by doing a PRA and thus figuring out with the residents what are their biggest problems and what are suggested solutions, some of which might be addressed by an interpretation centre of some kind . . .

For those who are unfamiliar with PRA methodology, in short, it's a workshop of 2–3 weeks on site where a team of outside tecnicos works with a local community team to investigate the site, discuss problems and solutions, and present them to the entire community who then discuss and prioritise them. A document results, done in part by the people and it is designed to promote autogestion on the part of the locals to use action plans and start their own development process.

In my case I will also be using the PRA results as a feasibility study for the establishment of an environmental interpretation centre which would provide economic, educational, and any number of other benefits.

PRA is an impressive methodology catching on in Latin America (while it is mainstay in India and Asia) which really puts into practice the principle of local participation that many talk about and few do.

I am currently recruiting tecnicos for the interdisciplinary team. . . . Requirements are proficiency in Spanish, some experience working with rural or local people, disciplinary skills that complement the yet unformed team, and a desire to work hard during the short stay. The offer is open to all students and professionals. I welcome recommendations of other people, especially latinos, as well.

(Jon Kohl, Yale School of Forestry and Environmental Studies, Environment in Latin America Network internet bulletin board, 21 March 1996)

While the techniques of local appraisal are well-intentioned by those who lead and conduct them, the critical questions concerning the balance of power are who leads them and to what ends. In general they are led, or at least significantly advised, by First World professionals, and the idea that a group of outsiders visiting for a short period of time can appreciate, let alone solve, the problems experienced by local communities is rather pretentious and patronising, and suggestive of neo-colonialist attitudes. It is no doubt exciting, and a little ramboesque, for the First World professional to be whisked off to help a community somewhere in Latin America, Oceania, South-East Asia or Africa, and will certainly add kudos to their curriculum vitae. But such approaches may not be appropriate for addressing the structural and long-term problems of community development. This is not to say that collaboration between First World professionals and local communities is not possible or desirable. But a crucial element in such collaboration might be to redress existing imbalances of power so that the outcome of the exercise represents the interests of local people rather than the interests and values of the 'outside tecnicos'.

Other participatory techniques

Another related technique which attempts to involve the notion of participation in the making of decisions is the Delphi technique, which is used to set threshold values or critical levels or standards of specific aspects of a development (such as pollution levels or maximum visitor numbers) or to identify positive and/or negative impacts of a development. It is a judgemental technique involving the subjective assessments of those who take part, although it is often seen 'as a means of collecting expert or informed opinion and of working towards consensus between experts on a given issue' (Green and Hunter, 1992: 37).

The technique uses responses of the participants to an initial questionnaire about the issue under study. The second stage compiles these responses and informs participants of the results (that is, the total responses). The third stage repeats the first but participants have the benefit of knowing all other responses. The stages can be repeated numerous times if a consensus is not close enough.

Although meetings between participants can take place as part of the process, one of the advantages of the technique is that it provides anonymity, or at least separation, for each individual participant, thereby reducing peer pressure in the formation of opinions and permitting more honest responses. Meetings and/or collection and dissemination of responses form an important part of the attempt to reach a consensus from all individual responses.

The example given in Box 8.4 illustrates the use of the technique. It highlights the general limitations on the depth of participation attained by the technique and the consequent suspicion with which its results may be treated by local people. Despite this, there is no doubt that such techniques may be useful in the field of tourism planning, as the example cited in Box 8.4 illustrates. It is not possible to be precise in a classification of this scheme according to Pretty's typology, largely because of the limited amount of information available and also because at the time of writing the exercise described had not taken place. One important issue concerns the adequacy of the representation of the local village population by the mayor – some groups may feel inadequately represented. On present information, however, its classification could range from Pretty's *passive participation* to *functional participation*.

Disadvantages of the Delphi technique include its method of selection of participants, the possibility of the dominant influence of particular personalities,

247

Box 8.4   An illustration of the Delphi technique

A landowner intends to lease an area of land close to the rim of an active volcano to a building consortium which has plans to develop the site for a hotel and tourist observation post. The hotel will incorporate a restaurant which will be open to non-residents. The site is in a national park area and is adjacent to a small wildlife reserve within the park. Other than the crater itself, which is devoid of vegetation other than a few mosses, the area outside the crater rim is covered by cloud forest and has all the rich and varied wildlife and plant life associated with that type of vegetation.

At present, there is only a rough track, just suitable for four-wheel drive vehicles. The scheme will necessitate the construction of a surfaced road. The Park authorities are seeking agreement from all interested parties about several factors concerning the development. These are:

- the width of the road;
- the capacity of the hotel;
- the height of the hotel;
- the numbers of tourists allowed into the reserve;
- the training and management of the tourist guides;
- the possibility of a minibus system from the village 3 kilometres away to the observation post;
- the possible need for a car park next to the observation post.

The people to be consulted are:

- the managing director of the building consortium;
- the landowner;
- the National Park director;
- the mayor of the nearby village;
- a biologist who works at the research station in the wildlife refuge;
- a vulcanologist (who works at the same place);
- the director of a tour company which is interested in running tours to the volcano from the capital city;
- a conservationist from a leading environmental organisation concerned particularly with tourism.

All these people are to be asked to answer the following questions as part of a Delphi process of approaching a consensus:

1. What should be the maximum height of the hotel?
2. How should sewage from the hotel be dealt with?
3. Should there be a maximum capacity fixed for the hotel?
4. What should be the maximum width of the road?
5. Who should pay for the road?
6. The hotel will have a car park for its residents' cars. Should a car park be constructed next to the observation post for other members of the public? Or should the public be encouraged to use a minibus service? (It is

Box 8.4
continued

possible to prevent cars, other than those of the hotel users, from using the road.)

7 As very few people visit the crater at present, there has been no need to restrict the numbers of people on the trails in the wildlife refuge. But many more visitors are expected once the hotel is built. What should be the maximum carrying capacity of a trail in the wildlife refuge?

8 Guides will be needed in the refuge. Who should train them?

its design by professional planners rather than those most affected by the plans, and the arbitrariness of the selection of its values. The selection of participants is normally made either by the professional planners or by the interested party who wishes to see the proposal go ahead; and is most unlikely to be made by those affected adversely by the plans. The anonymity of participants does not necessarily preclude the inordinate influence of a dominant personality over the outcome of the technique as a result of the power relationships between the participants. As with any subjective assessment technique, it is feasible that the same group could produce a different outcome at a different time and a different group could produce a different outcome at the same time. Moreover, it has the potential to be used as a means of ensuring that control stays with the 'experts' and out of the hands of the local people.

Analysis and development of the Delphi technique and other techniques are the subject of a considerable amount of work, references to some of which are given in Appendix 2.

Chapter 4 introduced the notion of carrying capacity for tourism destinations and illustrated its calculation in Box 4.8 (p. 107). Although Box 4.8 calculated the physical, real and effective carrying capacities of a trail in the Guayabo National Monument in Costa Rica, each of these measures refers to some tangible and physical factor pertaining to the area under study. Clark (1990), however, outlines the need to add a further dimension to these calculations by incorporating the social carrying capacity to measure the level at which tourist activity becomes a cause of social unrest and/or tourist discomfort. Clark expands the number of factors under consideration and the number of output measures or results, and makes it clear that he does not see carrying capacity purely as a negative idea which results in restriction, but relates it instead to the notion of sustainable development. Despite acknowledging difficulties in measurement and definition, he does assume that a scientifically rational balance can be reached between all these factors, resulting in an 'objective' measure.

Watson and Kopachevsky (1996), however, argue that the result of carrying capacity measurements will always depend on the context of the situation being measured and that this context will vary not just with the physical and social environments, but also with the values of those asking the questions and establishing the conditions for measurement: 'carrying capacities cannot be determined in the absence of value judgements that specify the type of experience a given area is attempting to provide . . . the establishment of target levels is fundamentally an exercise in human value judgement' (1996: 175). They identify different types of carrying capacity (see Box 8.5) and are adamant in their belief that values 'influence all phases and elements of social research' (177) and 'play a critical role in the choice and application of science' (177). Referring to the work of Thomas Kuhn (1962), they state that 'conceptual frameworks and paradigms rise and fall . . . as much on political grounds as on scientific ones' (177). In other words, human judgement will always be required in assessing appropriate threshold levels for a given activity, in this case tourism.

It is clear that carrying capacities may also vary with time, a point made by the Belize Centre for Environmental Studies (BCES): 'the physical carrying capacity of a road may decrease at night when visibility is less, or the environmental carrying capacity of Half Moon Caye, in terms of visitor numbers, may decrease when the boobies are nesting' (1994: 1). The way in which visitor

---

Box 8.5 Types of tourist carrying capacity

Ecological-environmental capacity

The level of tourist development or recreational activity beyond which the environment as previously experienced is degraded or compromised.

Physical-facility capacity

The level of tourist development or recreational activity beyond which facilities are 'saturated'; or physical deterioration of the environment occurs through overuse by tourists or inadequate infrastructural network.

Social-perceptual capacity

The level reached when local residents of an area no longer want tourists because they are destroying the environment, damaging the local culture or crowding them out of local activities.

Economic carrying capacity

The ability to absorb tourist functions without squeezing out desirable activities. Assumes that any limit to capacity can be overcome, even if at a cost – ecological, social, cultural or even political.

Psychological capacity

This is exceeded when tourists are no longer comfortable in the destination area, for reasons that can include perceived negative attitudes of the locals, crowding of the area (traffic jams) or the deterioration of the physical environment.

(adapted from Watson and Kopachevsky, 1996)

---

carrying capacity has had to be recalculated over time as visitor numbers have increased to the Galápagos Islands also illustrates the point about calculation changes with both time and perceptions or values. Jonathan Croall reports that the Galápagos Islands' 'sustainable' capacity of 12,000 visitors per annum set by the Ecuadorean government was soon raised to 50,000 for economic reasons (1995: 61).

This growing realisation, that the setting of limits is a normative process which cannot be divorced from the objectives of the exercise or from the values of those who set them, has recently led to inquiry into the techniques of setting limits of acceptable change, or LACs, which are to some extent developing out of the work on carrying capacity. One essential element that has been built into the development of LACs has been the involvement of different interest groups in the technique, on the grounds that the setting of limits based on value judgements would be more acceptable to users if they were involved in setting them. As Sidaway says, 'In LAC, the entire process involves the interest groups from the outset' (1994: 1).

Sidaway identifies the features which distinguish the LAC approach from carrying capacity and other management planning systems as 'its attempts to identify measurable aspects of quality, to monitor whether environmental quality is maintained and the degree of interest group involvement throughout the process' (3). The first two of these features show a similarity with another recent development in planning systems, that of sustainability indicators; and in its degree of interest group participation it resembles the Delphi technique discussed earlier.

Despite the recognition of LACs that the universal standards implied by carrying capacity calculations are not compatible with temporal and spatial variations and that the definition of standards of quality vary with time, space, interest group and value, the technique still attempts to define maximum or optimum thresholds applicable to a given situation. But as Munt (1992) asserts about carrying capacity, LACs also tend to provide 'both the ecological and social justification for forms of exclusionism – an objective carefully nurtured by a growing corps of eco-missionaries and ego-travellers in Third World countries' (213).

GOVERNMENT CONTROL/COMMUNITY CONTROL

In tourist-speak, suitable destinations are just as likely to be countries as they are to be specific small-scale resorts, towns or settlements. In fact a browse through the brochures of new forms of tourism shows that most are organised by country or even by groups of countries rather than by resort or community. But the countries that tour operators speak of are nation states and are run by governments which usually represent different interests and have different priorities from those of local communities. The Malaysian government, for instance, promoted 1990 as 'Visit Malaysia Year' with advertisements featuring images of indigenous peoples in colourful traditional dress while at the same time imprisoning members of these same groups for protesting against the logging of their land. Burma is a similar case (see p. 314), and there exist many other examples of government marketing policies being at odds with the same government's treatment of indigenous and other communities.

This raises the question of what is a community. Although a community can be defined by scale, sector, interest, level of power and by numerous other features which express its diversity and heterogeneity, it is taken here as an amorphous term over which there is considerable debate. For the purpose of discussion, community is not regarded here as an homogeneous construct; rather it is seen as something locational within which there are divisions of differing degrees of contrast according to many criteria. The formation and influence of local élites as a result of these divisions are discussed later. The definition of the community, then, differs according to the case study under question, and where divisions between sectors or groups within the community are significant these are pointed out and discussed if necessary and appropriate.

Gerardo Budowski has posed 'the possibility of a mass ecotourism towards natural areas' (1995) and asks whether this will affect the development of eco-tourism. He further asks whether the development of 'megahotels' will operate in opposition to the small rural and rustic hotels noted for the low impact of the tourism they promote. It is this last question which highlights the different courses of action available to national governments. On the one hand, they can promote relatively small-scale, locally owned, community-based tourism facilities (small hotels, pensions, restaurants and other facilities which form an integral part of the community in which they are located). On the other hand, they can attract transnational investment into the country in the form of large-scale, luxury hotels with all associated facilities integral to the hotel.

The former course of tourism development offers more chance that the economic benefits of the exchange will remain in the hands of local communities, although the government is still able to extract indirect revenue from taxes and from tourist contributions such as park entry fees, transport tariffs, banking charges and other enterprises which the tourists make use of and which the government either taxes or manages. The latter form of tourism development concentrates the economic benefits with the government, although it would be contested by the supporters of this course that the employment deriving from such schemes would also ensure some trickle-down of benefits to local workforces.

Of course there are more than two courses of action available to national governments and local communities. There are numerous ways in which revenue can be shared, both indirectly (through taxes) and directly (through allocation of a proportion of profits or takings – see the case of the Maasai cited later in this chapter), and there are combinations of these.

The cases of Belize and Costa Rica have been cited in a number of places throughout this book, partly because they have both built international reputations as destinations for new forms of tourism. They have also both publicised themselves as pursuing community-based tourism development. In 1994, Henry Young, then Belizean Minister of Tourism and the Environment, for instance, delivered a speech entitled 'Community-based tourism development in Belize: government policies and plans in support of community initiatives' in which he outlined plans 'to direct tourist dollars to . . . flow into and stay in local communities' (21). On p. 302 we give further details of how the Belizean government has used the language and rhetoric of sustainable tourism and its stated promotion of community-based tourism to further its international reputation and attractiveness.

It is noteworthy that the quotation above from Henry Young is found in a booklet funded by the United States Agency for International Development (USAID) and jointly published by the Ministry of Tourism and Environment and the Belize Enterprise for Sustained Technology (BEST). The parts played by BEST, USAID and the Belizean government in supposedly promoting community-based tourism development but actually undermining it in the region of Toledo are outlined below.

In 1991 the Toledo Ecotourism Association (TEA) initiated the construction of guesthouses in six indigenous villages (Kekchi, Mayan and Garífuna) in the district of Toledo in the south of Belize. Each guesthouse sleeps eight visitors, and is built in traditional style, using local materials. As extras, concrete floors, water tanks, screened windows and private shared bathroom facilities are also included.

Figure 8.1 The TEA guesthouse in Laguna, Toledo district, south Belize
Source: Martin Mowforth

Figure 8.2 Inside the Laguna guesthouse
Source: Martin Mowforth

The outside and inside of the guesthouse in the village of Laguna are shown in Figures 8.1 and 8.2. In the last two years, five more villages have entered the scheme, although in early 1996 there was a temporary lapse in activity in all but two of the villages because of the need to upgrade the facilities.

From the start the scheme encountered a number of difficulties. Local businesses, hotels and lodging houses in Punta Gorda, the district capital, opposed the scheme in the belief that it would take clients away from them. Cement for the construction was not easily obtained. Money and support from the government of Belize was slow in coming. Rivalries and local political squabbles between villages within the TEA hampered the smooth operation of the programme.

Time has helped overcome some of the local political squabbles and the government's suspicion of a scheme over which it did not have control. Recently, however, another problem has presented itself in the form of competition in a number of the villages. In a misguided attempt to improve an environmentally and socially tarnished image in the region, USAID is promoting the development of village-based and community-controlled tourism on condition that these developments also promote the benefits of competition. Through the organisation Belize Enterprise for Sustained Technology (BEST), the Agency has funded the construction of a new guesthouse at Laguna village. The BEST guesthouse is in competition for tourists with the TEA guesthouse.

In practice such developments are likely to cause rifts and rivalries between and within families in these villages. Some friction has already occurred. The villages are small and, despite individual effort, enterprise and family-based cultural development, many of the practices, customs and norms of village life have depended strongly on cooperation and community action rather than on the spirit of competition.

Additionally, the BEST promotion works directly against the aims of the TEA. The need to coordinate and plan the programme and to rotate visitors to different villages to ensure a fair distribution of the benefits has been at the heart of the efforts of the TEA. In one move, the USAID has created untold difficulties for the TEA and all the villagers who take part in its programme. And, as if not to be outdone, the United Kingdom's Overseas Development Administration (now the Department for International Development) has also funded the construction of a new guesthouse in a village which already has a TEA guesthouse.

The suitability of this type of promotion can be considered to be flawed. In the words of Chet Schmidt, an advisor to the TEA:

> These agencies showed what I consider to be no cultural sensitivity at all, not an overall idea of planning, nor an holistic view of the whole development process here. . . . All the ugly things that happen with uncontrolled tourism begin with this kind of thing – and it was done by foreign aid.
>
> (Schmidt, 1995)

Box 8.6 presents the case from the point of view of one of the villagers. The USAID and BEST efforts in this case would appear to be highly selective in their participation and somewhat arrogant in their willingness to override already existing community structures. On Pretty's typology their exercise could well classify as *manipulative participation*.

Because of the roles of transnational corporations (TNCs) and supranational institutions, the case of the Costa Rican government's promotion of both community-based, sustainable tourism at one moment and large-scale tourist condominia at another is presented later (see p. 310 and Box 9.10). It is no less relevant to the discussion here, however, and the gulf between the two policies is clearly demonstrated by the case. The arguments of those supporting community control of the development of tourism are aired regularly in the pages and letters' columns of the Costa Rican daily and weekly newspapers, and with tourism being the single most important foreign exchange earner in Costa Rica the debate is high on the political agenda and in the forefront of general discussion. The 1990s, however, have seen the government's approval of a number of large-scale, foreign-owned tourism development projects, despite the international renown Costa Rica has achieved through its earlier promotion of small-scale, community-based ecotourism projects.

In the same region, the policy divergence between small-scale, community-based tourism and large-scale, mass tourism is emphasised for Honduras by Ron Mader:

Box 8.6 A letter from Belize

RAINFOREST SOS

It has been very difficult for us to write what is happening here, probably because it is hard for us to understand . . .

With organised groups in 13 communities we have been told the TEA [Toledo Ecotourism Association] is the largest indigenous ecotourism conservation association in the western hemisphere . . .

The unfortunate fact is that even with this overwhelming evidence of the TEA's ability to help the local people to unite, to plan, operate, control and directly benefit from ecotourism in their areas, the major sources of funding in the country for conservation, sustainable national resource management and ecotourism development have refused to assist the TEA . . .

More difficult for us to understand is the fact that they are, on the other hand, helping to organise and are funding other independent individual groups to be involved with tourism in direct competition to the TEA groups . . .

These competing groups have divided our people and seriously threatened to weaken and destroy the unity the TEA has painstakingly developed over the years. This unity represents the only realistic hope and chance the rural and urban people of Toledo have to continue to control and benefit from responsible ecotourism.

We have been told that this is happening because there are other national and foreign interests who are not from our villages who want to take advantage of the opportunities for tourism and other developments . . .

The indigenous and other poor people know what this is all about. Before the coming of the TEA, 95 per cent of all tourism in Toledo was controlled by a small group of foreigners and wealthy Belizeans . . .

Many of them still resent and resist what they consider to be a loss of their exclusive political and economic powers. Those who oppose the real empowerment of the . . . people of Toledo are falling back on the old colonial system of divide and conquer. They have influenced the major funding agencies to use the money . . . earmarked to strengthen local community-based organisations for ecotourism and conservation development, to weaken the most successful association established for this purpose – the TEA.

. . . what about the major ecotourism organisations in the world and our government leaders who have endorsed this programme? It appears that they have been intimidated by the wealthy people . . .

Our only hope now lies with individual citizens abroad, who will write to the large conservation and ecotourism organisations to which they contribute, . . . They could demand to know why the TEA has been consistently denied this assistance, and to write to the Belizean Prime Minister and Minister of Tourism and the Environment to encourage them to continue to stand up for and encourage aid for the TEA.

(Tourism Concern, 1994b)

There are two strategies Honduras is pursuing simultaneously – though from different quarters. A World Bank consultant and officials from the Institute of Tourism tout the development of a handful of luxurious five-star hotels near various 'ecotourism' destinations – Cusuco, Tela, La Tigra, Celaque and Copán. . . . Another approach promotes community-based efforts. USAID has sponsored the creation of an ecotourism association – APROECOH – which has trained dozens of Hondurans. This type of grassroots, community development should be more successful (in my view).

(internet communication, 1996)

These examples highlight a number of points about this divergence. First, it is obvious that both governments and communities face a range of options and courses of action between the two extremes of government-inspired mega-projects and community-inspired local projects. While it is important that ideas for and control of tourism developments should come from within the community, it is also important that local communities can make use of and benefit from the assistance of national government resources to help establish and coordinate their ideas and schemes. This is especially so where management of protected areas is pertinent to the scheme and where the advice of specialist professionals may be helpful. It can also be crucial in enabling communities to gain access to the tourists themselves. Indeed, channels of communication and information should be an intrinsic part of the tourist system so that local communities can seek assistance when they deem that they require it.

Second, the examples cited here draw attention to the division between the rhetoric of national politicians and their actions. Governments which in public espouse the fine-sounding language of the sustainable and ethical high ground of local community tourism development may be subject to external pressures (from supranational lending agencies for example) which dictate a policy of economic liberalisation and foreign exchange maximisation. Such pressures and circumstances are more likely to lead them into the development of mega-projects which at best ignore and at worst trample on local communities, regardless of their rhetoric and stated aims.

Third, local communities may lack the base of resources, skills and finances required unless they have assistance from a higher tier authority such as provincial or central government. Hence, a partnership arrangement may often be more suitable than a community attempting to do everything entirely from within its own human, physical and financial resources. Partnerships do not necessarily have to include government departments or ministries, and indeed may be best advised to exclude them if they are perceived by the local community to be corrupt or likely to pervert the aims of the development. Industry or the academic community can also be involved in partnerships, especially where the tour operator's access to tourists is required or where survey or research work is required as part of the process of tourism development. Where the resources required are financial, however, then the assistance of governmental or international bodies may be necessary.

A number of aspects of partnerships are illustrated in Box 8.7, which outlines one community's development of the CAMPFIRE scheme (Communal Areas Management Programme for Indigenous Resources) in Zimbabwe. The CAMPFIRE initiative is designed to help rural communities manage their wildlife and natural resources for the benefit of the community as a whole, and treats wildlife as a resource rather than a threat and a problem. The CAMPFIRE scheme is heavily cited in tourism's academic literature, and more general details can be found in McIvor (1994), IIED (1994), Child (1996) and Murphree (1996). In all the CAMPFIRE examples and in others where partnerships may be more complex, there is a clear need for ultimate control to rest with the local community.

Box 8.7   A photo-tourism partnership in the CAMPFIRE scheme

The safari hunting contract provided the initial impetus for the programme, and still produces the bulk of the income. However, further schemes for generating revenue are now being developed. In 1993 a contract was set up with a photo-tourism safari operator, to run walking safaris from a tented camp. From the beginning, the local people were more involved in setting this up than they had been with the hunting contract, which was done purely at Council level. Still the business belongs to the operator, so the risks and responsibility lie with him – the joint venture is purely a financial one.

In 1994 a further step was taken in devolution, when one community began building their own small tourism camp. Limited funding was obtained from an outside donor, so that the local WWC [Ward Wildlife Committee] had to budget and plan the project properly. They have been the decision-making body at every stage. . . . When it is completed it will be staffed and managed by the WWC . . .

The money-making potential is not large, since the project is very modest, aimed at the budget traveller or weekenders from the city, but it has been, and will continue to be, an enormous learning experience for the people involved, as well as being a source of pride within the community. Having worked through this project, I can see the need for a centralised training facility to serve all such ventures. As well as initial training in project planning and budgeting, it should provide a forum for sharing ideas and experiences, so that each District can learn from the activities of others. This would require outside funding to set up and run . . .

The connection between the hunting, photo-tourism and small scale projects is community participation and responsibility.

(Bird, 1995)

Fourth, despite the need for partnership arrangements with outside bodies, there is also a need to balance the advice of the 'experts' or professionals with the advice of the community, where the term community is as all-inclusive as practice allows. Experts and professionals are as subject to values and the sway of competing interests as are local people. They may have greater knowledge of a particular field, they may have greater funds of money, they may have greater access to particular market sectors, but they all represent interests of some kind, whether it be market forces, a political dogma, a requirement for more research funding, or an ego; and it is important to balance the nature of their advice and expertise against the local interest.

Local élites

Notwithstanding the potential problems created by outside bodies, Krippendorf noted that: 'Some locals do, unquestionably, make a nice profit out of tourism, but they are usually a very small minority belonging to the propertied classes. It goes without saying that they are staunch advocates of a further development of tourism' (1987: 54–5). It should be acknowledged, then, that the local community itself is not immune to the divisions which may come from within its own number and which may expose either the influence of a dominant local élite or the need to balance the demands and wishes of different sectors of the community. Again, an analysis of the distribution of power, in this case within the local community, is essential for an understanding of the dynamics and effects of tourism developments.

The example of the Kuna Indians of Panama (Box 8.8) illustrates this point. In the main group of the San Blas Archipelago of islands on which the Kuna live (off the Caribbean coast of Panamá) there are only three hotels, one of which, the Hotel San Blas, is shown in Figure 8.3. It is generally perceived by the Kuna, for a number of reasons, as being in their own interests to keep it this way in order to prevent too great a number of tourists visiting their islands. But many of the Kuna manage to derive financial benefit from the tourists by selling their appliqué cloths (*molas*) (see Box 8.8). The greatest profits from tourism, however, undoubtedly accrue to the owners of the hotels and their families. Doubtless also these families gain respect from this position and in turn derive more than average influence in the development of their islands. In the case of the Kuna, this is not an influence which is much or generally begrudged on account of the general benefit derived from tourism and the decision-making systems in the community. Regardless of how this power and influence is used, however, the case serves to illustrate how local élites may be formed.

The case of the Toledo Ecotourism Association (TEA) discussed earlier illustrates the potential dangers of tourism benefits being divided among only a few. As the case makes clear, one of the essential objectives of the TEA is to distribute tourism benefits widely between and within the associated villages. In villages of up to 500 and 600 population, however, it is not possible for all

Box 8.8 *Cruising round the Choco and Kuna*

To visit the Choco tribe in the Darién jungle, we stepped off the ship into *cayucos* at 4 am and made the first two hours of our journey in these dug-out canoes in total darkness. We were amazed to see at first light the intricate palisades of mangrove on either side that showed our pilots knew precisely where to find the mouth of the Sambú River.

We powered upriver for another two hours in a convoy of eight vessels to . . . their little riparian village, about 40 houses on stilts built along a broad avenue that features centrally a basketball pitch. . . . Beyond are some clearings of light cultivation, but mainly they are hunters of meat, though their small stocky build suggests that over the centuries protein has been hard to come by.

They were waiting for us all along their riverside avenue, behind platforms and logs spread with the goods they had fashioned for sale, and were being continually reinforced by others who'd been alerted by messages in Coca-Cola bottles dropped from a plane by a dynamic American Mr Fixit who has lived in Panamá for 40 years.

The men make music; the women sell. They are the advance guard of a 'nation' of about 6,000 people. They carve beautifully in rosewood, imaginative little ornaments and earrings from ivory nut. The women make the dyes and the baskets. They used to make them large, but they've learned that tourists can only handle small ones. Their goods sold on merit too – 'Who would imagine that I'd get up at 3 am to do my Christmas shopping in the jungle?' said my schoolteacher friend – and I estimate that we spent $5,000. Our cruise director takes along a float of $3,000 to bankroll those who run out.

We left more like $10,000 with the Kuna on Acuatupu in the San Blas Islands. The Kuna is a larger nation, about 50,000, with a much greater exposure to tourists. They are not as good-looking as the Chocos, their features are sharper, but they're more together commercially . . . for the past 25 years they've had the rights and control for domestic purposes of the archipelago.

They don't carve, but they make brilliant 'molas' (a word for clothes that now means specifically their appliqué designs on squares which people have been known to buy for $10 and sell for $60 at Nieman Marcus). . . . The garment of one was overprinted, '*500 años de resistencia indígena . . .* '

And they make a dead set at the photographers, offering not only their own images but carefully contrived little tableaux. For instance, a little girl with an umbrella sitting on a bench, smoking a pipe and affecting to launder a brightly coloured shirt . . .

As the Kuna see the trade flagging, they rapidly pack up and jump into dug-outs, pulling plastic covers over themselves and their goods for the choppy journey . . . to a secondary outlet, the large cruise ship *Radisson Diamond*. I remarked to our ship's official photographer that we'd just been looking at what the anthropologists call 'staged authenticity'. He was shocked by this comment, and replied that that was surely an oxymoron. 'That's what they like about it', I said.

(Hamilton, 1995)

Figure 8.3 The Hotel San Blas, Panamá
Source: Martin Mowforth

families to participate in the activity of catering for tourists – not that all families would wish to do so. But inevitably, there is space for the rise of favouritism within the allocation of tourists to households for meals, for instance. In the TEA villages this has occurred in the past, but it is to the credit of the structure and operation of the TEA system of tourist allocation that the problem has thus far always been detected and corrected, in one case with the suspension of one family's involvement in the scheme. In the case of the TEA, though, the competing scheme promoted by USAID and BEST is likely to create divisions within the village which could lead to the formation of a local élite, especially as this competing scheme has access to far more funding opportunities than does the TEA.

Finally, it is worth pointing out that élitism may not relate solely to the financial aspect of tourism. Local district councils may develop an élitism of influence and decision-making without necessarily benefiting financially from it. In such cases a social distance and a communication gap may develop between the decision-makers and those they represent. Some district councils in the frequently acclaimed CAMPFIRE scheme in Zimbabwe during the 1980s (see Box 8.7), for example, have been described by McIvor as 'almost as remote as the central government in the minds of the people' (1994: 33). In such circumstances, representatives of local communities may take decisions on behalf of interests other than those of the people they represent. This is alluded to by O'Riordan, who states that 'participation on a mass scale is an idealistic dream. In a representative democracy, it is impractical and unnecessary; in a political culture with a tradition of élitism, it is out of the question' (1978: 153). In the prevailing economic system of capital accumulation and political system of representative democracy, nothing else should be expected.

DISPLACEMENT AND RESETTLEMENT

Of all the problems experienced by local communities facing tourism development schemes, the most harrowing involve accounts of people being displaced. Such events normally reflect the distribution of power around the activity of tourism and highlight the powerlessness of many local communities. And it seems to be rare that displacement and subsequent resettlement of displaced people result in a more even and equal development.

In the literature most case studies of displacement and resettlement illustrate a deteriorating situation for those displaced. This has been especially well-documented in cases where the development promotes mass forms of tourism: in Guatemala, 300 *campesino* families were evicted in June 1996 from land they claimed belonged to the state – police burned down their homes and arrested several of them – to make way for a Spanish businessman's plans to build a tourist complex (Flynn, 1996: 4); Chapter 9 gives details of the forced relocation of the Padaung communities in Burma for the development of tourist complexes and the creation of tourist attractions; and details of the displacement of groups in eastern Africa are documented in the following pages.

Around Third World countries, the list of such stories is endless and is charted regularly in the newsletters and publications of tourism campaigning groups, such as Tourism Concern, and human rights groups. One would assume, however, that the supposed ethical base of new forms of tourism would avoid such pitfalls. Unfortunately, much evidence appears to contradict this assumption, and displacement and resettlement have become frequent outcomes of policies aimed at conservation and protection. Two examples are cited here, both from Africa, together with others elsewhere in the book, in order to illustrate the ways in which the goals of tourist money, conservation and 'sustainable' development policies may be linked together in the dispossession of local communities and indigenous groups from their land. Neither example here, however, is entirely pessimistic and both include an element or two of a positive nature for local communities.

The Maasai in Kenya and Tanzania

Box 8.9 introduces the first case study of displacement, the Maasai in Kenya and Tanzania. It makes it clear that the displacement was not a single move, a one-off event; rather it has been a prolonged and systematic persecution of the Maasai by the Tanzanian and Kenyan authorities, although it should be stressed that these authorities were guided in their actions by First World conservationists and

---

Box 8.9   Displacement of the Maasai

Early 1900s European hunters eliminated some wildlife species and decimated others which had survived over 2,000 years of contact with Africans and their livestock.

Second World War Hundreds of thousands of wild animals killed to feed British troops.

1959 Serengeti National Park in northern Tanzania divided into the Ngorongoro Conservation Area (NCA) and smaller Serengeti National Park. The Maasai who lived in the latter (and had done so for over 200 years) were moved into the NCA so that the Serengeti would become a game park with no human interference. In the NCA, 'should there be any conflict between these interests, those of the latter [the game animals] must take precedence' (Governor of Tanganyika, 27 August 1959).

1960 Establishment of the Maasai Mara Game Reserve in Kenya, adjoining the Serengeti in Tanzania, further restricted the movements of the Maasai.

1974 Maasai forced to evacuate the two crater areas within the NCA on the grounds that their presence was detrimental to the wildlife and landscape.

1975 All cultivation within the NCA prohibited.

1976 Maasai prohibited from entering the Olduvai Gorge on the grounds that their 'livestock were detrimental to the archaeological value of the site' (Olerokonga, 1992: 6).

1980 Collection of resin, a source of cash for the Maasai, stopped. Burning grasses in the highland areas also restricted.

1987 Anti-cultivation operation mounted by the authorities against the Maasai, who were farming small plots: 666 people arrested; nine jailed for six months; 549 fined (Olerokonga, 1992: 6).

1994 Allegations of torture, false imprisonment, theft and corruption by the Kenya Wildlife Service against the Maasai (corroboration by Paul Ntiati, an African Wildlife Foundation representative, cited in Monbiot, 1994: 93).

---

operated not only in their own interests but also in those of developers in the tourism industry.

It is clear from Box 8.9 that after the Second World War, all the policies of exclusion and resettlement in this case have been pursued in the name of conservation, especially the conservation of wildlife. The mechanism for doing this has been the creation of national parks and wildlife reserves, and the impetus for creating them came largely from First World conservationists and scientists who suspected that pastoralism was responsible for environmental degradation and decreases in wildlife numbers. George Monbiot accuses some scientists of having 'maintained that local people have always been a threat to wildlife: that they hunt the game with destructive methods and over-graze the land. These arguments have been well-rehearsed among conservationists, and are known to many of the tourists visiting Kenya' (1995: 11).

Monbiot goes on to claim that the Maasai's activities did not threaten the wildlife and that the work of earlier scientists was clouded by 'colonial disdain' and 'genuine misunderstandings about savannah biology'. He summarises:

> What is incontestable is that a fantastic abundance of wild game continued to exist alongside the herds of the Maasai and other nomads up to and beyond the arrival of the British in East Africa. It was indeed because the Maasai had not destroyed the populations of game that the Europeans wanted to conserve their lands.
>
> (1995: 11)

Today, it is difficult to escape the realisation that wildlife and the pastoral activities of the Maasai have managed to coexist in the region for many centuries, and that the landscape (at least until recently) was the product of their grazing and burning practices. To Deihl, it seems 'ironic . . . that one of the first steps in establishing a national park is to rid the region of its original caretakers' (1985: 37).

As the Maasai have been excluded, so the tourists have been allowed access. Despite the exclusion of the Maasai from the crater areas within the Ngorongoro Conservation Area (NCA), for example, there are several camp sites for tourists on the crater floor, although casual camping is no longer allowed. Tracks and roads have been created to allow tourist vehicles easy access to wildlife, thereby destroying natural vegetation. Although the Maasai have been excluded from the Olduvai Gorge, many tourists enter it every day, some even removing stones as souvenirs (Olerokonga, 1992: 7). Tourists also gain access in hot air balloons (cost US$250 per person for a forty-five minute ride) gliding low over herds of wildlife, stampeding and disorienting them (Olindo, 1991: 37), although there is no recognition of this problem in Abercrombie and Kent's 1997 brochure in which they advertise 'A piloted hot-air balloon flight can be arranged for visitors to the Serengeti. Float feet above vast herds of game in the silence of a hot-air balloon' (29).

In 1991, Perez Olindo of the African Wildlife Foundation in Kenya and formerly director of the Kenya Wildlife Department outlined several of the plans of the Kenya Wildlife Service (KWS) which were aimed at addressing the tourism and conservation problems in the game reserves and national parks. These included road construction, a ban on the development of new tourist accommodation and on casual camping, minimum flight levels for balloons, and the promotion of ecological sensitivity in the tourists. There is no mention there of the guardians of the original environment, the Maasai, neither in terms of their land and grazing rights nor in terms of an acknowledgement that tourism and conservation have been largely responsible for their dispossession and displacement.

Although seen by some as the best disciplined conservation management force in Africa, Monbiot describes the KWS as a 'para-military sustainable tourism-conservation organisation' (1994: 121) and Fernandes says it 'has implemented sustainable market driven model growth strategies which deserve to be condemned' (1994: 11). Moreover, it is clear that much of what is done in the name of conservation is actually done to protect the profits obtained from tourism: 'Several times I was told by (tourism) conservation officials that the Maasai had to be kept out because the tourists did not want to see them there' (Monbiot, 1994: 119). And Olerokonga (1992) makes it clear that the tourist and the Maasai are currently alienated from each other.

Fernandes (1994) maintains that such strategies owe much to the Brundtland-inspired sustainable development approaches. It is the action plans and agendas which have emerged from this generalised approach (see Box 4.10) which Adams (1990) claims are 'firmly anchored with the existing economic paradigms of the industrialised North. This might be called the approach of "green growth" ' (67). The implication is clear: the power exercised here derives from the First World, is expressed by conservationists, acted upon by a powerful local élite, benefits First World tourists, and serves to increase the inequality of development by, first, preventing the local communities from conducting their traditional ways of life and, second, excluding them from the benefits of the activity of wildlife tourism. As Olerokonga explains:

> Since the mid-eighties many stand along the main road to Serengeti and the Ngorongoro Crater waiting for tourists and hoping that they will pay to take pictures of them. . . . For both, their only interest is to profit as much as possible from each other – the tourists by taking pictures of the Maasai, the Maasai by getting money from the tourist. They don't see each other as dignified human beings.
>
> (1992: 7)

In the first half of the 1990s, then, conservation in East Africa was still largely a matter of separating land from its traditional human inhabitants. New approaches are currently being explored, however, with the aim of attaining some redress of dignity and power. In 1996, for example, a small group of the Maasai opened the Kimana Community Wildlife Sanctuary covering over 6,500 acres in Kenya, having negotiated a deal with a British tour operator to construct a luxury lodge, from which a proportion of the tourist payments (approximately $12 out of each $80–100 paid per tourist per night) will be made to them. Having calculated their takings before the deal was signed, they now have plans for a school and clinic for their community. In November 1996, the British Guild of Travel Writers recognised the significance of this deal by the Kimana community with one of their annual awards. On the assumption that their plans come to fruition, it will be an important step for the Maasai – in the past such deals have directed benefits into the hands of a single individual or a single family rather than a whole group. In this case, all of the group have taken part in the plans and all will benefit directly from tourism. It may not make good their displacement, nor retrieve their former lifestyles, nor compensate them for all the losses and betrayals they have suffered in the past, but for one group of the Maasai at least it is a hope of improvement, even if it can hardly be said to threaten the dominant model of development which has been at least in part responsible for the displacement.

Such mild 'success' as this highlights a deep division among the Maasai over the wisdom of involvement in tourism projects such as this. The views of the Kimana community proponents of the sanctuary and lodge contrast sharply with the views expressed in the following statement from one of the Maasai:

> We know there is money to be made from tourism. We already have tourists staying on our lands in tented camps. And, yes, they bring us an income. We don't need the Kenya Wildlife Service to tell us that. But you can tell Dr Leakey [director of KWS until 1994][1] this. We don't want to be dependent on these tourists. We are Maasai and we want to herd cattle. If we stopped keeping cattle and depended on tourists, we would be ruined when the tourists stopped coming.
>
> (cited in Monbiot, 1994: 98)

Mountain gorilla conservation in Uganda, Rwanda and Zaire

The case of the international conservation agencies' attempts to save the remaining mountain gorillas from extinction in Uganda, Rwanda and Zaire illustrates an example of wildlife conservation that has had some success, although the onset of genocidal inter-ethnic strife in the region in the early 1990s has led to uncertainty about the long-term prospects for mountain gorillas. The case is included here because publicity about the success of the wildlife conservation and tourism revenue attraction measures has overshadowed the human displacement which these measures have caused.

Between 1960 and 1973 evidence showed that the mountain gorilla population in the area of the Virunga volcanoes had declined from 450 to 260, mainly through poaching (Weber, 1993). The gorillas were under pressure not just from poachers but also from local farmers who required more land to provide food for a growing population. As a result of this pressure, the Mountain Gorilla Project (MGP) was established in Rwanda in the late 1970s. The project had the linked goals of promoting ecologically sensitive tourism, improving park security and spreading conservation awareness. The ecologically sensitive tourism essentially took the form of gorilla watching similar to that described in Uganda's Mgahinga National Park by Melinda Ham (see Box 8.10).

For Rwanda's Parc National des Volcans, Lindberg and Huber (1993) show that the revenue from tourist fees to see the gorillas rose from US$7,000 in 1976 to US$1 million in 1989, a sum which far outweighed the cost of running the park (approximately US$200,000 in 1989). Demand was high and only

---

Box 8.10   *Cashing in on the silver-backed gorilla*

Holding their breath, the four tourists crouch in the grass, their long-lens cameras at the ready. Less than five metres away, their prey gently breaks off a piece of bamboo as thick as a human arm, tears off strips with his teeth and chews the soft green core.

The silver-backed gorilla stands proud. Beside him a female grooms her neighbour while a baby gorilla somersaults playfully beside her. They take little notice of the human presence . . .

The tourists have spent nearly two hours climbing up through the dense, muddy bamboo forest, accompanied by two trackers and a Uganda National Parks (UNP) ranger.

The trackers carry machetes to cut a narrow path through the forest. The ranger carries an AK47 assault rifle in case of a chance encounter with armed poachers.

Under the strict rules governing gorilla-tracking tourism in Uganda the tourists can spend only one hour with the gorillas once the trackers find them.

For this once-in-a-lifetime experience each tourist pays nearly US$150.

(Ham, 1995: 1–3)

---

twenty-four visitors per day were permitted entry. This allowed the fixing of a high individual fee – almost US$200 per person for a one-hour visit which, as Sherman and Dixon (1991) suggest, is around the highest charged anywhere in the world and may be near the upper limit that visitors are willing to pay. Moreover, it has been estimated that an extra US$3–5 million annually was paid into the national economy by foreign tourists (Weber, 1993).

The high fees have drawn charges of economic élitism, but the scheme was widely deemed to be successful. From the late 1970s to 1989, the gorilla population in this area rose from 260 to 320, which was due, according to William Weber, a specialist in primate conservation at the New York Zoological Society, to tourism revenues paying for more forest guards and in turn reducing gorilla poaching.

In Uganda a similar success in terms of wildlife conservation was achieved only after the displacement of over 1,300 people. Despite the success in Rwanda described above, it was clear that very considerable threats – civil unrest, poaching and loss of habitat to cultivation – still loomed over the primate population. Accordingly, in 1990 several international conservation agencies led by WWF formed the International Gorilla Conservation Programme (IGCP) in cooperation with the Ugandan, Rwandan and Zairean governments. Under the Global Environment Facility (see p. 169) the World Bank provided US$4 million to fund gorilla conservation projects in Bwindi and Mgahinga National Parks in Uganda.

Uganda has only begun to realise and exploit the potential tourist revenue from its gorillas in the last decade. The Mgahinga National Park covers only

33 square kilometres in the Ugandan part of the Virunga Mountains and shares the gorillas with Rwanda and Zaire. One of the effects of the IGCP, the eviction of local people, is described in Box 8.11, again by Melinda Ham, who points out that Mgahinga was already 'being eyed by major tourism companies for investment' (Ham, 1995: 3).

Such displacement, resettlement and later compensation certainly qualifies the success of the conservation measures. Ham describes the Mgahinga case as an 'ideological battle between international and local conservationists' and cites Jaap Schoorl, a Uganda National Parks advisor seconded from the international aid agency CARE, as saying 'We have learned that if you don't get the cooperation of the local people you can forget about gorilla conservation altogether. People have to see the benefits of having a national park on their doorstep' (1).

After the eviction in 1993 the Ugandan government established an advisory committee for the region which included representatives from the local community.

> The committee decided that the whole community could still have access to the park, to collect water, gather plants or place their beehives in the forest. But access would be limited and subject to detailed agreement between all groups. The committee also agreed that the community would receive ten per cent of the revenue generated from the park entry fees paid by tourists.
>
> (Ham, 1995: 3)

Box 8.11 Of gorillas and people

In the Mgahinga National Park [Uganda], 1,318 peasant farmers had cleared ten square kilometres of bamboo forest illegally, planting peas, potatoes and wheat on terraced plots. Every year, the villagers pushed the gorillas further up the volcanoes' slopes and across the border into Zaire and Rwanda.

The villagers . . . used the trees and bamboo for building materials, plants, bark and fibres for medicinal purposes and placed their beehives in the forest. Their water supply also came from streams in the reserve.

Originally Mgahinga . . . was only a forest reserve. But in 1992, the IGCP [International Gorilla Conservation Programme] encouraged efforts to turn it into a protected National Park. . . . In mid-1992 the UNP [Uganda National Parks] evicted the peasants from their illegal plots. However, after a year of bitter hostility between National Park authorities and the community, the villagers were compensated with funds provided by the US Agency for International Development . . .

Didas Mutabazi, one of Mgahinga's six park rangers, says: 'Even though the villagers have received money, they are still very angry about losing their land.' . . . Only two farmers were offered employment in the new national park as trackers. Both refuse to talk about the loss of their land.

(Ham, 1995: 1–3)

This example shows up the potential conflicts between wildlife conservation and human survival strategies and serves to highlight the importance of tourist revenue in these conflicts. Tourism and conservation can be the cause of displacement and are factors which serve to increase the unevenness and inequality of development. But potentially they also offer a partial solution to some of the problems which they create in the first place.

The Ugandan example further highlights the gap between two opposing conservationist views widely discussed in Chapter 6 – *parquismo*, as Lorenzo Cardenal calls it, and an integrated approach. The former is the policy of excluding humans from the area to be protected and conserved (the *parque*), and the latter refers to the integration of human activity with flora and fauna conservation. In the case of the Mgahinga National Park, the difference is emphasised by the words of Jaap Schoorl above. In a different context, Jean Carrière of CEDLA (Centro de Estudios y Documentación Latinoamericano) describes US-influenced environmental institutions as tending 'to see environmental protection in isolation from the social context, and [to] soon convert Costa Rica's forests into fenced-off green museums surrounded by starving peasant families' (1991: 198).

Colchester (1994) has pointed out that conservation NGOs have generally derived their funding from the establishment and have attempted to use the power of the state to impose their visions and goals. In search of funding for their conservation programmes, international conservation organisations tend to form alliances with supranational organisations such as the World Bank, First World governments, or even transnational corporations. In so doing they find their programmes becoming geared to the kind of economic justifications – hence, the need to raise tourist revenue – which fit with the dominant economic paradigm promoted by these organisations. With such justifications, and with the moral high ground of ecological sustainability on the side of the conservationists, it is relatively easy to treat local people and communities as inconvenient and to confirm the need to displace them.

VISITOR AND HOST ATTITUDES

Tourism is widely touted as a beneficial cultural exchange for all parties involved, as having contributed to the general well-being of peoples around the world, as having stimulated economic development and as 'fostering peace and understanding between nations' (World Tourism Organisation, 1991). With such descriptors one might be forgiven for believing that the relationships between tourists and local people in the visited destination areas are always rosy. Many friendships are indeed made between the visitors and visited. But the relationships between visitors and hosts give rise to a range of other outcomes. In the following pages we discuss three possible outcomes of the relationships created by tourism to Third World destinations.

Transculturation

One of the most protracted criticisms of tourism in the Third World has concerned its impact on local cultures. Critics have consistently rounded upon cultural 'bastardisation', 'trinketisation', the destruction of indigenous cultures, and so on. Tourism, they contend, is a process of acculturation through which Third World cultures are assimilated into materialistic First World lifestyles.

In the first part of the book, it was suggested that advocacy of the need to protect cultures finds strong resonance in colonialism and the romanticism of the past, a kind of institutional racism that celebrates primitiveness. As Robins describes, in a process of unequal cultural encounter, 'foreign' populations 'have been compelled to be the subjects and subalterns of western empire, while, no less significantly, the west has come face-to-face with the "alien" and "exotic" culture of its "Other" ' (1991).

Similarly, in writing about travel writing, Pratt refers to 'contact zones', 'social spaces where disparate cultures meet, clash, and grapple with each other, often in highly asymmetrical relations of domination and subordination – like colonialism, slavery, or their aftermaths as they are lived out across the globe today' (1992: 4). Contemporary tourism, and particularly new tourism in the Third World, is staged in these so-called contact zones, which serve to emphasise that tourism is experienced in sharply differentiated ways by visitor and host. Pratt continues:

> A 'contact' perspective emphasizes how subjects are constituted in and by their relations to each other. It treats the relations among colonisers and colonised, or travellers and 'travelees', not in terms of separateness or apartheid, but in terms of co-presence, interaction, interlocking understandings and practices, often within radically asymmetrical relations of power.
>
> (1992: 7)

However, this asymmetry of power between host and guest tells only half the story. First, this has been encapsulated in the ideas of Edward Said in *Orientalism* (Chapter 3) – and the images portrayed in the collage in Figure 3.2. More often than not, it is charities, social movements and tourists that talk about the rights, cultural practices and uniqueness of Third World cultures, as if these people do not have a voice (which of course many do not) and are unable to represent their own views. Hence, Said's satirical 'quotation' of Marx: 'They cannot represent themselves; they must be represented.'

Second, through a process termed transculturation, Pratt (1992) attempts to encapsulate the way in which marginalised or subordinated groups select and invent from materials transmitted to them by dominant 'metropolitan' cultures. Hall (1995) refers to this as a 'cultural strategy' which operates between previously sharply differentiated cultures which are forced to interact. It is this process of change that those engaged in the promotion and undertaking of new tourism find difficult to accept. It is a feeling that we are somehow being cheated of 'authentic experiences', that this is no longer the real thing. This search for authenticity lies at the heart of much new tourism activity, as noted in Chapter 3, where it was suggested that authenticity might be understood as a part of the desire for (cultural) sustainability. It is an aspect of new tourism that is sharply reflected in trekking, an activity about solitude and distance from other tourists, but also about contact with 'real' cultures.

Boxes 8.12 and 8.13 outline some characteristics of the activity of trekking. On the basis of his observations of trekking in Thailand (1974 and 1989), Erik Cohen modified the idea of *staged authenticity* identified by MacCannell (1973, 1976) into the idea of *communicative staging* (see Box 3.6). Central to this debate has been the way in which local populations adapt to tourism. The last example cited in Box 8.13 particularly reflects the notion of transculturation in its description of those manifestations of First World influence which have been 'selected' for use by the younger Sherpa men.

Box 8.12 Focus on trekking

Trekking is the visiting of off-the-beaten-track locations and involves walking, often but not always in organised parties accompanied by a number of porters. The names of some of the small independent tour operators (High Places, Himalayan Kingdoms) which organise trekking tours testify to the rugged mountainous areas often visited by trekking parties and individual trekkers, but gentler highland areas such as Thailand, Kenya and Tanzania are also favourite trekking destinations.

The phenomenal growth of trekking in South-East Asia, Latin America and Africa (Brockelman and Dearden, 1990) underlines its importance in new middle-class travel. In Nepal the number of trekking permits issued has risen from eight in 1966 to approximately 13,000 in 1976, 47,000 in 1987 and 61,000 in 1988. The Annapurna area attracts over 38,000 trekkers annually, roughly equivalent to the area's year-round population; the Khumbu (Everest) region, home to around 4,000 people, receives 8,000 trekkers a year. (Source: Tourism Concern, 'Trekking in the Himalayas', information sheet, undated.)

The environmental effects of trekking in the Himalayas are documented in many papers and articles. Nepal's forest area is believed to be decreasing at a rate of 3 per cent per year with higher rates in lowland areas and heavily trekked routes in the hills. One hectare of cleared forest loses 30–75 tons of soil annually. Also regularly cited is the problem of litter, with a special topic of interest appearing to be the ugly and unhygienic streamers of used toilet paper, which are the exclusive contribution of westerners.

In Nepal trekking also provides an estimated 24,000 full-time jobs with as many as 70,000 people employed as porters on a freelance basis.

As Tourism Concern make clear, the cultural impacts of trekking are almost impossible to quantify, but Box 8.13 gives anecdotal evidence of some of the subtle cultural impacts of trekkers upon local populations.

Box 8.13 *The cultural effects of trekking*

We wait on a path in the Hinku Valley as another weather-battered group creaks towards us from Mera, one of Nepal's 20,000 feet trekking mountains. Their strained, peeling faces contrast with their porters' clear complexions and bored expressions. One scabby Lancastrian in hi-tech gear gasps through wind-cracked lips, 'It's amazing. But now I'm shattered; I'm emotionally and physically drained.'

Twenty over-burdened, under-clad porters rush past, anxious that nothing should interrupt their journey home to the comfort of a wood fire. It is just as well they hurried away, as they might not have cared for the parting words of one of our number. . . . 'They have to learn,' he said. 'They can't keep chopping down trees.'

His remarks epitomised one of the chief ambiguities in what has come to be known as 'sustainable tourism' – tourism that, according to its proponents, should do minimal harm to the environment and tries to put something back.

(Gordon-Walker, 1993)

After he [Sir Edmund Hillary] had climbed and succeeded, he was so grateful to the Sherpas because without them he couldn't have done it. He wanted to do something . . . and decided that what these people needed were schools. After the school had been built, as elsewhere in Nepal governmental authorities took over. They wanted to use the schools to homogenise the country's diverse ethnic communities. So they're using the Hillary schools to teach the Sherpa children Hindu ways and Nepali, and English and Mathematics as well, so that they can serve the tourists and bring tourist currency to the country.

In these schools they teach nothing about the Sherpa-Tibetan culture – and nothing about their own 1,200 year old written language, which is classical Tibetan, of course. Most of the Sherpa children growing up in Kathmandu . . . do not learn a word of Sherpa. If you get off the aeroplane at Lukla, . . . you are met by a whole group of youngsters who speak 'Hillary School English' to serve you. Over the years they have been completely incorporated into the tourist economy. Most of the younger inhabitants of this area don't know how to run their farm any more.

(Kvalöy, 1993)

Younger Sherpa men, of all Nepalis, are the most 'hip'. They dress in expensive jeans, tracksuits, baseball caps. They have Walkmans, tend to speak good English and smoke designer cigarettes. The women, however, have on the whole kept up the traditional – and hard – way of life . . .

. . . Western influence, which followed expeditions to Everest . . . , has greatly affected the Sherpas' way of life. Schools, hospitals and clinics, postal services, air transport and radio communication changed a semi-nomadic life to one much more dependent on tourism, trekking and expeditions.

(Klatchko, 1991)

One advantage of the concept of transculturation is that it allows us to explore possibilities that lie beyond the often repeated, even slavish, charge that tourism distorts, disrupts and bastardises Third World cultures. In some instances this is, of course, exactly what happens, especially where the power of the tourism industry is intense, as described in the next few pages. But it is a view that debars us from considering how the visited actually adapt and borrow from cultural practices and in turn modify their own cultural practices or ways of making a living, even in circumstances where their power is differentially distributed.

Tribal peoples and zooification

Survival International, an INGO which supports indigenous groups, has in recent years adopted the term 'tribal peoples', reflecting their representation of people who live by tribal norms, customs and practices rather than those of mainstream society. The term indigenous groups includes tribal peoples, while the term tribal peoples excludes members of indigenous groups who live by the norms and practices of mainstream society. In the tourism industry not all host communities are tribal peoples but tribal peoples which have had contact with the 'civilisation' of the First World are potential host communities. This chapter has already cited a number of examples of tribal peoples and their interaction with the tourist industry: the Maya, the Choco and Kuna, the Sherpas, and the Maasai. This section highlights one particular point about their experience as 'hosts' in the tourism industry: namely, their treatment as objects to be viewed, a process which might be called the zooification of tribal peoples.

The nineteenth-century and earlier Christian fundamentalist and explorer's view of tribal peoples as savages whose souls needed saving may not be as pervasive a public viewpoint now as it used to be in the early part of this century. But it helped to shape the common perception of tribal peoples as 'noble savages'. The characteristic of nobility owes something to the development of this early First World view by conservationists and environmentalists in their interpretation of wilderness as inclusive of the peoples indigenous to it. ' . . . these images [of godless, natural, wild and blameless savages] are retained to this day and lie behind conservationist policies of "enforced primitivism", whereby indigenous people are accommodated in protected areas so long as they conform to the stereotype and do not adopt modern practices' (Colchester, 1994: 3). The build-up of area protection policies and an associated conservation ethic (see Chapter 4) have clearly been important in the promotion of the common perception of tribal peoples as natural and wild. It is this perception which, in some cases, dominates the relationship and exchange between tourists and tribal peoples and which confirms and strengthens the already prevailing prejudices and can, at times, lead to the process of zooification.

Visiting tribal peoples' settlements is an activity especially associated with new forms of tourism. Such visits are advertised as small group tours, implying low

273

impact; tribal communities are described as 'almost untouched' or 'totally unchanged', implying an authentic experience; conditions are referred to as 'primitive', implying an experience with a difference, thereby conferring status on the tourist; and some tours to tribal settlements mention the possibilities for artefact purchases, implying that the exchange will assist in the development of the settlements and peoples visited.

In general, 'small group tours' is an accurate description, but these can hardly be claimed to be low in impact. Even the lone traveller can have an insidious and disruptive effect on local culture, especially if the host community has had little contact with mainstream society. Descriptions of the supposedly primitive nature of tribal peoples by tour operators are used to emphasise the 'otherness' of the experience. Again, examples have been cited where this cultural authenticity is often falsely enacted, such as the divesting of T-shirts solely for the sake of the visiting tourists (see the first example given in Box 8.14). The economic exchange involved can of course be significant, as the Choco and Kuna cases illustrate. Equally, however, it can contribute to the 'trinketisation' of a culture and can lead to conditions of near-slave labour, as in the case of the Yagua in Peru in which traders or middlemen take all the profit (see Box 8.14).

The authenticity of the tourist experience is put into question by many of the examples cited in this chapter. Even in those cases where control of the visit is in the hands of the tribal people, the nature of their ceremonies or products is altered for the sake of the visiting tourists. The Kuna, for instance, carry out rehearsals of dances and songs on the night before a cruise ship is due to arrive. As Box 8.8 outlines, the Kuna are adept at staging authenticity. Many other cases, a few of which are given in Box 8.14, show that the exchange is much more unequal. One example actually refers to the treatment of the Penan (in Sarawak) as being 'like animals in a zoo', but most of the other examples also hint at this kind of treatment. Knowledge of the fact that the tourist experience is staged often fails to deter the tourist from wanting to experience it, a reflection perhaps of the new tourist's need to collect cultural capital (see Chapter 5).

The zooification process involves turning tribal peoples into one of the 'sights' of a rainforest expedition or a trek. As a 1995 Survival International background sheet says, 'All too often tour operators treat tribal peoples as exotic objects to be enjoyed as part of the scenery.' But the following example, from the Green Travel bulletin board on the internet (see Appendix 1), shows that this attitude is just as likely to come from professionals such as archaeologists and anthropologists as it is from tour operators.

> There is a group of archaeologists and anthropologists who are travelling the jungles of Mexico, Belize, Guatemala and Honduras totally self-contained on X-country bikes. They are uploading journals and graphics via a satellite linkage of their adventures for all to see on a World Wide Web site. . . . The purpose of the expedition is to explore remote and nearly inaccessible Mayan pre-Columbian ruins and to study the present

day Mayas in their native habitat. These travellers are all young people
. . . and their adventures make . . . interesting reading for anyone.

(email communication, 1995)

Perhaps the process of zooification is best summarised by Rigoberta Menchú,
a Guatemalan Quiché Indian and Nobel Peace Prize winner: 'What hurts
Indians most is that our costumes are considered beautiful, but it's as if the
person wearing it didn't exist' (quoted in Survival International, 1995).

Box 8.14 Tourists and tribal peoples

The scene is a Bushman camp in a remote part of Botswana. In the distance
a plume of dust shows the arrival of a jeep. The people drop whatever they are
doing, quickly pull off their T-shirts, trousers and cotton dresses, and begin to
dance.

In the Himalayas, fields lie uncultivated. The men who once farmed them
have become porters to climbing expeditions.

In the Peruvian Amazon, women of the Yagua tribe make bags, hammocks
and jewelry for the tourists. They work hastily because they are being paid
almost nothing. The traders make all the profits. Decoration is crude, the finish
careless – 'Well,' the women say, 'these people don't know any better.'

In Tuareg camps around Tamanrasset in Algeria, the tents of the drought-
stricken refugees are normally covered with plastic sheets – only when the
tourists arrive are the old coverings of animal hide brought out.

After an Amazon trip, Mick Jagger told an interviewer: 'What the tour guides
do is take all the Indians' clothes off and put their little skirts on them, hand
them a spear which they hand back at the end. The Indians dance a little
around you.'

A Survival member writes to us in disgust: 'In the Mulu National Park
(Sarawak) I realised how the Penan are being treated like animals in a zoo.
Almost every tourist group that visits the park is taken there to walk around and
look at the Penan "way of life". A longhouse is now in the process of being built
for them, which as you know is not the way the Penan live. The atmosphere is
one of despair.'

Another Survival member describes how her party were taken to stay in a
village of the Meo tribe in the hills of Thailand. She paints a picture of the tragic
gap in understanding that remained between the hill people and their well-
intentioned visitors, who were quite unprepared for the 'primitive' conditions
they found. A young German, looking bewildered, asked where the toilet was.
Our guide laughed, and exclaimed 'Everywhere!' As the initial exhilaration of
our adventure gave way to fatigue, relations with the villagers became strained
and our interaction with them was reduced to stares and picture taking.
Occasionally, as an inadequate response to our own guilt and to demonstrate
our gratitude we made them gifts of whatever we had. A soap box or a second
hand toothbrush.

(Survival International, 1991)

The process of zooifying tribal peoples leads inevitably to a position of power-lessness as well as a complete loss of human dignity. The key to avoiding such situations is control of and participation in the tourism activity, which do not necessarily mean simply a greater share of the financial profits. As Pretty's typology suggests, it also implies control over all the conditions of the tourism development. Again, as was noted in the discussion on local participation, one of the most important elements in the success of a tourism scheme is that the idea and impetus for it should come from within the community itself.

It is of course too simplistic to demand a single-minded, blanket policy of total control to the tribal groups involved in any tourist development. There are dangers, as Marcus Colchester points out, in making 'an assumption that once an area is under indigenous ownership and control the problem is solved. . . . This is patently not the case' (1994: 57). Notwithstanding these dangers, it can be argued that the community has to own and control the development if it is to avoid the pitfalls associated with external control.

Doxey's levels of host irritation

Broadening out the analysis of relationships between hosts and visitors to include local communities from the Third World rather than just tribal peoples, it is a helpful starting-point to use Doxey's index of irritation, first put forward in 1975 and 1976. This is often referred to as Doxey's Irridex and is illustrated in Table 8.2.

The Irridex is a causal model of the effects of tourism developments on the social relationships between visitors and the visited. Beginning with a state of very little tourism development and only the occasional passing visitor, the model's four stages describe different states of tourism development and the ways in which tourists and local people perceive each other in these stages. Its final stage is that of antagonism in which the stresses and tensions between the visitors and visited, resulting from high levels of development for the tourists, are at a peak and are likely to lead to a deterioration in the reputation of the destination.

Clearly, this is a highly generalised model, and its sequence and relevance will be subject to a wide variety of factors which differ with time and space. Its original application was in a First World, mass tourism context, but it is feasible that the relationships between Third World communities and the new tourists who visit them will follow a similar sequence to that of the relationships in First World resorts. (Tourist motivations may be somewhat different, but tourism effects are not likely to be dissimilar.)

Community control of the developments from the outset may go some way towards breaking what may appear from the Irridex to be the inevitability of the sequence of worsening social relationships. The Irridex relates the type of social relationship (euphoria, apathy, annoyance, antagonism) directly to the level of development of tourist facilities and infrastructure. The last two stages indicate

Table 8.2 Doxey's levels of host irritation extended

Doxey's Irridex	Social relationships	Power relationships
Euphoria	Initial phase of development; visitors and investors welcome	Little planning or formalised control mechanism; greater potential for control by local individuals and groups in this phase
Apathy	Visitors taken for granted; contacts between residents and outsiders more formal (commercial)	Planning concerned mostly with marketing; tourism industry association begins to assert its interest
Annoyance	Saturation points approached; residents have misgivings about tourist industry	Planners attempt to control by increasing infrastructure rather than limiting growth; local protest groups begin to assert an interest
Antagonism	Irritations openly expressed; visitors seen as cause of all problems	Planning is remedial but promotion is increased to offset deteriorating reputation of destination; power struggle between interest groups forces compromise

Source: adapted from Doxey, 1975 and 1976

that a level of change to local lifestyles above what is considered acceptable by local people has been reached, and especially in the final stage has been surpassed. This may come about as a result of dimensional changes, such as overcrowding (in which case planning and visitor management techniques may be able to provide solutions) or structural changes (such as the outside influence of foreign investors or national politicians pursuing goals different from those of the local community). The latter cause especially implies that local control of development may act as a solution, and it is interesting to speculate on the association between Doxey's levels of irritation and the degree of local control. This association may be a pointer for worthwhile future research, a starting-point for which we have provided by extending the Irridex in Table 8.2 to include speculation on the power relationships implied by the level of irritation as well as the social relationships.

It should be stressed that the Irridex is offered here only as a loose framework for considering the relationships between visitor and visited. It should be noted that its applicability will be compromised by circumstance. It should also be noted that Doxey's is not the only attempt to characterise the different stages and features of these social relationships. Butler (1975) and Murphy (1983), for instance, offer more detailed models and descriptors of social interaction that make explicit allowance for a number of variable factors. Both acknowledge that communities can adjust their lifestyles in order to overcome stresses caused by uneasy social relationships between visitors and visited. But perhaps rather than looking for remedial action to counter the inequalities and unevenness of tourism developments, it might be more suitable for local communities to control developments from the start, as has been pointed out throughout this chapter.

CONCLUSION

The relationships of power, one of this book's key themes, are central to a consideration of the role of local communities in tourism. In this chapter these relationships have been examined with the assistance of Pretty's typology of participation, a seven-point range of types describing differing degrees of involvement and control by local people over (tourism) developments in their communities. The case studies presented in this chapter have been related to this typology and it is suggested that reference to the scale of participation is something that could profitably happen at the outset of all tourism development schemes. Likewise, techniques for assessing the degree of local participation in schemes need themselves to be subjected to a consideration of who is doing the assessing and for what purpose.

A note of caution needs to be sounded, however, for the general assumption that the greater the degree of local control and participation the greater the scheme's supposed sustainability and the wider the distribution of benefits within the community does not always hold true. First, as has already been well established, sustainability differs according to the interests of those who are defining it, and the interests of the local community will not necessarily coincide with those of others; nor is it likely that the interests of the local community will be the same for all within the community. Second, local power relationships within the community can be as factional as those which include players on a broader stage such as national governments, INGOs and supra-national institutions. Thus, the emergence of local élites is as likely to produce inequalities within the community, just as these other players produce disparities of benefits at a different level.

At its worst, tourism and the conservation measures which have been used to support it have been responsible for the displacement and resettlement of local communities. The case studies from eastern Africa used to illustrate these phenomena are disturbing, but it is noteworthy that they are not entirely negative examples. A number of other case studies highlight the important point that only where the impetus for tourism development comes from within the community is the prevailing inequality of development likely to be challenged.

The notion of transculturation, the process by which local communities adapt themselves to the cultural mores and habits of those with whom they interact, was used to demonstrate that this interaction is not purely a case of the imposition of one set of cultural values upon another, as it is often represented in tourism analyses. The term 'zooification' was introduced, however, to illustrate that in the cases of some tribal peoples this dehumanisation of local peoples does indeed take place, leaving them no power or dignity. Less extreme cases of First World new tourist–Third World local community interaction may be appropriately analysed according to Doxey's Irridex, which can be extended to describe the relationships of power as well as the social relationships between the visitors and the visited.

9

GOVERNMENTS AND TOURISM
What can we sell off?

The central governments of nation states are the last of the key players in this analysis of tourism. It is governments that possess the potential power to control, plan and direct the growth and development of tourism. And it is largely through governments that tourism-related international investments and loans and overseas aid are agreed and channelled.

It has already been argued that tourism has come to represent a considerable attraction to many Third World governments. It has been widely promoted both within the Third World and by First World 'experts' as a means of economic diversification and an important mechanism in producing foreign exchange. There can be little surprise that in Third World countries characterised by indebtedness and by primary industries (such as agriculture and mining) adversely affected by world market prices, tourism has come to represent something of a panacea.

In this chapter we take 'governments' as a further way of exploring the politics of Third World tourism development. As has already been argued, the consideration of politics has been a poor partner in tourism analysis. The assessment of government activity has tended to focus on strategies, policies and programmes of national level planning. While such analysis is an essential part of assessing the approaches of Third World governments to tourism, it tends to minimise broader consideration of the inherently political nature of tourism development. In other words, the pressures that governments face from external influences such as lending policies, First World foreign policy or the activities of international NGOs tend to be bypassed; similarly internal forces may also be overlooked. In short, the politics of tourism is another way of exploring unequal and uneven development.

In the first half of this chapter, politics is established as a critical factor in Third World tourism analysis. Most importantly, the need to acknowledge that tourism is not a neutral factor that can be assessed as just another government economic activity is established. Tourism is used for a variety of political purposes and there is a wide range of external influences. The structural adjustment policies of supranational institutions such as the World Bank and IMF, for instance, are considered along with the policies promoted by the World Trade Organisation and World Travel and Tourism Council. This provides the opportunity to reflect on the political globalisation outlined in Chapter 2 and in particular to consider the nature of sustainability promoted by such institutions.

The second half of the chapter uses a number of case studies to illustrate the development of new forms of tourism and the ways in which these are intimately related to sustainability. In particular, we analyse sustainability and sustainable tourism as part of a political discourse. Throughout this chapter, the ways in which politics is reflected through the power jigsaw (Figure 2.5) are considered.

THE POLITICS OF TOURISM

We are constantly reminded by the media of the hazards of travel in the Third World and by implication the critical nature of the politics of tourism. At its most dramatic are the newsworthy events, the killing or kidnapping of western tourists often by political factions eager to further their causes. For these factions, tourism becomes a means to an end: access to the system of global communication. To First World populations, such events provide a succession of countries – Algeria, Cambodia, Colombia, Egypt, Indonesia, India, Morocco, Peru – which, we are told, are plagued by fundamentalist groupings and where terrorism is rife. Such presentations have been one of the principal ways in which our geographical imaginations are filled out, and show how tourists and the tourism industry decide where, and where not, to visit and invest; for example, in Chapters 3 and 5 it was suggested that 'dangerous' regions may actually act as a bonus for adventurous 'new tourists'.

As Richter (1994) demonstrates, the USA has dozens of countries for which *DON'T GO* warnings are provided, of which over 80 per cent are in the Third World. But such warnings, Richter continues, appear to be based as much on ideological reasons as on potential security risks to US tourists. Countries to which the USA is more kindly disposed, such as Israel, Brazil, Mexico and Egypt, appear to require much higher levels of danger to tourists before travel advisories are made against them. For example, up to 1994, although tourists had been signalled as potential targets for attack by Islamic groups in Egypt, this large-scale receiver of US foreign aid had not been mentioned in State Department advisories (Richter, 1994). Similarly, despite the seeds of a civil war in southern Mexico, this country too has escaped travel advisories: a reflection, some would argue, of Mexico's 'integration' into the North American Free Trade Agreement (NAFTA).

At the other extreme, countries to which the USA considers itself ideologically opposed, namely Cuba, North Korea and Vietnam, remain off-limits to US tourists. The notion that *tourism is politics* is forcefully expressed by Conrad Hilton (of Hilton hotels), who is reported as saying 'each of our hotels is a little America' and 'we are doing our bit to spread world peace, and to fight socialism' (quoted in Crick, 1989: 325). Conversely, tourism is used by some Third World governments to gloss over glaring social inequalities and in some cases, as discussed later in this chapter, the systematic abuse of human rights.

Of course, for many Third World countries, political stability is one of the principal keys to securing a steady stream of First World tourists. The case of The Gambia (Box 9.1) shows how profoundly political instability, or just the First World's judgement of such instability, can affect a country's tourism fortunes. Following political instability in 1994, and subsequent British Foreign Office travel warnings, the tourism industry in The Gambia all but collapsed. Teye (1986, 1988) has argued that the perception of political instability arising from the liberation wars and frequency of coups d'état after decolonisation in sub-Saharan Africa has seriously stunted tourism development in some states. Clearly then, it is not only, or even mainly, the actual occurrence of political instability that is of critical importance, but rather the way that it is perceived, constructed and represented in the First World. Taking this reasoning a little further, it can be argued that it is also the perception of sustainability by First World agencies that is crucial to how some forms of tourism in Third World countries are considered.

These are perhaps dramatic examples, but they help to emphasise the inherently political nature of tourism development and the way in which power is transmitted through tourism. Countries promoting tourism (both mass and new) can be influenced by a range of factors, from the decisions of First World institutions and tourists, to the way in which particular Third World destinations are perceived in the First World. Consequently, Third World governments are also eager to control the way in which their tourism image is projected. We must therefore ask whether the messages we receive reflect truly what is happening in reality.

ASSESSING THE POLITICS OF TOURISM

Although the size and scale of the global tourist industry would seem to suggest that tourism is a hot political issue, so far we have argued that the politics of tourism has remained relatively underexplored (Richter, 1983). In terms of assessing the activities of Third World governments, the majority of analyses have been policy studies focusing upon national tourism policies, including both individual case studies and comparative policy analyses between countries (see, for example, Richter and Richter, 1985; World Tourism Organisation, 1994). A good deal of this work is also focused upon the identification of relevant and appropriate methodologies for planning and implementing tourism policy (for example, Gunn, 1994). Clearly such discussion is of interest and use, and has a wide appeal to practitioners.

Its weaknesses, however, lie in its often prescriptive nature and its formulation from within the First World and, in many cases, the failure to set policy and action in a broader and more critical framework which acknowledges that there are competing interests. In Chapter 3 it was argued that the political economy

---

Box 9.1   The Gambia

Thompson *et al.* summarise the economic context of The Gamb

> Since the mid-1980s, faced with falling prices for grouɪ\u...
> increasing debt repayments, The Gambia has undergone a period oɪ
> economic restructuring under IMF stipulations. The two major economic
> reform programmes embarked upon by the Gambian Government since
> then, namely the 1985 IMF-inspired Economic Reform Programme (ERP)
> and the 1990 Programme for Sustained Development (PSD) have
> attempted to diversify the economy including the agricultural, industrial
> and service sectors. Both programmes advocated that the government
> should institute economic and political frameworks within which develop-
> ment can take place, and that economic progress should be dictated by
> market forces. Both programmes emphasise the important role played
> by tourism in the development process.
>
> (1995: 573)

Tourism makes a substantial contribution to the economy, estimated to be 12 per cent of gross domestic product (GDP) by the end of the 1980s and has become the principal source of foreign exchange.

In 1994, Captain Yahyah Jammeh seized power in a coup d'état and established the Armed Forces Provisional Ruling Council (AFPRC). A failed counter-coup in November 1994 resulted in the British Foreign Office issuing Travel Advice not to visit The Gambia. As a result of this advice, air charter arrivals (December to March, the peak of high season) dropped from 45,733 in 1993/94 to 8,363 in 1994/95. There was an estimated 60 per cent reduction in the contribution of restaurants and hotels to the GDP and an estimated direct job loss of around 10,000. The indirect or knock-on effects are estimated at least double this impact. Overall, tourism contracted by 60 per cent and the economy shrank by 6.2 per cent.

Despite the lifting of the formal travel advice for the 1995/96 season, the Economist Intelligence Unit identified what appeared to be further pressure applied by the British Foreign Office: 'the UK appears determined to maintain pressure for an early return to full democracy and civil freedoms, and this position is likely to influence other trading and aid partners' (EIU, 1995: 17). 'Tourism, accounting for around 12 per cent of GDP, has yet to pick up after the most recent season (November 1994 – May 1995), ruined by European governments' advice to their nationals to stay away' (EIU, 1995).

(Information and extracts from Economist Intelligence Unit, 1995; Evans, O'Hare and Thompson, 1994; Thompson, O'Hare and Evans, 1995)

---

approach is one of the few attempts to illustrate the way in which control of tourism development is held firmly in the First World. As Britton argues:

> The World Tourism Organisation, International Monetary Fund, United
> Nations, World Bank and UNESCO, among others, set the parameters of

tourism planning, promotion, identification of tourism products, investment and infrastructure construction policies often in conjunction with
metropolitan tourism companies.

(Britton 1982: 339)

Of course, it is not just external influences that affect tourism policy and
development, but internal factors as well. This includes the conflicts between
governmental and non-governmental interests, and conflicting interests and
priorities within governments themselves. The latter are particularly prevalent
where responsibilities over tourism resources overlap more than one government
department. In Uganda, for example, there are three separate government
departments with responsibilities for national parks, the focus of Uganda's
tourism. Box 9.2 illustrates a similar problem of different outlooks adopted
towards its national parks by different government departments in Guatemala.
Clearly, it is necessary to avoid the crude suggestion that government policy and
action is an undifferentiated whole.

A non-critical approach also tends to assume that there are rational policy
decisions that will lead to the surmounting of potential problems. As Hall and
Jenkins argue, while 'rationality' represents an influential approach to policy
analysis, it is also 'extremely misleading as it fails to recognise the inherently
political nature of public policy' (1995: 65). Take, for example, the approach of
Edward Inskeep, a tourism planner who has worked with the WTO, World
Bank and United Nations Development Programme (UNDP) in many Third
World countries:

> [the] planning of tourism is necessary not only for scientific purposes and
> to conserve the environment for the benefit of residents, but also for the
> protection of long-term investments in tourism infrastructure, attractions,
> facilities, services, and marketing programs.
>
> (1987: 119)

What is important here is to question who benefits from planning and policy
formulation and who controls these decision-making processes. If anything, the
'benefit to residents' is a poor partner of the need to protect 'long-term investments' and is couched in terms of conserving the environment for the local
people. The suspicion that environmental conservation is not really about a
concern for local people is confirmed: 'Increasingly, tourists are demanding that
their environments be high-quality and pollution-free as well as inherently interesting, and some tourists will change travel patterns if environmental quality
expectations are not met' (Inskeep, 1987: 119). The results and effects of the
Tourism Master Plan for Bali, outlined in one of the case studies given later in
this chapter, are interesting to note in this regard.

Ultimately then, we must ask who controls government policy and planning
and what forces drive the processes of its formulation and implementation.
Sustainability is a critical consideration here, for Agenda 21 designates national

Box 9.2 Guatemala's protected areas

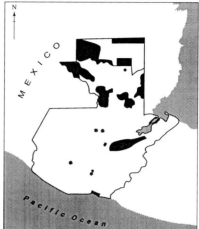

Figure 1 INGUAT's version   *Figure 2* CONAP's version

Figures 1 and 2 above show the protected areas of Guatemala – according to two different arms of the Guatemalan government. Box figure 1 gives the version according to INGUAT, the Guatemalan Institute of Tourism; Box figure 2 gives the version according to CONAP, the National Council of Protected Areas, which is charged with overseeing the protection of these areas.

The differences between the two are technically due, in the main, to INGUAT's inclusion of all the areas whose status as a protected area is only proposed rather than actual. But these differences also reflect the roles and outlooks of the two departments. INGUAT is in the business of enhancing the country's image, of making it look attractive to potential visitors; so it covers as much of the national area as possible with the status of a protected area, with the greenwash of pristine natural areas. CONAP, on the other hand, is more realistic; it is aware that it does not have the resources necessary to oversee those areas which are already designated as protected, still less all those which are proposed.

governments as having most of the responsibility for leading the process. The overall aim of developing a sustainable tourism programme has been set out by the WTTC, which defines Agenda 21 as the process 'to establish systems and procedures to incorporate sustainable development considerations at the core of the decision-making process and to identify actions necessary to bring sustainable tourism development into being' (1995: 38). We must also ask in whose interests such actions are undertaken and explore how the uneven and unequal nature of development is reflected through this.

The notion of 'interests' is a useful one in focusing attention on which groups, organisations and institutions have influence and power in decision-making. As Table 9.1 implies, such influences can vary from the structural adjustment policies of the World Bank and IMF to the campaigns against visiting certain countries (due, for example, to the infringement of human rights within that country – see the case study of Burma outlined later in this chapter).

Box 9.1 on The Gambia helps to summarise the discussion so far and identify different ways in which power is expressed through political processes. It is illustrative of the typical problems faced by Third World countries in their efforts to develop tourism. In this case, the expansion of the sector has been encouraged by the internationally inspired structural reform programmes which have clearly pushed The Gambia along a particular development path. But First World countries also retain the potential to control the tourism industry (especially when much of the industry caters for mass packaged charters, as in this case).

Table 9.1 Tourism interest groups

Scale	Industry groups	Non-industry groups	Single interest groups
International	World Tourism Organisation, WTTC, World Bank, regional development banks, World Trade Organisation	Environmental and social organisations, e.g. IUCN, WWF, ECTWT, TEN, FOE, Ecotourism Society, Friends of Conservation	Occasional environmental or social issues, often location-specific, e.g. End Child Prostitution in Asian Tourism (ECPAT), Global Anti-Golf Movement (GAGM), Burma Action Group (BAG)
National	National tourism industry associations (e.g. Belize Tourism Industry Association – BTIA), trade unions, national professional, and trade associations	Environmental and consumer organisations, e.g. National Trust, Tourism Concern, Wilderness Society, Sierra Club, Audubon Society, ANCON (Panamá)	Single-issue environmental groups, e.g. those opposing airport development, Surfers Against Sewage
Local	Chambers of commerce, regional tourism business associations, local area promotion partnerships	Rate-payers' and residents' associations, e.g. Talamanca Association for Eco-tourism and Conservation	Groups opposed to tourist development in a specific location, e.g. anti-resort development groups

Source: adapted from Hall and Jenkins, 1995

TOURISM AS POLITICS

Chapter 3 showed how radical critiques have revealed the shortcomings of Third World tourism and how tourism is an additional element in uneven and unequal development. However, this has not prevented countries which pursue socialist strategies and ideologies from dabbling in tourism, and in some cases arguing for forms of tourism that are not subordinating and are responsive to the issues of power and control exercised by the First World so heavily criticised in the political economy framework. As Crick (1989) argues, in some countries this has resulted in seemingly glaring contradictions to the anti-colonial ideologies adopted by newly independent countries. Indeed, the discussion in Chapter 3 suggested that tourism may help to maintain systems of neo-colonial power and control.

Tanzania is a frequently cited example where an emphasis on tourism development in the 1970s ran contrary to the anti-colonial socialism adopted by Nyerere in the wake of independence. The Tanzanian intellectual, Shivji, was very critical of the separation of politics from economics: "The justification for tourism in terms of it being "economically good" though it may have adverse social, cultural and political effects, completely fails to appreciate the integrated nature of the system of underdevelopment' (1973: quoted in Crick, 1989: 321).

Other independent countries pursuing socialist ideals have also explored the development of tourism. In Cuba, for example, as much effort has been put into widening the appeal of Cuba as radical chic, with solidarity and study tours, as has been devoted to stemming the foreign exchange leakage characteristic of other destinations in the region. In order to reduce the leakage of foreign exchange, Cuba has established a tourism industry that carefully, if uneasily, separates tourism spending from the national economy, and forces tourists to buy goods and services in foreign currency (mainly US dollars).

Following a successful coup d'état in 1979 in Grenada, a regional neighbour of Cuba, Maurice Bishop, head of the People's Revolutionary Government (PRG), acknowledged the shortcomings of mass tourism as it existed in the Caribbean and started to identify the basis of an alternative type of tourism. Outlining the parallels of tourism with colonialism which we discussed earlier, the socialist Prime Minister argued:

> in the early days and even now, most tourists are white. This clear association of 'whiteness' and 'privilege' is a major problem for Caribbean people just emerging out of a racist colonial history where we had been so carefully taught the superiority of things white and inferiority of things black.
>
> (Bishop, 1983: 69)

Bishop elaborated his ideas further at a Caribbean regional conference on the impact of tourism, where he identified the need to replace 'old tourism' with what he termed 'new tourism'. Along with Jamaica under the socialist leadership of Prime Minister Michael Manley, Bishop attempted to chart the development of a tourism designed to escape the problems that had been associated with 'old' tourism, bound up (as argued in Chapter 3) with 'colonial and imperialist connotations'.

> It was foreign-owned and controlled, unrelated to the needs and development of the Caribbean people, and it brought with it a number of distinct socio-cultural and environmental hazards such as the race question and undesirable social and economic patterns such as drug abuse and prostitution.
>
> (Bishop, 1983: 71)

Bishop's vision was for a 'new' tourism to escape these characteristics. It was to involve all people (as both 'guests' and tourists); it would seek to create linkages between the different sectors of the economy; and above all it was conceived as a tool for development. The construction of Grenada's airport, a key part of the country's attempt to attract tourism (see Box 9.3), encapsulates and symbolises the political nature of tourism development. Bishop was murdered in October 1983 and with him died this vision of new tourism. Of course, it is highly questionable whether such a vision was realisable, given the context of uneven and unequal development in which Third World tourism operates.

A rather different aspect of the inherently political nature of tourism is the way in which the USA sought to destabilise both Grenada and Jamaica, under Bishop and Manley respectively, through negative publicity and foreign policy statements that were hostile to both countries (Thomas, 1988). In the case of Grenada this is particularly poignant. Grenada quickly gained an extraordinary international reputation, with Fidel Castro assessing that in conjunction with Nicaragua and Cuba they were representative of 'three giants rising to defend their rights to independence, sovereignty and justice on the very threshold of imperialism' (Clark, 1983, quoted in Thomas, 1988). As Thomas concludes from the radical agendas pursued in these countries and the reworking of tourism that they advocated, 'they confirmed that any attempt at radical social reorganisation in the region would be met by insistent efforts on the part of the US to destabilise the process (or worse), especially if the proposed reorganisation involved new options for foreign policy and external relations' (1988: 246).

GLOBALISATION AND THE POLITICS OF EXTERNAL INFLUENCES

Supranational institutions

Chapter 2 identified a number of key interests in tourism. Supranational institutions involve a degree of political integration between states but are not

---

Box 9.3 *Grenada's airport*

The absence of an international airport was a clear weakness in Grenada's tourism plans and a brake on its development, a shortcoming that Bishop readily acknowledged in describing the airport as 'the gateway to our future' (Marcus and Taber, 1983). The majority of capital (US$60 million) and technical assistance for the airport's construction came from Grenada's ideological neighbour, Cuba.

President Reagan, however, claimed that the construction involved a new naval base, a new air base, storage bases and barracks for troops, and training grounds. 'And, of course, one can believe that they are all there to export nutmeg. . . . It is not nutmeg that is at stake in Central America and the Caribbean, it is the United States' national security' (President Reagan, March 1983, quoted in Ferguson, 1990).

As Ferguson points out, despite USAID's hostility to PRG [People's Revolutionary Government] policies and projects, it was deeply ironic that USAID were financing and administering the completion of the international airport, especially given the initial announcement that the USA had no plans to assist a project that was never intended to promote the development of Grenada's tourism.

Post-revolutionary Grenada provided US technocrats an opportunity for a structurally adjusted blueprint – a 'USAID inspired restructuring and free market deregulation' (Ferguson, 1990: 114). Grenada was subsequently a laboratory and showcase of the US ideological superiority. Much USAID/NNP [New National Party] policy and finance was targeted on environmental and infrastructural improvements designed to attract foreign investment and tourism. Indeed tourism has been one of the principal factors in the US-approved development strategy for the Caribbean. In many other places too, tourism is alluded to as a peace industry, and a way of allowing the expansion of capitalist relations of production in a stable and peaceful environment. 'The overwhelming majority of USAID funds have . . . gone towards dismantling the state sector, encouraging private enterprise and wooing foreign capital – with negligible success' (Ferguson, 1990: 39; see also McAfee, 1991).

As Ferguson argues, the failure of the model is all the more significant given the symbolic importance placed on Grenada's development by the US government.

(Information and some extracts from Ferguson, 1990)

---

subject to the laws of national governments. Such organisations operate in a global sphere of activity and are able to influence, and in some cases determine, economic and political activities in the national sphere.

Supranational organisations exist in many fields of activity. Those described as socio-environmental organisations were discussed in Chapter 6. While these may in certain instances influence the policies pursued by national governments, this section deals with those organisations dedicated to direct and deliberate

Box 9.4   The World Bank and IMF

The World Bank and IMF were born at the Bretton Woods conference in the USA in 1944. Their purpose was to provide an order to the global economy to prevent the collapse of international trade and the development of isolationist economic policies, which it was believed had led to the Depression of the 1930s and the rise of fascism. The conference rejected the proposals of John Maynard Keynes to establish a world reserve currency administered by a central bank. Instead, it 'opted for a system based on the free movement of capital and goods with the US dollar as the international currency' (New Internationalist, 1994: 14).

Initially, a more specific purpose was to provide funds to assist in the rebuilding of war-torn Europe. The European countries objected to the severe loan conditions, so the World Bank worked through the Marshall Plan which provided US finance to rebuild Europe mainly through grants rather than loans. Third World countries did not receive the same treatment. Instead, they were pressured to keep their economies completely open to foreign goods and capital; and in exchange for financial assistance, countries were expected to adopt structural adjustment policies (see Box 9.6).

Partly as a result of the debt crisis of the 1970s and 1980s (see Box 9.5), both institutions now exercise enormous leverage over the policies pursued by most governments which require financial assistance. 'That influence is enhanced by donor insistence upon an IMF and World Bank seal of approval as a condition for aid and debt relief' (Oxfam, 1995: 8).

As Duncan Green points out, the two organisations are not democratically run: 'decisions at the IMF and World Bank are taken on the basis of "one dollar, one vote", guaranteeing the dominance of both by the US government' (1995: 34). Voting rights in 1995, for instance, for a selection of countries at the World Bank were as follows:

	% of total votes
United States	17.18
Japan	6.31
United Kingdom	4.68
Saudi Arabia	3.02
Total vote for the following 23 countries (Benin, Burkina Faso, Cameroon, Cape Verde, Central African Republic, Chad, Comoros, Congo, Côte d'Ivoire, Djibouti, Equatorial Guinea, Gabon, Guinea-Bissau, Madagascar, Mali, Mauritania, Mauritius, Niger, Rwanda, Sao Tomé and Príncipe, Senegal, Togo, Zaire)	1.58

involvement in the formulation of national economic policy through money-lending activities. The two most powerful and dominant of these are the World Bank and the International Monetary Fund (IMF), although regional finance and development organisations, such as the European Bank for Reconstruction and Development (EBRD), the Inter-American Development Bank (IDB) and the US Agency for International Development (USAID) also wield very considerable power over the policies pursued by Third World countries. A brief background to the formation and role of the World Bank and IMF is given in Box 9.4.

The economic system promoted by these institutions is based on the free movement of capital and goods with the US dollar as the international currency. In such a system, tourism can be seen as much the same as any other cash crop, and pressure exerted by the development banks upon the government of a country to increase its export earnings by boosting production of export crops is as applicable to tourism as it is to sugar, coffee, cotton, bauxite, or other products. Foreign exchange earnings can be boosted by attracting more international tourists and the foreign exchange which they bring with them.

The debt crisis of the 1980s (see Box 9.5) and the First World's reaction to it increased the power of the IMF and the World Bank. Spurred on by the prevailing ideology of President Reagan and Prime Minister Thatcher, the two institutions embarked on a mission to adjust structurally Third World economies so that they could meet debt obligations and fall into line with the right-wing economic orthodoxy sweeping the globe.

From the point of view of our analysis, what is crucial here is that structural adjustment programmes (SAPs) imposed by the IMF and World Bank on Third World governments effectively force those governments to pursue specific policies not of their own design. These policies make living conditions for the population harsher and more difficult while supposedly enabling the government to increase its foreign exchange earnings and thereby pay off its debt. The kind of policies associated with SAPs are listed in Box 9.6. It is worth noting, however, that most Third World countries already had little effective control over the direction of their policies before the debt crisis began. Colonial powers of the First World and transnational corporations (TNCs) dictated the main thrust of economic policy throughout the Third World. The IMF, World Bank and other supranational lending agencies along with the TNCs took over the mechanism of power from the former colonial powers. Moreover, proponents of SAPs point out that once the debt had become a reality, there were no policies other than those given in Box 9.6 which could realistically be imposed in a world of capital accumulation.

The loans which form part of the SAPs are made only on strict condition that the national economy is restructured according to these policies (see Box 9.6). And this conditionality has meant that the economic and fiscal policies of most African and Latin American countries are now dictated from Washington. In the words of Green:

Box 9.5   Origins of the debt crisis

The early 1970s brought huge profits for members of the Oil Producing and Exporting Countries (OPEC). These profits were surplus to requirements within the OPEC countries – that is, more than could be spent – and could not all be spent at once in any case.

Much of the money was therefore banked and invested in the commercial banks of the First World. Awash with petro-dollars, the banks sought to make the most effective, that is, profitable, use of this money. One of the ways in which they recycled it was by financing development projects in Third World countries, including tourist resort schemes.

In the rush for 'development', many of the projects financed were ill-considered and poorly planned. Some of the money was spirited away by Third World élites with access to power brokers and decision-makers, and more was paid direct to First World companies taking part in the projects. A fair proportion of it therefore found its way back into the same First World banks that lent it in the first place.

In the late 1970s and the first half of the 1980s, interest rates on these loans rose sharply, an event that had not been anticipated by the lenders or borrowers in the early 1970s. Many borrowing countries found themselves unable to pay back even the interest on the loans, quite apart from the principal loan itself. First World bankers and politicians became concerned 'that the sheer volume of unpayable loans would undermine the world financial system. They turned to the World Bank and the IMF, who were to restructure Third World economies so they could meet their debt obligations' (New Internationalist, 1994: 15).

Hence, the rise of structural adjustment programmes (SAPs).

In the aftermath of the debt crisis, the IMF and later World Bank's use of conditionality allowed the powerful industrialised nations to revamp one Third World economy after another along free-market lines. Critics believe that in the process, the IMF has systematically put the powerful nations' self-interest before the welfare of the Third World poor . . .

(1995: 37)

The initial, rhetorical hope that SAPs might be successful in delivering widespread prosperity through highly dubious mechanisms such as the 'trickle-down' effect and belief in the benefits of economic globalisation are confounded somewhat by the evidence. As Ayesha Imam states:

In most countries in Africa, SAPs have not led to increases in production or investment. . . . In fact, according to the UN Economic Commission for Africa, production has decreased over the decade since SAPs have been implemented. Investment as a proportion of GDP has fallen too. Budget deficits have grown and a greater proportion of export-earned money is now used solely to service debts. . . . The general standard of living has fallen. There have been painfully deep cuts in the provision of health, education and other social services and in the subsidies on basic necessities.

(1994: 13)

Box 9.6 *Policies and conditions associated with structural adjustment programmes (SAPs)*

A Increase earnings of foreign capital

- *Boost export production*
  One implication of this is that the production of goods for local and national needs is relegated in importance as a result of the drive for export production. Many Third World countries which used to be self-sufficient in foodstuffs now have to import basic grains and other foodstuffs. In 1996, for instance, Honduras began to import beans for the first time in its history. 'It is important to remember, in this context, that tourism is an export' (Gonsalves, 1995: 34).
- *Devalue the national currency*
  Devaluation makes the country more attractive to foreign investors seeking a cheap place to produce their goods. The cheapness of its currency against other currencies also increases its attraction to foreign tourists.
- *Reduce and/or abolish import tariffs*
  Import tariffs are designed to protect the price of the production of goods for home consumption against the entry of cheaper imported goods. In practice, removing them leads to the flooding of the country with expensive luxury goods and durables with high value added in production from the First World.

B Reduce state involvement in the economy

- *Privatise state-run enterprises*
  Tourism is a flagship of the private sector.
- *Cut public spending on activities which cannot be privatised*
  For example, social services, health provision, education – although increasingly attempts are being made to privatise these services too.
- *Deregulation*
  Reduction of all forms of controls, especially economic regulations, which inhibit industry's ability to maximise profits.

C Fiscal measures

- *Reduce inflation*
  This increases security of investment, including investments in tourist developments.
- *Cut interest rates*
  This increases investment, including investment in the tourism industry.
- *Encourage investment rather than saving*

In 1994, the UNDP's *Human Development Report* showed that in terms of control of total world income, the ratio between the richest 20 per cent and the poorest 20 per cent had increased from 30 to 1 in 1960 to 61 to 1 in 1991 (UNDP, 1994); in other words, in 1991 the richest 20 per cent of the world's population received an income 61 times greater than the poorest 20 per cent.

Even within the World Bank, there is awareness of the effects of these policies. Herman Daly, a renowned and respected economist in the Bank's Environment Department, has identified some major conflicts between the ultimate goal of international free trade, at which the Bank's policies aim, and national policies pursued in the national interest. These conflicts include: 'getting prices right; moving towards a more just distribution; fostering community; controlling the macroeconomy; and keeping scale within ecological limits' (1993: 124).

In the world of tourism, as in many other industries, the process of structural adjustment effectively delivers control of development to the TNCs and consultancies, most of which are based in the First World. This relegates the role of the national government to one of providing the necessary infrastructure. For this, it must seek more loans, leaving it yet more indebted than before. Some countries, such as Honduras and Panama for example, are now even offering free trade zones for tourism companies wishing to build and operate resorts.

An earlier quotation from Paul Gonsalves on pp. 191–2 illustrates the way in which TNCs based in the First World can operate to maximise their own profits at the same time as ensuring that as little as possible accrues to the national government of the visited country. And Gonsalves concludes that 'the economies of the so-called Third World are in effect mere extensions of the economic priorities of the First World' (1995: 39).

The list of development disasters caused by these supranational organisations in funding such schemes as large-scale dams, roads through sensitive ecosystems, monocultivation practices, infrastructural developments for the promotion of tourism, and many others, is long and well documented. And they have not just been confined to the IMF and World Bank. During the 1980s, the USAID also gained notoriety for its arrogant and callous disregard for the environments and people affected by its development projects. Organisations such as World Bank Watch, Survival International, the World Development Movement, Friends of the Earth International and many others have documented these effects, exerted heavy pressure against them, and publicised the events and examples to such an extent that the start of the 1990s saw a growing recognition by the supranational organisations themselves of the effects of their policies. Both the IMF and World Bank increased their staff who examine the social and environmental impacts of their programmes. The USAID also made strong efforts to green its image and to promote, for instance, community-based tourism schemes rather than grandiose resort projects. A tourism-related example of the misdirection and failure of their efforts was given on pp. 253–5. That these efforts to foster a new environmentally friendly image for themselves have failed should come as no surprise, given that their *modus operandi* of forcing national governments to pursue a dictated development path has remained unchanged.

With the same increasing awareness of the need to demonstrate that they are changing, even if they are not doing so in reality, the IMF and World Bank have acknowledged the role of non-governmental organisations (NGOs) and have made attempts to enlist their advice. Despite this, Pierre Galand, Oxfam-Belgium's general secretary, resigned from the NGO working group on the World Bank and its Steering Committee. In his letter of resignation he said to the World Bank:

> You have stolen speeches from NGOs concerned with development, eco-development, poverty and popular participation. . . . The Bank has learned to do excellent analyses and is capable of saying what is important: popular participation – particularly of women – , the popular struggle against poverty, and the necessity to protect the environment. . . . Why are so many beautiful speeches accompanied by such scandalous practices?
>
> (1994: 12–13)

In even more unequivocal terms, the Permanent Peoples' Tribunal in their 1994 evaluation of IMF and World Bank policies with respect to international law and the right to self-determination, concluded: 'There is overwhelming evidence that neither the World Bank nor the IMF have changed their socially and ecologically destructive philosophy' (1994).

Agenda 21 for the travel and tourism industry

Chapter 4 examined the applicability of Agenda 21 to tourism. The role of different tourism interests within Agenda 21 has been elaborated in a document co-authored by the World Travel and Tourism Council (WTTC), the World Tourism Organisation and the Earth Council, entitled Agenda 21 for the Travel and Tourism Industry: Towards Environmentally Sustainable Development (1995). Table 9.2 extracts the duties placed upon central and local governments (and national tourism authorities and representative trade organisations such as the WTTC and WTO) in facilitating 'sustainable' tourism development.

There are a number of observations that can be made about the priorities set out in this document. First, it is clearly set within a framework of environmental sustainability. This is indicated in both the report's rubric, 'Towards Environmentally Sustainable Development', and in the respective forewords of the collaborating bodies:

> A key conclusion of the Earth Summit was the importance of harnessing the entrepreneurial drive of the private sector in the cause of environmentally compatible development. . . . The environment is our core asset, the key component of product quality, and an increasing priority for our consumers.
>
> (Geoffrey Lipman, President, WTTC)

Table 9.2 Agenda 21: responsibilities of governments

Priority area	Priorities and objectives	Responses and examples of success
i	Assessing the capacity of existing regulatory, economic and voluntary framework to bring about sustainable tourism, and developing policies to facilitate sustainable tourism	Charge negative externalities to producer; provide cost incentives to companies minimising waste; encourage responsible entrepreneurship through codes of practice. More than 100 codes now exist
ii	Assessing the economic, energy, social, cultural and environmental implications of an organisation's own operations	Ranges from a more efficient use of water to encouraging staff to use environmentally benign transport and ensuring the destination's character is truly represented in marketing. Green Globe was established – a worldwide environmental management and awareness programme; the Belize Ministry of Tourism is a member
iii	To train and educate all stakeholders in travel and tourism about the need to develop more sustainable forms of tourism	Organisations should work with government departments. The WTO has established a worldwide Network of Education and Training Centres to promote tourism education and training
iv	Develop and implement effective land-use planning that maximises the environmental and economic benefits of travel and tourism and minimises environmental and cultural damage	Includes working with government planning authorities on sustainable tourism planning, i.e. transport planning and coastal zone management. Bermuda is offered as one of the world's most affluent countries because of planning regulations
v	Facilitate information exchange between 'developed' and 'developing' countries	Includes developing partnerships with 'less developed' countries, advising them on sustainable tourism. A good example is the WTO's Guide for Local Planners
vi	Provide opportunities for all sectors of society to participate in sustainable tourism	Promote and ensure the participation of women, indigenous people, etc., and develop appropriate training courses. A key example would be the CAMPFIRE project in Zimbabwe
vii	Design new tourism products that are sustainable: social, cultural, economic and environmental	Ranges from defining what makes resorts sustainable to using local materials in construction and

296

Table 9.2 cont.

		ensuring government departments realise the benefits of tourism in conservation. Examples include the game parks of Kenya, South Africa and Zimbabwe, and the investigations of governments in Belize, Brazil and Costa Rica into how tourism can be used in environmental conservation.
viii	Measure progress in sustainable development by setting realistic indicators at national and local level	Ranges from assessing which data are appropriate to exchanging experiences with other organisations. Examples include the participation of national tourism authorities (NTAs) in Argentina, Mexico and Turkey in the WTO's programme to develop indicators
ix	Develop partnerships to facilitate sustainable tourism development and responsible entrepreneurship	Includes government departments providing a coordinating mechanism for those responsible for sustainable tourism development and ensuring the necessary infrastructure (e.g., sewage treatment, recycling facilities, etc.) is in place. Trinidad and Tobago tourist board has joined Green Globe

Source: WTTC, WTO and Earth Council, 1995

A good deal of our Travel & Tourism activity relies on . . . fragile natural or cultural resources, so it is in our interests to protect them for the future . . . Travel & Tourism will inevitably continue to increase. Meeting this growth in a responsible, sustainable way, that preserves and enhances the beauty of the attraction, is the challenge we all face.

(Antonio Enriquez Savignac, Secretary General, WTO)

The overriding concern for the industry must be to seek out ways to enhance rather than degrade its core product, the environment.

(Maurice Strong, Chairman, Earth Council)

Reflecting for a moment on some of the words used here to describe the environment: 'core asset', 'core product', 'product quality', 'consumers', 'resources', 'preserve' – the treatment of the environment as a marketable product is clear. It is also clear that the drive 'towards environmentally sustainable development' is a commercial necessity, in order to maintain the marketability and attractiveness of this commodity. More noticeable is the way in which much, if not all, the discussion bypasses the interests of local people. Where social considerations do appear they are an adjunct. Social development, it is assumed, will be achieved at the same time as improving the environment; the benefits will magically 'trickle down' to local people in the same way as the World Bank and IMF suggest. The ability for local people to choose whether they wish to engage with tourism is not on offer. Rather, it appears to be an agenda for removing the 'growing resentment of residents in some destinations' (3) and allowing for the expansion of the industry.

As is indicated in the quotations above, the report fails to reflect the inherent inequality expressed through global tourism. It is written from a thoroughly First World perspective, where the majority of the tourism industry, tourists and tourism interests are located. The forewords talk of 'our Travel and Tourism activity', 'our consumers', 'our interests'. The supporting statement for priority (vi), for example, extolls the participation of all sectors of society in sustainable tourism and identifies the promotion of the participation of women and indigenous people in 'appropriate forms of tourism development'. But, of course, these remain the receivers of tourism and the report is silent on the gross inequality of who is, and who is not, able to travel. The Third World, so it would appear (and as argued in Chapter 3), is there for our entertainment. It exists to serve us and see to our needs. As the report states, the 'challenge . . . is to ensure that tourists don't "love nature to death"' (49); in other words, to ensure that the tourism product is retained.

Even 'peace' is invoked as an essential prerequisite to successful sustainable tourism development, further reflecting the overall intention of protecting the tourism product while allowing for overall global expansion. Priority (vii) supporting statement urges the use of 'the potential of the tourism industry to promote peace between nations and people' (50), acknowledging that 'Peace is a pre-requisite for a sustainable Travel & Tourism industry and, by facilitating contact and understanding between cultures, well-designed tourism products can make a significant contribution to world peace.' (Similar exhortations have been made elsewhere, particularly by business commentators and by international conventions, for example, the 1995 World Conference on Sustainable Tourism and its product the Charter for Sustainable Tourism – see below). While peace is of course a noble goal, once again there is an insipid suggestion that we need to move towards one happy global community, 'In recognition of economic and social cohesion among the peoples of the world as a fundamental principle of sustainable development.'

Not only are Third World populations the receivers of visitors, according to this report they are also receivers of (First World) information. Table 9.2 reflects this bias. Priority area (v), for example, involving the facilitation of information exchange, more accurately involves advising Third World countries on how to make tourism sustainable. Far from exchange, this is information delivery and represents a First World perspective on how certain goals are to be achieved.

Similarly, priority (viii) supporting statement encourages the 'exchange' of indicators with other organisations 'especially those in developing countries' (51). Cynically, many of the priorities and objectives advocating 'training' courses are a further indication of the cycles of dependency and development aid-giving – a 'jobs for the boys' syndrome – that such frameworks maintain.

The participatory processes which the report anticipates are very passive, and it may be of interest to consider the contents of Table 9.2 in relation to Pretty's typology of participation given in Table 8.1. Participation, so the report suggests, is to be undertaken after key initial decisions have been made. In the wake of the discovery of Mayan ruins near the city of Tekax (in south-east Mexico), and the realisation that this had significant tourism potential, it was agreed that the 'people of the area should participate in the development in order to maximise economic benefits'. The critical point here is that the decision to develop the site and to maximise its economic benefits had already been taken.

Finally, the report offers a Framework for Action, although on the subject of the means of implementation the report is baffling and also reflects the First World perspective. It is difficult to conceive of the processes and procedures set out in the report – that is, securing the commitment of top management to the concept of sustainability, communicating this to all staff, establishing realistic and achievable targets, implementing action programmes, monitoring progress, and so on – working in practice. In short, it amounts to top-draw management-speak: many windy words within a highly ambiguous and inappropriate framework. We are served up the familiar dish of 'ensure', 'promote', 'advise', 'develop', 'work with', 'reduce', 'minimise', 'establish', 'secure', 'prioritise', and so on. It is very difficult to believe that many grossly underresourced and under-staffed government departments would be in a position to undertake such a grand scheme. How the report's recommendations are to be implemented in practice represents a serious failing.

The report and its Framework for Action are in effect just one more in a pile of conventions, codes of conduct, declarations and the like weighing down on the governments of Third World countries in respect of tourism development (see Figure 9.1). All these pressures are full of fine words and lofty principles, but few if any of them offer the resources required to enable Third World governments to consider sustainability in the development of their tourism industries.

The Charter for Sustainable Tourism

A further example of how international conventions and organisations have attempted to set a framework for sustainable tourism development is given in Box 9.7. The Charter for Sustainable Tourism was the product of the World Conference on Sustainable Tourism held in the Canary Islands in 1995 – a conference attended by delegates representing a wide range of interests including government, industry, academia and NGOs.

Figure 9.1 Squashing the Third World

Box 9.7 *The Charter for Sustainable Tourism*

April 1995: World Conference on Sustainable Tourism

Reflecting the problems and opportunities offered by tourism and mindful of recommendations from Agenda 21, the Universal Declaration of Human Rights, the Manila Declaration on World Tourism, the Hague Declaration, Tourism Bill of Rights and Tourist Code, the conference formulated a Charter for Sustainable Tourism.

Extracts from the Final Resolution of the Charter illustrate its importance.

1 The Conference recommends state and regional governments to draw up urgently plans of action for sustainable development applied to tourism.

2 The Conference agrees to refer the Charter for Sustainable Tourism to the Secretary of the United Nations, so that it may be taken up by the bodies and agencies of the United Nations system . . . for submission to the General Assembly.

The following are from the principles and objectives of the Charter:

1 Tourism development shall be based on criteria of sustainability, which means that it must be ecologically bearable in the long term, as well as economically viable, and ethically and socially equitable for local communities. . . . A requirement of sound management of tourism is that the sustainability of the resources on which it depends must be guaranteed.

2 Tourism should contribute to sustainable development and be integrated with the natural, cultural and human environment. . . . Tourism should ensure an acceptable evolution as regards its influence on natural resources, biodiversity and the capacity for assimilation of many impacts and residues produced.

9 Governments and the competent authorities, with the participation of NGOs and local communities, shall undertake actions aimed at integrating the planning of tourism as a contribution to sustainable development.

10 Governments and multilateral organisations should prioritise and strengthen direct and indirect aid to tourism projects which contribute to improving the quality of the environment . . . it is necessary to explore thoroughly the application of internationally harmonised economic, legal and fiscal instruments to ensure the sustainable use of resources in tourism.

11 Environmentally and culturally vulnerable spaces, both now and in the future, shall be given special priority in the matter of technical cooperation and financial aid for sustainable tourism development. Special treatment should be given to zones that have been degraded by obsolete and high impact tourism models.

12 The promotion of alternative forms of tourism that are compatible with the principles of sustainable development . . . represent a guarantee of stability in the medium and the long run.

301

Box 9.7
continued

> 13 Governments . . . should promote and participate in the creation of open networks for research, dissemination of information and transfer of appropriate knowledge on tourism and environmentally sustainable tourism technologies.
>
> 14 The establishment of a sustainable tourism policy necessarily requires the support and promotion of environmentally compatible tourism management systems, feasibility studies for the transformation of the sector, as well as the implementation of demonstration projects and the development of international cooperation programmes.
>
> (extracts from World Conference on Sustainable Tourism, 1995)

Like Agenda 21 for the Travel and Tourism Industry (and most other action frameworks of this nature), the Charter for Sustainable Tourism is devoid of ideas about how the grand scheme is to be implemented. Representatives of First World NGOs who form TEN (the Third World Tourism European Ecumenical Network) welcomed this publication, in part due to its emphasis on governments ensuring more just and sustainable forms of tourism, and, critically, because of the importance it will have when adopted by the General Assembly of the United Nations. But, as TEN argue, there is in fact little to distinguish the Charter from 'the many other codes that have been published in the past 30 years'.

More invidiously, the Charter is also reflective of the potential for coercion and the manifestations of unequal power that are often apparent in international conventions and agreements. Declarations 10 and 11, for example, provide clear indications of how certain countries could be forced to adopt given strategies and policies that would lead tourism development along specific lines. In this respect, the Charter's Declaration appears to be as much about a new round of development and the opportunities for 'ecocrats' to assess the sustainability of certain tourism products. It represents an opportunity to identify new measures that can be agreed 'internationally', new avenues for donor funding and new opportunities for 'international' cooperation (see Declaration 14).

Within the overall context of a global response, the Charter provides a superficial acknowledgement of the aspirations and choices of local people. Notwithstanding the effort to drop in the importance of so-called local communities, with oblique references and hints that these people too must have a say and a chance of employment, in effect such declarations are about preserving the tourism product.

SUSTAINABLE TOURISM AS POLITICAL DISCOURSE

Lessons from Belize

'Beautiful, multi-faceted jewel' (*Financial Times*, 22 September 1992); 'To sea is to Belize' (*Guardian*, 3 April 1993); 'Mounts, mountains and Mayans' (*Observer*,

27 March 1994); and 'Too good for tourists' (*Independent*, 3 November 1992) – this is how four UK broadsheet newspapers referred to the small Central American country of Belize (see Box 9.8). Up to the late 1980s few would have heard of the former British colony of British Honduras, which gained independence in 1981. In the post-independence period, tourism has become an attractive strategy for diversifying the economy away from the export of agricultural produce. Belize is an excellent example of how a country uses the notion of sustainability to promote the development of new types of tourism and how a government's use of the language of sustainable tourism is often sufficient to win international accolade, even among academics and environmentalists. In the early 1990s it seemed as if Belize had emerged as the ecotourism capital of the world, hosting a number of international conferences (see pp. 232–4). The experience of Belize provides an example of sustainability as a political discourse.

Unlike its CARICOM (Caribbean Community and Common Market) partners, Belize recognised the need to avoid the negative externalities, foreign exchange leakages and the foreign control of the mass tourism industry developed throughout much of the Caribbean. It also acknowledged the potentially disastrous impact that large-scale tourism development would have on the relatively fragile sub-tropical environment.

In their first term of office, the United Democratic Party (UDP), a right-of-centre and pro-US free market economics party, was the first Belizean administration to focus upon tourism. It approved an *Integrated Tourism Policy and Strategy Statement* in its last year in office (Government of Belize, 1989). This acknowledged the significance of tourism to the government, which had made it the second economic priority next to agriculture. In this way, significant emphasis was placed on the overall potential of tourism, 'its economic stimulus and linkage with other subsectors of the economy' (1989: 1), and the guiding objectives stressed the maximisation of economic benefits. This included a particular emphasis on 'long stay and upper income travellers'.

In the 1989 general election the victorious People's United Party (PUP – a nationalist party and like its increasingly indistinguishable rival, a pro-US and free market oriented party) had spelled out clearly the advantages of ecotourism in the party's manifesto (*Belizeans First 1989–1994*) in what it also termed *Tourism with Dignity.*

> Belize has in abundance what few countries in the world still have left – nature. Ours, for the most part, is still intact. And nature is rapidly becoming tourism's Holy Grail. Belize has so far escaped the most brutal blows of man's destructive hand. . . . But, we must ask, for how long? The answer is, forever if we safeguard them. The protection of our national environment and development of eco-tourism are compatible goals which will be given high priority. The new PUP Government will place the portfolios of Environment and Tourism in one Ministry.
>
> (PUP, 1989)

Box 9.8   Focus on Belize

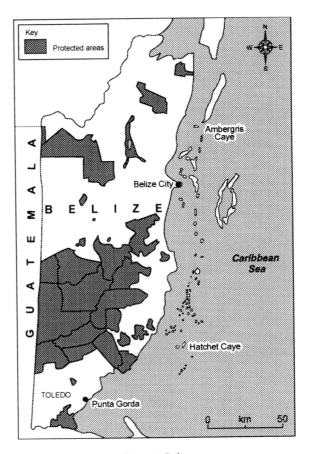

Figure 1 Belize

Virtually all tourist sources on Belize cite large tracts of 'pristine' rain forest, numerous Mayan archaeological sites, and the second longest barrier reef (after Australia).

Writers point to a number of successful community-based and conservation projects. These include the Community Baboon Sanctuary, Sandy Beach Lodge (Women's Cooperative), the Toledo Eco-tourism Association (see p. 253), the Hol Chan Marine Reserve, the Cockscomb Basin Wildlife Sanctuary (Jaguar Preserve), and Crooked Tree Wildlife Sanctuary.

On paper, much of the surface area of the country enjoys some form of designated protection from development (see Figure 1).

Growth rates in tourist arrivals and receipts are shown in Figure 2.

(Source: Simons, 1988; Cater, 1992)

Box 9.8
continued

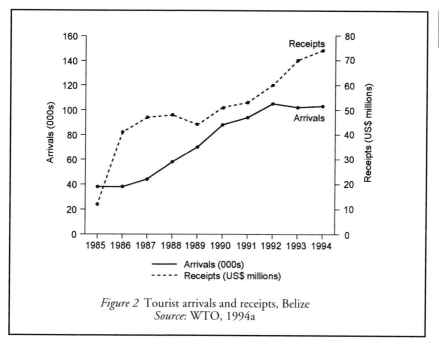

Figure 2 Tourist arrivals and receipts, Belize
Source: WTO, 1994a

In the wake of its election victory, the PUP did indeed combine these portfolios and Belize has subsequently enjoyed popular international support and praise for its stand on the environment and promotion of sustainable tourism development and has hosted a number of conferences on ecotourism (see p. 232). As we argued earlier, the conferences reveal the serious deficiencies and contradictions in ecotourism (see Box 9.9), not least that the conference fees, half the average monthly Belizean wage, ensured that the only local representation was from the educated non-governmental organisations; hardly a wide spectrum of views.

The Belizean government has used its commitment to sustainable tourism to full effect. At an ecology and tourism symposium in Costa Rica for example, the then Tourism and Environment Minister, Glenn Godfrey, emphasised the relationships between environment, conservation and ecotourism and the fundamental requirement of local control. 'Conservation and therefore ecotourism thrives best where the sunlight penetrates to the lowest levels of autonomous local and community government.' It was a stance he later reiterated at the Rio Summit, assuring the audience that the Belizean government was committed to 'community-based ecotourism'.

Eco-terrorism

But the government's statements have not always matched their actions and a succession of examples from Belize begin to provide a rather less optimistic picture of the use of sustainable tourism.

Box 9.9   Like any other business

Hosted in a new luxury hotel – the construction of which necessitated stripping the protective mangrove cover – on the ecologically fragile outskirts of Belize City, two recent major international conferences on ecotourism highlighted the conflict between environmentally conscious activists and entrepreneurs who have seized upon ecotourism as a convenient marketing tool.

The keynote speaker at the First Caribbean Ecotourism Conference in 1991, Voit Gilmore from the American Society of Travel Agents, promised 'millions of Americans just waiting to come'. Howard Hills, a US investor speaking at the 1992 First World Congress on Tourism and the Environment, promised money (see p. 228). Ecotourism, Hills argued, is just like any other business. He boasted that the Overseas Private Investment Corporation would lend up to $50 million to any 'environmentally sound' tourism development, conditional upon a 25 per cent US stake in the project.

Environmentalists consider such sentiments anathema to the very ethos of ecotourism. They argue that a small scale, locally controlled and ecologically sensitive industry can neither sustain many visitors, nor be a big money-maker. In contrast to the Belizean government's view of ecotourism as a way to develop economically and to earn foreign exchange, environmental activists see ecotourism as justification for ecological and cultural conservation. An incredulous anthropologist from Scotland's Edinburgh University could only express dismay at Hill's remarks. 'The whole idea of ecotourism,' he said, 'is that it is like no other business.'

(Higinio and Munt, 1993, 8–10)

In 1992 the popular Belizean weekly *Amandala* carried the headline 'Eco-terrorism at Hatchet Caye', following a US resort owner's attempt to blow up part of the ecologically fragile coral reef in order to make his resort more accessible to visiting boats. The developer, Donald McKenzie, was subsequently arrested, fined US$2,000 (for 'dynamiting without a permit') and fled the country (Otis, 1992). Of course similar incidents have been reported elsewhere in the world where government rhetoric extolling the virtues of the controlled growth of tourism has been contradicted in practice. In some cases this is a direct result of weak and ineffectual legislation controlling environmental impacts, particularly the inability to enforce legislation. In other cases it reflects the complete absence of necessary controls or the government's willingness to turn a blind eye.

More than environmental damage, however, the Hatchet Caye incident further underlined the unease with the degree and growth of foreign ownership of tourism resources in Belize. *Amandala* reflects these concerns in statements that clearly resonate Steve Britton's political economy of tourism – refer back to Chapter 3.

As Belize struggles to find the balance between the development and conservation of our natural resources, more and more we will run into the Hatchet Caye syndrome: foreign nationals and corporations who do not believe that they are answerable to us.

. . .

Belize and Belizeans are beginning to be trampled in the rush by those who regard our laws and traditions as inconvenient at best, and our sovereignty as theirs, bought and paid by their money. They wheedle, bribe and connive to extract their fortunes. And when all the above fail, they will resort to terrorism; covert campaigns, destabilisation and destruction, to have their way.

(Amandala, 1992: 6)

A devil's bargain

The concerns of Belizeans, especially the government's interpretation of what is and what is not ecotourism, were heightened by the announcement of a joint venture with a Mississippi-based company for a US$1 billion development to cover two-thirds of Ambergris Caye, the biggest of Belize's offshore islands. The Ambergris Caye controversy has raised serious doubts about the scale and nature of tourist development. Presented as a major contribution to re-investing in the Belizean people (a reflection of the PUP's 1989 election soundbite 'Belizeans First'), the government bought back the northern two-thirds of the 20,000-acre Ambergris Caye from its US owner. It was expected that a newly established Ambergris Caye Planning Authority (ACPA), would have its jurisdiction and planning powers extended over this area of land, known as the Pinkerton Estate.

Without the knowledge of ACPA, however, the government set up a semi-autonomous development corporation, appointed by and answerable to the Tourism and Environment Minister, Glenn Godfrey (who was also the attorney general). It was planned to earmark approximately half the 20,000 acres for conservation and 2,500 acres for Belizeans. But it is on the remaining 7,500 acres that the corporation proposed a US$50 million 'sustainable development'. In the relatively fragile ecology of Ambergris Caye, it is difficult to reconcile Minister Godfrey's statement that 'in Belize we must keep tourist developments small' (address to First World Congress on Tourism and the Environment, April 1992) with the claim that the proposed development represents 'an integrated and ecologically sound resort development' when the development was to comprise at least one international hotel, two 'all-inclusive spa hotels', three to five up-scale lodges, two golf courses, town houses and villas, a thousand luxury homes, polo fields and stables.

Although the scheme has not progressed, it is illustrative of the 'seductions of mass tourism' as Godfrey refers to it, and the manner in which local control is compromised, if it is not entirely absent. Just two days before the contract was to be signed between the developer and the Belizean government, Godfrey presented the document to ACPA for the first time in spite of discussions having proceeded over the two previous years. A furious member of the ACPA encapsulated the mood of local people arguing that if the proposed agreement of North Ambergris Caye was as good as claimed, why had the minister not let the people know: 'When the government acquired the Pinkerton Estate, we were told in no uncertain terms that we were getting back control of the land and meaningful participation. To me, 75% to foreigners and 25% to Belizeans is not fulfilling the promise.' As Otis reports in *The Nation* newspaper, one Belize-based diplomat commented, 'There is a latent resentment of foreign ownership. But Belizeans made a devil's bargain. They sold the place' (1992).

A further case in point was the proposed multi-million dollar Belize City Tourism District Project, which would have involved large-scale foreign capital investment or large-scale borrowing from US-dominated lending institutions such as the World Bank. New Orleans-based architects completed a USAID-sponsored initial study for a waterfront-style regeneration project. The planners hoped to build an exclusive downtown area complete with new hotels and shopping facilities, a promenade with seafront restaurants and cafés, an ethnic crafts and food market and marina for small yachts cruising the Caribbean. It was, in other words, a mini Miami (the consultants even used Baltimore's Haborplace as a comparison), designed to transform the seafront district into a vacation *pied-à-terre* for would-be ecotourists, who would presumably feel protected from the urban problems that have become legendary in guidebooks to Belize. The scheme was also underpinned by the philosophy that the benefits would, somehow, trickle down to the rest of the city's population.

However, not only would ordinary Belizeans gain little from this up-scale project, but local representatives were not consulted in the planning process. In a bizarre twist, the city's Planning Department first learnt of the scheme when a three-dimensional model appeared in the window of the city's supermarket. In a city of 60,000 where the city council is severely strapped for money and poverty is on the increase, the consultants also lauded the scheme for allowing the authorities to 'concentrate its resources' (and by inference cutting back spending in other, poorer, neighbourhoods) in one area. In this way, it was argued, the district would be a showcase with cleaning, street lighting and road maintenance, for example, maximised. Such recommendations were regarded by many as deeply insulting.

Paradise sold

Much of this expressed concern over the control of tourism development in Belize is also reflected in the ownership of tourism resources and facilities, a point already emphasised by the *Amandala* quote above. Some commentators have already observed that much of the tourism industry is already in the hands of the country's small, but powerful expatriate community (Everitt, 1987; Cater, 1992). (Box 5.8 ('Too good for tourists') especially reflects this fact in the last paragraph.) Cater, a delegate at the ecotourism conference in 1991, found that 25 per cent of the registered delegates were US citizens. More significantly, among the 43 per cent registered as Belizeans were many expatriate Americans. This is a phenomenon witnessed in many countries with expatriates running bars, restaurants, tour operations, lodges and hotels. Updating these figures by one year, Mowforth, attending the 1992 First World Congress on Tourism and the Environment in Belize City, found that more than 40 per cent of the 300 delegates came from the USA and Canada. Again, many of the 29 per cent of delegates who were Belizean were expatriate Americans. (As already noted, Belize makes something of an industry out of hosting conferences on ecotourism, sustainable tourism, environmental issues relating to tourism, or some combination of these terms. In this context the unintended irony in the title – First World Congress – is of interest.)

In Belize the expatriate community is a resolute and well-represented lobby. Expatriate riverside lodge owners from the western district of Cayo, for example, demanded strict zoning laws to protect their interests from further tourist development. The depth of expatriate ownership is further reflected through the USAID-initiated Belize Tourism Industry Association (BTIA), an umbrella group representing private interests. In 1992, BTIA was thrown into turmoil when during the annual election of officers, a dispute erupted over whether seats on the committee should be restricted to Belizeans, even though 65 per cent of the membership were expatriates.

It appears, therefore, that the Belizean government's stated commitment to community tourism development is becoming steadily more elusive as foreign control of the industry takes over. It is not just the large multinational companies and developers cited in Britton's political economy, but small-scale, alternative tourism businesses as well. As Everitt (1987) comments of the post-independence period in Belize, with foreign (mainly Canadian and US) interests in control of the country's most successful tourist facilities, tourism has emerged as a symptom or contributory factor in the creation and aura of neo-colonialism. And it is difficult to escape the impression that Belize is little more than a piece of real estate. As an advertisement for Ambergris Caye's Club Caribbean in the US publication *Belize Currents* claims: 'own your own piece of paradise . . . Prices start as low as $9,950 . . . Values are starting to soar.'

With the importance of tourism growing in the national economy, and the lure of perceived gains from the growth and development of the industry increasing, one of the most positively cited laboratories of successful ecotourism must be critically challenged. The government's need for foreign exchange is often in direct conflict with the need to control the scale and content of tourism development (Simons, 1988). Expensive new infrastructure projects such as airports, roads and ports require a return on investment. Locally based and controlled tourism projects, while worthy, do not provide the necessary return. Indeed, a government's aspirations and expectations of tourism development often stand in contradiction to the aspirations of local people.

For many, Belize's problems are couched within the narrow confines of the threat of too many tourists and the effect they may have on the environment. As Sloan concludes 'Belize is fast becoming one of the most popular ecotourism destinations in the hemisphere. But there's a hitch: All those ecotourists are threatening to trample the place to death – literally' (Sloan, 1993). For others, the problems are considerably more deep-rooted. As Everitt concludes of tourism, it is 'clear that economic imperialism is rife in Belize despite independence, and ranges throughout the economy which also sells its produce to the United States' (1987: 51). Like citrus fruit or bananas, tourism has emerged as another cash crop in the national economy, one that is steadily eroding the image of small-scale, locally owned projects.

Belize is also a perfect example of the complex interaction between a range of interests and why it is problematic to examine individual interests in isolation. In this case the interests of local people, multinational environmental NGOs, aid agencies and the Belizean government are intimately intertwined.

The second and third case studies provide further examples of government policy and the discourse of sustainable tourism development. The second case study is from another Central American country, Costa Rica, and the third is from Bali.

Lessons from Costa Rica

Over the last thirty years Costa Rica has developed a system of national parks and other protected areas which now cover just over a quarter of the country's land area. Despite severe underfunding for the protection of these areas (Blake and Becher, 1991: 82–3; McNeil, 1996: 45–6), this system has formed the base resource of a tourism industry that has relied heavily on the attraction of the country's natural diversity of flora, fauna and landscape.

Consequently, it has become renowned as a destination for ecotourists; outside San José, the capital city, its tourism industry has been largely based on small-scale, locally owned lodges and hotels which form an integral part of both the communities and natural environments in which they are located. Even within San José, there is a good supply of small hotels and only a limited supply of large-scale, luxury hotels.

Partly through the efforts of its ministers speaking at international conferences, Costa Rica also gained an international reputation as a leader in environmental conservation. Whether this was deserved or not is rather debatable given its high rate of deforestation, especially during the 1980s, outside its protected areas.

In recent years, however, international admiration of the Costa Rican government's shunning of large-scale, mass tourism developments has begun to fade as numerous contracts have been signed with international consortia to build tourist condominia offering the 'four Ss' of mass tourism but also claiming to be environmentally sensitive and offering nature as part of the attraction. A number of features of these schemes are briefly outlined in Box 9.10.

Box 9.10   Recent tourism developments in Costa Rica

In *Tambor* on the Pacific coast peninsula of Nicoya, a subsidiary of the Barceló Group, a Spanish holiday firm which owns 4-star and 5-star hotels in Tunisia, Mexico, Spain, the Dominican Republic, Costa Rica and the USA, built a 400-room hotel complex within 50 metres of the highwater mark. In so doing they stripped a hillside, filled in a swamp, extracted sand from a nearby river, destroyed a hillside to quarry its stone and threatened local species, all without the necessary building permits from the ministries of Housing, Health, Public Works and Transport and Natural Resources. Additionally, it has been reported that white sand was removed from a nearby beach and used to cover the original black sand beach at the complex. The hotel opened in 1992 and has been accused of depositing its sewage in the Río Pánica.

Just to the north of *Jacó* on the Pacific coast of Costa Rica, construction has begun on the Los Sueños resort hotel, condominiums and marina at the once tranquil village of Playa Herradura. Apart from its negative aesthetic effect on the village, the project threatens an extensive tidal estuary and a marsh that stretches along the entire beach. These wetlands are an essential habitat, food source and nursery for a wide range of wildlife.

The massive *Papagayo* project includes the construction of 1,144 homes, 6,270 condo-hotel units, 6,584 hotel rooms, a shopping centre and a golf course. It forms a large part of the Costa Rican Institute of Tourism's development of Culebra Bay, also on the Pacific coast. The scheme goes under the title 'Papagayo Ecodevelopment', but despite the planning which has obviously gone into it, the 'eco' more appropriately refers to the *eco*nomic wealth it will generate for its investors rather than to the local *eco*logy it will save.

Of the Papagayo project (Box 9.10), Jeff Marshall, a seasoned commentator on the Costa Rican tourism scene, says:

> The enormous scope of this project is entirely inconsistent with the concept of sustainable and socially responsible ecotourism, which forms the very foundation of the highly successful tourism industry in this country. . . . It is nothing more than a high-profit real estate scheme designed to make a bundle of money for a few Costa Rican insiders and their foreign corporate allies.

(1994: 2)

The link between the burden of international debt and government capitulation to the pressure of transnational companies to develop large-scale tourism projects in order to gain much-needed foreign exchange may be denied in some quarters. But while it may be difficult to prove the existence of the link, it is also difficult to deny.

The examples outlined in Box 9.10 illustrate the subversive power of the pressure exerted by international consortia – baying at the heels of the international lending agencies which stipulate the need for national governments to raise foreign exchange to service their debt. As Michael Kaye, president of Costa Rica Expeditions tour company, has stated, 'the short-term temptations of the fast and easy money from mass tourism development in the context of an economy the size of Costa Rica's should not be under-estimated' (1994).

Under such pressure, it should be no surprise that the government of Costa Rica has felt obliged to abandon its previously cautious support for a sensitive and responsible type of community-based tourism development from which it, the government, gained relatively little financial benefit as most of this was received directly by Costa Ricans.

In 1992, Deirdre Evans-Pritchard, an English academic who used to run the Eco Institute of Costa Rica, pointed out that the Costa Rican Chamber of Tourism (CANATUR) was under the same kind of pressure as the government. ' . . . the 1992 CANATUR congress championed noble causes and concerns in the name of small scale sustainable tourism development and then, in the final ceremony, awarded prizes to the four largest, most resort-like hotels in the country' (1993: 779). Under such pressure, perhaps the government of Costa Rica is as unpleased about the decisions it has to take as are most Costa Ricans. But it must have been particularly galling in the first half of 1995 to see the decline in the numbers of incoming tourists. Accommodation capacity was barely 50 per cent used, and it was clear that tourists were seeking other destinations. Developments such as those listed in Box 9.10 were intended to increase tourist receipts, not to make them fall.

At that particular time, however, the Costa Rican government suffered from another loss of power over its tourism industry which may account, at least in part, for the reduction in tourist numbers. In 1994, the Minister of Environment, René Castro, increased the entry fees to most national parks for foreigners by a factor of ten, ostensibly to further the government's sustainable development programme

> by limiting the number of tourists using the country's natural resources. In one sense, his policy 'worked' – between '94 and '96, the nation's parks recorded an overall 26.5 per cent decrease in visits. Castro may have been motivated by more than the principle of sustainable development. The majority of foreign visitors to Costa Rica are 'shoestring' travellers, low-budget backpackers from the US, Europe and Australia. Both Castro and Figueres [the President] . . . may have hoped that the tourist industry could recover any losses incurred by the decrease in backpackers through its promotion of Costa Rica as a travel destination for affluent visitors.
>
> (quoted in Mesoamérica 1996: 9)

In response to the widespread complaints arising from the price increase and the significant decline in tourist receipts, from 1 April 1996, the entry fees were lowered from $15 to $6. Additionally, hotel and car rental companies lowered their daily rates by 15 per cent. CANATUR president, Mauricio Ventura, asked 'Who is the winner in this?' and answered, 'The tourist' (Mesoamérica, 1996). Another clear case of power resting with the First World.

The Costa Rican Institute of Tourism's claim that the country is a model of sustainable development now rings a little hollow. As Jeff Marshall asks:

> Instead of promoting massive centralised tourism projects tied to foreign capital, why doesn't the ICT concentrate on the careful planning and development of a network of quality, locally owned and operated, small scale tourism businesses that are compatible with the natural environment and integrated within the existing structures of local communities and economies?
>
> (Marshall, 1994: 36)

Lessons from Bali[1]

The island of Bali, Indonesia's main cultural tourist draw, has long attracted foreign visitors. Increasing consumer demand (see Figure 5.2, p. 152) has coincided with the Indonesian government's own targets for tourism, which have used Bali as a pivotal marketing tool. At the same time, the special nature of Bali's attractions and the assumed fragility of its cultural life have made it one of the earliest testing grounds for theories about cultural sustainability in the context of tourism development.

National policies to increase tourism were accompanied by a conscious attempt to maximise economic benefits and minimise unwelcome impacts on the island's culture and environment. A Tourism Master Plan drawn up by French consultants and released in 1971 recommended that 'any changes should reinforce not detract from traditional Balinese social structures' (Noronha, 1979: 195). These were based on *banjars*, neighbourhood associations which controlled virtually every aspect of people's lives. As part of the strategy, tourism development was to be confined to the area around the airport, on the southern peninsula. A new resort, Nusa Dua, was constructed there to cater for the luxury package market and the conference trade. From there tourists are taken to visit famous temples, see craftspeople and artists at work or watch dance performances, with the proceeds going into the treasury of the *banjar* involved. Elsewhere, the largely agricultural life of the islanders was planned to continue.

The Plan's formula for sustainable tourism certainly increased visitor numbers and limited spatial tourism development. But it 'gave rise to undisguised criticism in Bali. . . . The Master Plan was a plan for tourism, but it clearly was not a plan for the development of Bali' (Picard, 1991: 83). In practice, the new hotels needed skills and capital not available locally, thus bringing in outside investors, both Indonesian and foreign, and limiting the projected local employment benefits. Moreover the planners had not reckoned with the huge popularity

313

of Bali among independent travellers and young people booking flight-only holidays – the 'new' travellers. By 1985, ad hoc, unregulated tourist enterprises of all kinds were flourishing alongside the planned new hotels and packaged cultural attractions.

With tourist numbers exceeding forecasts, international assistance from the UNDP was sought to help with the management, planning and control of tourism on the island. Meanwhile, concern at the growing imbalance of prosperity between the south and other parts of the island led to an apparent *volte face*. The authorities discarded the original vision of a tourist ghetto and instituted a Spatial Arrangement Plan in 1989 which identified sixteen centres for tourism development around the island as part of a dispersal strategy.

This move could be seen as reflecting the agenda of local government, answering local needs, rather than national government, intent as the latter was on preserving the tourism product. In effect, however, the Spatial Arrangement Plan's wider distribution of tourism's economic benefits served to reinforce the kind of power structure that is most damaging to Bali's cultural continuity.

In 1995, an Australian report highlighted disturbing patterns of ownership in new resort developments around the island. It confirmed that these are no more likely to be controlled by local Balinese than the resorts in the southern enclave. Powerful Javanese and foreign investment interests are now firmly entrenched in the island's tourism economy (Aditjondro, 1995: 15–21). Sustainability for these investors is strictly a matter of sustaining profits and their concerns are more about the availability of land for golf courses and condominia than about local needs and the continuation of a flourishing cultural community. The provincial authorities appear helpless to prevent the construction of new attractions alien to Balinese culture; new hotels are sited on sacred land despite local protests; and rural land is transformed into tourist playgrounds.

The Bali Sustainable Development Project undertaken jointly by academics at Canada's University of Waterloo and Indonesia's universities of Gadja Meja and Udayana (Wall and Dibnah, 1992; Wall, 1993) has attempted to demonstrate an alternative vision to the process of relentless outside manipulation, setting out proposals for integrating tourism into the wider Balinese economy. Some of its conclusions have been incorporated into the Bali Planning Department's latest five-year development plan, which recognises the importance of maintaining a viable agricultural sector and of supporting village institutions. Implicit in this is the need to maintain local autonomy. But its effectiveness will depend on the nature of its challenge to the prevailing power relationships, in which outside investors now appear to be dominant.

Human rights and tourism in Burma

Aung San Suu Kyi's National League for Democracy (NLD) won 81 per cent of the parliamentary seats in elections in Burma (Myanmar) in 1990. The response by the generals of the State Law and Order Council (SLORC) was to imprison

314

most of the newly elected parliamentarians and to turn the country into what Amnesty International has called 'a prison without walls'.

Since that time, the United Nations Commission on Human Rights has described the country's violations of human rights as:

> extremely serious, in particular concerning the practice of torture, summary and arbitrary execution, forced labour, including forced porter-ing for the military, abuse of women, politically motivated arrests and detention . . . important restriction on the exercise of fundamental freedoms and the imposition of oppressive measures directed, in particular, at minority groups.
>
> (1993)

The issue of tourism in Burma highlights a number of points of interest in this chapter. These concern not just the relationship of the Burmese government to the tourism industry, but also the relationship of First World governments and tourism-related TNCs to the Burmese government.

Since 1990, 'The SLORC has been wooing new investors – especially in the area of tourism and hotel development – with offers of ten year tax breaks and full repatriation of profits' (Mahr and Sutcliffe, 1996: 28). As Mahr and Sutcliffe also point out, such investment from TNCs based in the First World lends the regime political legitimacy, which is confirmed by the arrival of First World tourists. And tourists have been arriving, with a tenfold increase between 1992 and 1995, when approximately 100,000 visited the country.

The SLORC declared 1996 the 'Visit Myanmar[2] Year' and increased its efforts to profit from the tourism industry by aiming at a target of half a million visi-tors for 1996. To prepare for the expected increase in tourists, the SLORC put its people to work – forcibly – as described by Mahr and Sutcliffe:

> Whole communities have been forcibly relocated from their homes to make way for foreign hotels and to 'clean up' areas for the eyes of tourists. . . . Thousands of people have been forced to work without pay on tourist sites – restoring the moat around Mandalay Palace, for example. Plans have been put into action to relocate the more 'picturesque' ethnic peoples to special villages where they can be visited by tourist groups in what amounts to a human zoo. Model villages for this purpose are under construction near Rangoon and some Padaung people have already been resettled to Inle Lake. There tourists can visit – for a charge – and take photos of 'long-necked' Padaung women.
>
> (1996: 29)

These human rights violations have been documented in detail by the London-based Burma Action Group (Sutcliffe, 1995) and reported by Tourism Concern and increasingly in the western media. No British tour operators can currently be unaware of the debate around the ethical grounds for not visiting Burma. New hotels are being built by TNCs from Hong Kong, Japan, Malaysia,

Singapore, South Korea and Thailand, and French and Swiss interests are also involved. Companies such as Voyages Jules Verne, Explore, Bales, Coromandel and David Sayers Travel all run cultural or trekking tours to Burma. Explore describes Burma as 'half-hidden from foreign eyes, a remote and little known country. Yet it presents travellers with a kaleidoscope of vivid impressions. Places like Rangoon, Pagan, Mandalay and Inle Lake enchant us with a magic all of their own' (65).

Burma represents an attractive Third World destination for the new middle-class tourists from the First World, as Explore's description makes clear. But more importantly than this, from the standpoints of First World corporate decision-makers, it represents an attractive investment opportunity for First World tourism businesses. And First World governments can and do assist in these processes of tourism investment and associated human rights violations.

The British government's stated policy is not to promote investment in Burma because of the SLORC's human rights violations. In 1996, however, despite its stated policy, the government's Department of Trade and Industry (DTI) allocated £50,000 travelling expenses to trade missions to Burma for the year. After the first mission at the end of February, in the words of Tourism Concern, the

> Shadow Foreign Affairs Minister [Derek Fatchett] asked a parliamentary question about the DTI's plans for future trade missions and the reply was 'we intend to evaluate the results of this mission before deciding on supporting the second.' Civil service speak for 'oops – we've been found out.'
>
> (1996b)

It can be assumed that First World governments would not try to cover up public knowledge of such trade missions if they believed that the development to be promoted by these missions could be portrayed as being ethically sound or in the interests of sustainability. The example of Burma is indicative of the 'reality' to which governments and the industry work. It is a reality which they use to justify their actions and investments. If ethical considerations contradict this reality, then they must be hidden from view. And if the contradiction is exposed, then weasel words must be found to justify the investment.

Tourism in Burma provides an illustration of the way in which investment is used to serve specific interests, those of the Burmese government, First World governments, First World tourists, and tourism-related businesses based in the First World. Other interests, those of the visited, the resettled, the majority of Burmese people, are decidedly not served by these investments and developments.

But whereas in many Third World destination countries the governments have to be coerced by First World governments, international lending agencies and TNCs, in the case of Burma the SLORC actively enlists the assistance of these bodies. Power is still very much based in the First World, but with a willing First World bastion in exotic Burma's military and business élite.

Despite the clarity and documentation of the link between Burma's human rights violations and its exploitation of nature and cultural tourism, the brochures of many new tourism companies still extol its virtues. In such circumstances, then, it is hardly surprising that for many tourists the relationship between travel, tourism and human rights has remained rather nebulous. Indeed, as was suggested in Chapters 3 and 5, the occurrence of human rights abuses and their coincidence with perceived danger may actually act as a fillip to tourism in certain circumstances and among certain types of tourists.

CONCLUSION

In this chapter it has been argued that an understanding of issues of power is essential to an analysis of tourism developments and the role of governments in this. The necessity for such an understanding is as great for analyses of new forms of tourism to Third World destinations as it is for mass tourism developments.

It is commonly perceived, especially from local community level, that power rests with the national government. This may indeed be the case in some instances, but we have suggested that the actions and policies pursued by national governments are often circumscribed at best and are sometimes dictated by the influence of external organisations and forces. This includes both the inevitability that tourism must increase as well as the contest over which form of tourism development should take place. It is especially so for Third World governments weighed down by a burden of debt – a burden that was largely foisted upon them by First World banks and governments as well as their own incautiousness resulting from the unseemly rush for western-style development.

The outcome of this process of First World influence on Third World development has been government policies which increase the unevenness and inequality of development within national boundaries and which further widen the gap between First and Third Worlds. The policies pursued for the development of tourism are those most suitable to maximise the profits and enjoyment of First World investors and tourists rather than those most beneficial to Third World communities and governments.

We have tried to show how, even where governments have attempted to follow environmentally benign, socially and economically beneficial and culturally sensitive tourism developments, forces beyond their control have subverted the policies and have served only to emphasise seemingly ever-widening gaps. In the process, government attempts to satisfy outsiders' views of the content of tourism have turned sustainable tourism into an item of political discourse.

10

CONCLUSION

This book has sought to provide a broad debate as a means of critically assessing the important developments in contemporary tourism. It has attempted to explore the ways in which new forms of tourism are interrelated with notions of sustainability and are reflected in the Third World. This approach has enabled us to view and analyse tourism in a broader context and to stress that tourism is not the sole focus of our discussion. Instead, we set out to explore the way in which socio-cultural, economic and political processes operate *on* and *through* tourism; in other words, tourism is a mirror of these wider processes. In addition, it was inappropriate to provide further analysis and evidence of the now well-documented environmental, economic and socio-cultural impacts *of* tourism.

KEY THEMES AND KEY WORDS

In order to achieve this broader analysis, a number of key themes (uneven and unequal development, globalisation and relationships of power) and key words (tourism, the Third World and sustainability) were identified. Table 10.1 summarises these positions and seeks to illustrate the relationships between the key words and key themes by drawing some examples from the book. Clearly the relationships summarised in the cells of the table are not exclusive and there is considerable overlap between both the key words and key themes.

Sustainability, new tourism and the Third World, it was argued, were best thought of as key words as each is open to competing meanings and interpretations. While to a backpacker the Third World may represent exotic cultures and environments and an endless world of adventure, to mass tourists in the USA, Europe or the Pacific Rim, the notion of the Third World may represent a threat, and a place either to avoid when holidaying or where a holiday needs to be carefully controlled, through all-inclusive enclave tourism, for example.

Arguably, the most important purpose for this book was to establish that these key words are hotly contested. This helps us understand, for example, the way in which new tourism seeks to differentiate and distance itself from mass forms of tourism and how sustainability does not have a fixed and agreed definition that is shared by everyone.

Table 10.1 Key themes and key words

Key words		Key themes			
		Uneven and unequal development	*Relationships of power*	*Globalisation*	*Tourism and geographical imagination*
	New tourism	Majority of tourists from the First World, especially new tourists	First World owns majority of tourism resources	Places drawn into global tourism, and new tourisms spread	'Benidorm' v. 'Himalayas' Independent v. mass tourism
	Third World	Third World is structurally disadvantaged – global inequality	Third World required to adjust structurally for First World institutions	Third World increasingly interdependent with the First World	Third World as poverty stricken Third World as environmental paradise
	Sustainability	First World prescribing solutions: 'Think globally, act locally'	Third World governments and NGOs clash at the Rio Summit over First World	'Think globally, act locally' Planet Earth – global environmental interdependency	Third World environments as sizeable tourism assets Third World environments need saving

The first of the key themes, uneven and unequal development, was adapted as shorthand for indicating the inequality faced by certain nation states, regions, communities and individuals. The principal focus was on the way in which inequality is expressed globally through the First and Third Worlds, but clearly unevenness is also expressed, for example, in the way certain Third World destinations become tourist magnets while others do not. The second and closely associated theme, relationships of power, builds upon this inequality by offering a more nuanced and extended discussion of how power is transmitted in and through tourism. This varies from the power wielded by private business consortia such as the WTTC, to the priorities of environmental sustainability, to the way in which new tourism draws in and controls new Third World destinations. The final theme, globalisation, seeks to provide an overall framework for understanding the way in which the world has seemingly shrunk as both a cause and consequence of the expansion of tourism; and new tourism has played an interesting role in this relative shrinkage.

In Table 10.1 we have added 'tourism and geographical imagination' to the themes. This is a subtheme, so to speak, and it is timely to review its relationship to the key words. The notion of tourism and geographical imagination seeks to emphasise that we are dealing with the nature of representation, how the context and meanings of words such as tourism, the Third World and sustainability are socially constructed by host communities, governments, TNCs, smaller operators, environmental organisations and tourists. They are words whose meaning is open to interpretation and are continually contested between different interest groups – a point that is reflected in the examples provided in the final column of Table 10.1.

NEW FORMS OF THIRD WORLD TOURISM

The new forms of tourism were summarised in Chapter 4. As the early chapters of the book argue and as Figure 10.1 and Table 10.1 portray, they have arisen as a result of a wide variety of economic, socio-cultural and environmental factors. Factors such as a structural economic shift from a Fordist to a post-Fordist mode of production, accompanied by cultural shifts characterised as moving from modernism to postmodernism, and a growing environmentalism, help explain the increase in the number of new forms of tourism. Independent travel has sought to distance itself from mass tourism, and a variety of benevolent terms (appropriate, alternative, acceptable, responsible, sustainable, and so on) have been employed in an attempt to assert that it is independent travel that is morally and ethically acceptable.

We have attempted to demonstrate a tentative link between the economic and cultural shifts in contemporary capitalism and the growth of the new middle classes. In Chapter 5 we suggested that it is these class fractions in particular (though not exclusively) that can be linked to the growth of new tourism. Furthermore, these new groups and their motives are more formally represented

UNEVEN AND UNEQUAL DEVELOPMENT

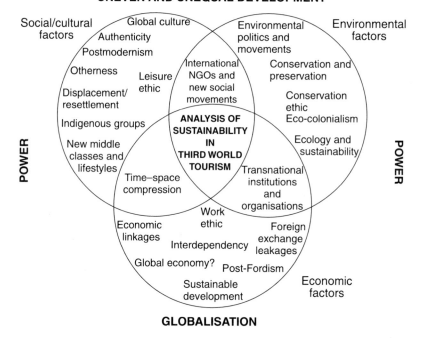

Figure 10.1 Tourism, sustainability and globalisation

by the international NGOs which campaign around issues of conservation, ecology, pollution, human rights and cultural identity to which many of the new middle classes subscribe. These organisations represent the vanguard of this movement to gain a moral base and a justification for their own definition of sustainability, which is often constrained by a restricted association with ecology. The subliminal promotion of a conservation ethic and the assured certainty of one form of sustainability are responsible for the emergence of attitudes which in some cases have been characterised as eco-colonialist.

Chapter 7 examined the ways in which the new tour operators, as well as the more established and conventional ones which also lay claim to sustainability, are defining the notion to fit conveniently with their own practices and purposes. This can be achieved through a range of techniques. The chapter also showed how supposed 'business realities' tend to subvert all these efforts and to turn them into relatively ineffective palliatives and marketing ploys. Thus the claims to sustainability can be seen as having little or no effect on business as usual. Despite this, the new forms of tour operation have brought into existence a variety of new posts within the industry to promote its claims to sustainability.

The new forms of tourism have changed the relationships of power and exchange for many destination communities and these were discussed in Chapter

8. The roles of destination communities in Third World countries may not have changed so drastically as a result of the emergence of new forms of tourism, but local participation in the decision-making and running of tourism schemes is highly variable. For destination communities, particular problems have arisen as a result of the conservation movement and its penchant for promoting area protection measures (such as the designation of national parks) which have often excluded local populations and prevented them from continuing 'life as normal'. This has been particularly so for indigenous groups who find themselves fulfilling the role of 'zooified' objects for the entertainment of new tourists. Host populations have found themselves drawn into the new tourism activities as local entrepreneurs, as local élites, as applicants for government funds, as local service providers and local operators, rather than simply as the objects to be viewed. These changing relationships have occasionally had positive effects, but in other cases have led to the perpetuation of unequal divisions.

Finally, that governments have begun to adapt to the new forms of tourism was demonstrated in Chapter 9. At this level the power relationships are often dominated by external influences upon nation states and the whole process of 'development' is often determined by the requirements of international finance.

GLOBALISATION, SUSTAINABILITY AND TOURISM

Figure 2.4 presented a graphic illustration of the interplay of the key factors involved in an analysis of sustainability in Third World tourism. In the chapters that followed we attempted to broaden the notion of sustainability and to analyse the new forms of tourism from different perspectives using an inter-disciplinary approach. Figure 2.4 was a much-simplified version, presented to show the framework for the analysis in the subsequent chapters.

In Figure 10.1, the same illustration is 'fleshed out' with some of the detail of the analyses of these relationships, using examples from throughout the book. The figure links the key themes and key words discussed above and expands the notion of sustainability in Third World tourism by presenting relationships between the key themes and the factors which feed into an analysis of tourism.

The core of the analysis has been an examination of the notion of sustainability in Third World tourism. This analysis is affected by socio-cultural, environmental and economic factors. Figure 10.1 provides examples of the forms that these factors take, which were discussed at appropriate places throughout the book. The economic factors, for example, range from the influences of post-Fordist modes of production giving more flexible forms of tourism (discussed in Chapter 2) to the global economic linkages exploited by transnational companies (Chapters 7 and 9). It is not just the effects of tourism on the economies of Third World destinations that should concern us, but the manner in which an understanding of global and economic forces bears upon an understanding of contemporary tourism.

Similarly, the discussion of socio-cultural factors also seeks to set the debate within a broad analysis. Factors range from postmodernism and the leisure ethic to the problems of displacement and resettlement caused by these holiday pursuits for those at the receiving end (Chapter 8).

Linking the socio-cultural and environmental factors are the new international socio-environmental movements (Chapter 6) whose interests and concern for environmental issues show a considerable overlap with those of the new middle classes. Again, an understanding of both environmental politics at a global level and the effects and impacts of environmentalism at a local level are required.

The examples drawn into Figure 10.1 are by no means exclusive and you may wish to consider other examples which illustrate these factors and their inter-relationships. Overall the figure serves as a reminder that a deeper understanding of sustainability and Third World tourism can be gained from a critical assessment of an array of global and local factors within an interdisciplinary framework. Arguably of most significance is the way in which power and uneven and unequal development are manifested through these processes and reflected through tourism. The core of the diagram in Figure 10.1, where the three sets of factors and the key themes overlap, is the focus of Chapters 5–9 of the book, in which different aspects of sustainability in new forms of tourism were examined.

SUSTAINABILITY AND POWER

A key characteristic of our argument, as Figure 10.1 emphasises, is the need to understand sustainability in a broad context. We have argued that such a con-text provides a richer basis for understanding and analysing the emergence of these new forms of tourism; in a nutshell, sustainability and new tourism are intimately related. It is also suggested throughout the book that sustainability is open to a wide variety of competing meanings. At its most basic, the perception and understanding of sustainability for the residents of the leafy suburbs of Vancouver, London or Melbourne is considerably different from that of local tourist destination communities in Bali, Bolivia or Uganda. Moreover, as has been suggested, sustainability is a conduit, as it were, of power; and power is reflected in and transmitted through notions of sustainability, as both Figure 10.1 and Table 10.1 seek to illustrate through examples.

Chapter 2 provided a background discussion to concepts of power and focused on the usefulness of these in critically assessing the role of sustainability. In the first place it was suggested that sustainability is ideological in the sense that it is from the First World that the notion has emerged and, in the main, it is First World interests that are served through the promotion of sustainability. But it is also ideological in the way in which notions of sustainability are forced upon the Third World and enforced by First World interests and institutions. While this is a considerable generalisation, it helps capture the inequality felt by Third World countries at the Rio Summit. It is also a point underlined by the power (and pervasiveness) of global environmentalism; in Box 3.2 Sachs captures this sense of First World environmental domination.

Sustainability was also referred to as a discourse, a question, as Eagleton suggests, of asking 'who is saying what to whom and for what purposes' (1991: 9). Sustainability, as the later chapters of the book stress, is used by a variety of interests in a variety of ways as a means of supporting and enhancing their basis of power. This ranges from the activities of the tourism industry to the interpretation adopted by or foisted upon Third World governments or the discourse adopted by the myriad of environmental organisations. In short, sustainability and sustainable tourism reflect a discourse that is contested and through which power circulates.

The final concept of power discussed was hegemony, and again (as with discourse) this relates usefully to the contested meaning of sustainability, and the manner in which it must be continually renewed, redefined, defended, and so on. This is demonstrably the case with the range of new tourisms that have emerged and the protracted debates over the most 'appropriate' way to holiday (as Figure 3.3 suggests). For example, Chapter 5 sought to demonstrate the practical strategies adopted by new tourists and the way they both seek to sustain lifestyles and cut out a tourism niche for themselves; a similar analysis was applied to the new tourism operators. Another way of illustrating the hegemony of sustainability and its relation to tourism is through a continuum of meanings adopted by the socio-environmental movements (Table 6.1). Overall we have tried to demonstrate why notions of power and contest are fundamental to an understanding of sustainability and especially the emergence of new forms of tourism.

In Chapter 4 we made reference to what some see as a guiding spiritual or philo-sophical maxim which is often used to explain dominant behaviour patterns and lifestyles: the ethics. It might be more appropriate to see these as descriptors of lifestyles rather than explanations. The abstract notions of the work ethic and the leisure ethic – not mutually exclusive – were briefly discussed with respect to the prevailing patterns of holidaymaking. We also introduced the idea of the emer-gence of what might be called a conservation ethic as a limited descriptor of the holidaymaking behaviour of some sectors of the new tourists. All of the 'ethics' were considered to be rather weak, although they do have a degree of relevance, at least as an aid to understanding. Doubt was especially focused upon this last ethic, the conservation ethic, and whether it can or will come to explain the behaviour patterns of anything more than an insignificant number of tourists.

NEW TOURISM, NEW CRITIQUES, NEW RESEARCH

The emergence of new forms of tourism begs many questions. We have attempted to identify some of these and answer others. Tourism is a vast and complex industry and field of study, and we have provided a partial analysis of some of its facets.

We have adopted a *thematic approach* in order to re-contextualise the analysis of Third World tourism. In part this seeks to avoid a case study approach, analysing examples of new tourism development and extrapolating lessons from these. There are many excellent accounts of individual cases – of Annapurna in Nepal, CAMPFIRE in Zimbabwe or the Toledo village guesthouse scheme in Belize – to name three of the most oft-quoted examples. This is not to say that a thematic approach is better or more correct than case study analyses – both have an important part to play. But we have been conscious of the need to place the analysis of local tourism development within a global context; as suggested above, an understanding of new tourism must start with an assessment of the broader forces at work.

The analysis of new tourism should also embrace a much fuller understanding of the tourism process itself and the *theoretical frameworks* that can help us make better sense of new developments. It is worth repeating the spirit of Britton's observation and criticism on this count:

> Although oversimplifying, we could characterise the 'geography of tourism' as being . . . dealt within descriptive and weakly theorised ways. . . . This problem is of fundamental importance as it has meant an absence of an adequate theoretical foundation for our understanding of the dynamics of the industry and the social activities it involves.
>
> (1991: 451)

Taking these sentiments on board, we have attempted to develop a critique of new tourism by extending and applying a political economy approach and engaging in a *multidisciplinary* discussion. Far from heralding the death of dominance and control of Third World destinations by foreign-owned mass tourism interests, the emergence of new forms of tourism represents a supplementary, if considerably more nuanced way in which relationships of power are transmitted and circulated.

As the discussion above implies, there is a growing need for analysis of new tourism to supplement the existing studies. Arguably, new tourism is currently a western phenomenon. Indeed, those Third World countries experiencing major increases in numbers of visitors from South-East Asian countries are affected far more by the impact of conventional mass tourism. It would nevertheless be interesting to ask whether the South-East Asian middle classes will resemble the new middle classes elsewhere in their cultural lifestyles and the subsequent production of new forms of tourism. If this were to be the case, it would dramatically increase the number of people who desire a holiday with a difference.

RESEARCHING NEW TOURISM

Clearly, then, there is a considerable need for research to clarify these trends and patterns, and there are ample opportunities for you to conduct your own research into new tourism. It is not necessary to make expensive and time-consuming research trips overseas. Arguably one of the most important research sources is on our doorstep, in the form of tourists. For example, an analysis of the journeys, aspirations, expectations and lifestyles (to name only a few research-able factors) of new tourists would provide some interesting and worthwhile data and analysis in its own right.

Similarly, we are also surrounded by new tourism providers and operators (and by their brochures, advertisements, and so on), that allow for some inter-esting considerations of the way in which such providers feed into the broader processes of differentiation in the tourism industry.

Quite apart from the brochures and newsletters of tourism companies, other valuable research sources include the annual reports of a range of governmental and non-governmental organisations, statistics from the World Tourism Organisation, the World Bank and the UNDP's *Human Development Report*, and a whole bank of travel writing in newspapers and magazines. The reports of organisations such as Tourism Concern and the Campaign for Environmentally Responsible Tourism (CERT) (see Appendix 3) will be of particular help. Indeed, an excellent introduction to tackling undergraduate research dissertations is Tourism Concern's *Writing Your Dissertation on Sustainable Tourism* (1996c). The point to emphasise here is that research into tourism is 'do-able' and enjoyable, and we hope that some of the approaches we have used here will stimulate further thinking and further research projects.

WHITHER NEW FORMS OF TOURISM?

At the end of Chapter 4 we briefly speculated on the question of 'Whither sustainability in tourism?' The need for a political analysis of the tourism industry in order to reveal the broader context in which it is set was emphasised. Without such analysis it is unlikely that the currently prevailing power structures which hold up their own definitions of sustainability in tourism will be seriously challenged.

We conclude here with speculation on whether new forms of tourism are here to stay. The future development of new forms of tourism are considered from the standpoints of each of the players in the industry. It is important to stress here that this brief discussion covers only what we consider to be likely and does not examine what might be considered desirable.

The growth in the demand for new forms of tourism to Third World destinations will probably continue. In the First World, the appeal of an experi-ence that is adventurous, authentic and alternative is unlikely to decline, even if it may be subject to temporal fluctuations arising out of First World recessions or spatial fluctuations due to local circumstances such as war. Moreover, it is likely that the diversity of forms which are demanded will continue to grow – there are still more hidden corners of the planet and underexposed cultures for the new middle classes to see and 'tourisms' with which the new middle classes are anxious to differentiate themselves from others. Indeed, as a group, the new middle classes of the First World look likely to grow in size and proportion; and differentiation of fashions and tastes within this group is likely to increase, leading to ever greater demand for holidays with a difference. This may be less likely in the case of the tourist patterns of the South-East Asian new middle classes, who, as has already been noted, appear to have pursued, in large part to date, conventional mass tourism activities rather than the new forms.

The differentiation of fashion in holidays is no less likely to be associated with sustainability and the conservation ethic in the future than it is now. And this association is likely to be fostered, successfully, by the international environmental and conservation NGOs. Some of these organisations, such as the Ecotourism Society, have recently begun to produce manuals on 'how to do ecotourism right', as Whelan suggests (1991: 4). Such assistance effectively offers local communities and others guidelines on how to make the prevailing system work for them, in particular how to work within the system of current power relationships to promote their own project or development. It is likely that this will only strengthen and perpetuate existing power relationships rather than bring about change.

The industry is unlikely to change its spots or its *modus operandi*. There seems no prospect of changes to the dominant imperatives of capitalism and capital growth. It is likely, however, that the industry will become the major proponent of 'sustainability'. The industry's definition of sustainability is unlikely to become much clearer than it is at present, but the techniques of conveying it, especially to tourists, will become more refined, more complex and more sophisticated. This applies to operators and service providers in both the mainstream industry and the new forms of the industry and at all scales and sizes. As the large operators diversify their activities, they are likely to associate their new offerings increasingly with the notion of sustainability; and as new operators emerge to cater for new, alternative forms of tourism, so they too will increasingly deploy links with conservation, ecology and matters ethical, to their own ends.

Local communities will continue to adapt themselves to offer the type of holiday that the new tourists demand. This will not drastically alter the relationships of power that operate between the relevant interest groups for, in doing so, they will fall into line with dominant practice. There may, however, be a development of at least local significance – the possibility that some Third World communities will take a degree of control over their own exploitation of tourism, and particularly new forms of tourism, which will represent, at least for them, a re-balancing of power. Although the number of communities in such a position is likely to increase, it is likely to happen in only a small way (albeit significant for the communities concerned). Some examples of this possibility were discussed in Chapter 8. Overall, however, change is unlikely to be significant or substantial.

National governments are also unlikely to promote change that will alter the balance of power between all the players in the field of tourism. We have tried to show throughout this book that even countries which are held up as models of sustainable tourism development, such as Costa Rica and to a lesser extent Belize, are just as likely to end up promoting the development of mass tourism enclaves. More significantly, where they do genuinely promote small-scale, sensitive and 'low impact' tourism, they are no more likely to break out of the existing power structures. Of course this is not always a result of a desire to stay within the confines of 'acceptable' economic development processes; but the external pressures of the transnational companies, First World governments and supranational lending agencies are generally too great for Third World countries to withstand. It has also been suggested (Chapters 7 and 9) that where these supranational organisations paint themselves as green or ethical, the paint is likely to represent only a thin gloss, over a body of policy which remains largely intact and unchanged.

If this round-up appears rather pessimistic, short on hope and delivering the message that we can expect 'more of the same', then we should also point out that the viewpoint was a global one: in other words, while we have used local case studies to illustrate the book, we have not organised the arguments around them. As Chapter 2 stressed, too local an analysis would tend to ignore the wider global forces and too global an analysis would tend to ignore local differences and examples. There is plenty of scope for communities and others to take control of the development of the industry around them and a few examples given in Chapter 8 illustrated this, even if they all had their imperfections. Notwithstanding imperfections, there are successes and the obstacles to change are not insurmountable. The possibilities for change, however, are unlikely to come from the top or from the middle, where the power of vested interest is too great; it is more likely to come from below, where the need for change is the greatest. At this level, the resources of nature, power and finance and the control over them are small in dimension and change is unlikely to be significant on anything but a local scale.

NOTES

1 INTRODUCTION

1 Equally, this critique of First World package tourism and its effects on the Third World has not accounted for the phenomenal growth of package tourism demanded by the burgeoning middle classes in South Asia (including Indonesia and Malaysia: see *Economist*, 1995; Hitchcock *et al.*, 1993). However, this is not the focus of debate in this book.

2 Where tables are used in this book we adopt the standard definitions given by the World Tourism Organisation (*Tourism Market Trends* (annual), Madrid: WTO) as follows:

Visitor: Any person who travels to a country other than that in which s/he has his/her usual residence but outside his/her usual residence but outside his/her usual environment for a period not exceeding 12 months and whose main purpose of visit is other than the exercise of an activity remunerated from within the country visited.

Tourist: A visitor who stays at least one night in a collective or private accommodation in the country visited.

Same-day visitor: Visitor who does not spend the night in a collective or private accommodation in the country visited.

Arrivals: All data refer to arrivals and not to actual number of people travelling. One person visiting the same country several times during the year is counted each time as a new arrival. Likewise, the same person visiting several countries during the same trip is counted each time as a new arrival.

3 It should be noted that the growth of the middle classes in some countries, especially those in South Asia and the Pacific Rim, is also of increasing significance (*Economist*, 1995). Most of the patterns of tourism arising from these countries resemble conventional mass forms of tourism rather than the new forms discussed in this book and are characteristic largely of the new middle classes of North America, Europe and Australasia.

2 GLOBALISATION AND SUSTAINABILITY

1 Some authors argue that the new middle classes are inseparable from postmodernism. Lash argues that 'economic growth and cultural change (post-Fordism and postmodernism) constitute . . . the two sides of these new post-industrial urban middle classes' (1991: 252); and Savage *et al.* refer to a growing convergence that postmodernism is 'some kind of middle class phenomenon' (1992: 1).

3   POWER AND TOURISM

1 Sex tourism has emerged as a major activity in a number of countries, especially in South-East Asia. Edward Said (1991) depressingly catalogues the 'formidable structure of cultural domination' (25) in his study of *Orientalism*. It represents such a 'system of truths', he argues, that it has 'rarely offered the individual anything but imperialism, racism, and ethnocentrism for dealing with "other" cultures' (204). The most persistent value of western cultural life, and one which is remarkably persistent, is sex: sexual experimentation and fantasy, promise, desire, delight and unlimited sensuality. It is this key feature of Orientalism (carefully nurtured by the nineteenth century's colonial wayfarers) that has been so cleverly refined in the world of mass communications, but remains a standardised cultural stereotyping of 'the mysterious Orient' – smiling, servile and sexy.

2 Susan Sontag (1979) makes this point about the 'asethetising tendency of photography' (109), that photography 'develops in tandem' with tourism (9) and that, ultimately, the 'medium which conveys distress ends by neutralising it' (109). Or, in other words, photographs preserve the status quo, and 'aestheticize the injuries of class, race and sex' (178).

4   TOURISM AND SUSTAINABILITY

1 The motor car, however, has an extremely powerful lobby and many supporters who would pit the arguments of personal freedom and inevitability against environmental sustainability.

2 Details of other techniques in this category and their application are not given here, although a number of examples are discussed in Chapters 5–9. Instead, see Green and Hunter (1992), Briassoulis (1992), Fletcher (1989), and Witt and Moutinho (1994).

5   A NEW TOURIST CLASS

1 For this chapter we have drawn heavily on our field notes and on an analysis of tour brochures offering holidays to a new middle-class clientele.

2 Reference is sometimes made to the *lifestyle* movement. This was especially associated with yuppies and was encapsulated in the idea that your 'style of consumption could both indicate and influence [your] position in a new social and cultural order' (Shurmer-Smith and Hannam, 1994).

3 Baudrillard is poignant here in his exposition of the 'into', being the 'key to everything': 'Into your sexuality, into your own desire . . . The hedonism of the "into"' (1988: 35). It is within this context that control and self are most eloquently expressed. Not only are the new middle classes able to map their futures in time, but they are able to extend this control spatially. 'Done it', 'Chill out', 'the World's a breeze!'

4 The representations of tourists and tourism in a number of tourist brochures and an analysis of travel reviews appearing in UK broadsheet newspapers are of particular significance in reflecting the characteristics of contemporary travel. As a number of authors (Massey, 1995; Munt, 1994b) have argued, these are a significant medium through which 'geographical imaginations' are formed and reflected (Walter, 1982). In particular, the manner in which the *Independent* mirrors the cultural and 'political' lifestyles of the new middle class is of especial interest. It is characterised by distaste with 'yuppie' culture, a veneer of classlessness and political independence, and an adoption of fractured non-party political issues (new socio-environmental movement): the campaigns on Bosnia, dog mess and café street life are illustrative of this.

Similarly, tourism has become a focal point, with the appearance of the *Independent Traveller* section and its campaigning stance on more sustainable forms of travel. In this respect, the *Independent* is a mouthpiece of new middle-class travel and frequently has contributions from organisations such as Tourism Concern.

5 Interestingly, the brochure says 'Essentially our tours are holidays; we have fun and although we focus on the crafts and textiles of a country, we do not miss out on any of the general sightseeing. So in India, we will still visit the Taj Mahal, in Peru we will explore the ruins of Machu Picchu and so on. . . . We fly with comfortable scheduled airlines. . . . Our accommodation is always of a good standard and is chosen for its ambience and cleanliness whether in colonial houses, castles, hotels, pensions or even tents.'

6 This argument is strongly reflected in Sontag's arguments regarding photography, another principal means of recording and translating the experience of tourism: 'To take a picture is to have an interest in things as they are, in the status quo remaining unchanged . . . to be in complicity with whatever makes a subject interesting, worth photographing – including, when that is the interest, another person's pain or misfortune' (1979: 12). And in so doing this process aestheticises reality, a tendency in photography that Sontag observes as 'the medium which conveys distress ends by neutralising it' (109).

6   SOCIO-ENVIRONMENTAL ORGANISATIONS

1 'Largely' is an important qualification here, for clearly it would be to overstate the case considerably to suggest that environmentalism and the middle classes are inseparable and mutually inclusive: not all middle classes are supportive of environmentalism.

2 This latter position is reflective of Dobson's attention to ecologism. It is interesting to note Dobson's call for a separation of environmentalism – which is easily integrated into other ideologies, such as socialism or feminism – from ecologism which he argues is an ideology in its own right. Much of the debate over tourism is about the absorption or integration of environmental concerns into existing approaches, whether they be political ideology (eco-socialism) or hegemonic strategies (the adoption of green concerns by major transnational tourism companies, for example).

7   THE INDUSTRY

1 Council Directive of 13 June 1990 (990/314/EEC), *Official Journal of the European Communities* L158/59, 23 June 1990.

8   'HOSTS' AND DESTINATIONS

1 David Western's succession of Dr Leakey as director of the KWS has brought about a change of attitude to the wildlife/human conflict in Kenya since 1994. Most Maasai await evidence that this change will shift the balance of power in their favour.

9   GOVERNMENTS AND TOURISM

1 This section was prepared by Alison Stancliffe, and was specially commissioned for this book.

2 The SLORC renamed the country 'Myanmar' in 1989.

APPENDIX 1
TRAVEL-RELATED BULLETIN
BOARDS AND WEB SITES ON
THE INTERNET

BULLETIN BOARDS FOR THOSE INTERESTED IN
TRAVEL AND TOURISM

Explorer

'For the Soft Adventure and Ecotourism Industry. This list was established to provide a forum for the discussion of industry issues, needs and concerns.' (email communication, 1996)

To subscribe, send an email message to

majordomo@newnorth.net

and in the message, put

subscribe explorer

Green-Travel

This is 'a moderated[1] mailing list for the sharing of information about culturally and environmentally responsible, or sustainable, travel and tourism worldwide'. (email communication, 1996)

To subscribe, send an email message to

majordomo@igc.apc.org

and in the message, put

subscribe green-travel

---

1 A 'moderated' list means that communications to the list are vetted, so that those which do not support the stated purpose of the list can be rejected.

Tourism

A discussion group aimed at academics and researchers working in all areas of tourism. Issues include the teaching of tourism at undergraduate and postgraduate levels and research into tourism.

To subscribe, send an email message to

mailbase@mailbase.ac.uk

and in the message, put only the following (substituting appropriately)

join tourism *<firstname>* *<lastname>*

Trinet

(Tourism Research Information Network). Intended to facilitate exchange of information on research projects, references, conferences, grant proposals and tourism curricula and education. 'The use of TRINET will be limited to research-oriented and scholarly exchanges.'

To subscribe, send a request to

psheldon@uhmtravel.tim.hawaii.edu

WEB SITES ON TRAVEL AND TOURISM

The GNN Travellers' Centre

A collection of tourism-related information. Its goal is to use the internet as a tool for travel research, as a vehicle for travel journalism and as a forum for information exchange.

To explore it, go to

http://www.gnn.com

The Lonely Planet site

Information about the Lonely Planet series of publications.

To explore it, go to

http://www.lonelyplanet.com.au

The Mountain Forum

Information about 'mountain conservation issues and equitable and ecologically sustainable development in mountains' (email communication, 1996).

The site is located at

http://www.mtnforum.org

Tourism Concern

Tourism Concern is a membership network of people with an active concern for tourism's impact on community and environment. It advocates tourism that is just, participatory and sustainable.

Its web site is at

http://www.oneworld.org/tourconcern

APPENDIX 2
REFERENCES FOR SELECTED
TECHNIQUES OF
SUSTAINABILITY
MEASUREMENT

Stated Preference surveys

Fowkes, A., Nash, C. and Tweddle, G. (1989) *Valuing the Attributes of Freight Transport Quality: Results of the Stated Preference Survey*, Leeds: University of Leeds Institute for Transport Studies.

Freeman-Fox and Associates (1976) *Gloucestershire: A Survey of Public Attitudes*, part 2, *General Methods and Tables*, London: Freeman-Fox Associates and Gloucestershire County Council.

Jones, P., Ferguson, D. and Bradley, M. (1986) *Combining the Household Activity Approach with Stated Preference as a Tool for Policy Making: The Case of Peak Spreading in Adelaide*, Oxford: Transport Studies Unit, Oxford University.

Jones, P. *et al.* (1990) *Trondheim Toll Ring: Stated Preference Study: Pilot Survey Assessment: Interim Report on the Full Study*, Oxford: Transport Studies Unit, Oxford University.

Marks, P. and Fowkes, A. (1986) 'Stated Preference experiments concerning long distance business travel in Great Britain', Leeds: University of Leeds Institute for Transport Studies, Working Paper 219.

Pearce, D. (1993) *Economic Values and the Natural World*, London: Earthscan.

Polak, J.W. (1992) 'Recent developments in the use of Stated Preference techniques for the valuation of travel time, Transport Studies Unit, Oxford University, paper presented to seminar on "The Value of Travel Time"', Institute of Transport Economics, Oslo, May 1992.

Propper, C. (1988) *Estimation of the Value of Time Spent on NHS Waiting Lists Using Stated Preference Methodology*, York: University of York.

Wardman, M. (1987) *The Distribution of Individual Values of Time: An Empirical Study Using Stated Preference Data*, Leeds: University of Leeds Institute for Transport Studies.

Contingent Valuation

Bateman, I., Willis, K. and Garrod, G. (1993) *Consistency between Contingent Valuation Estimates: A Comparison of Two Studies of UK National Parks*, Newcastle: Department of Agricultural Economics and Food Marketing, University of Newcastle upon Tyne.

Garrod, G. and Willis, K. (1990) *Contingent Valuation Techniques: A Review of Their Unbiassedness, Efficiency and Consistency*, Newcastle: Department of Agricultural Economics and Food Marketing, University of Newcastle upon Tyne.

Giles, N. (1995) *A Socio-economic Valuation of Grasslands Rich in Wildlife Resources: Using Contingent Valuation Methodology*, Plymouth: University of Plymouth.

O'Doherty, R. (1993) 'Public participation in the development of local plans and the role of Contingent Valuation', Working Papers in Economics 4, Department of Economics, University of West of England.

O'Doherty, R. (1993) 'Using Contingent Valuation to enhance public participation in local planning: empirical evidence on the value of open space', Working Papers in Economics 7, Department of Economics, University of West of England.

Pearce, D. (1993) *Economic Values and the Natural World*, London: Earthscan.

Pearce, D. and Moran, D. (1994) *The Economic Value of Biodiversity*, London: Earthscan.

Willis, K. and Garrod, G. (1991) *Landscape Values: A Contingent Valuation Approach and Case Study of the Yorkshire Dales National Park*, Newcastle: Department of Agricultural Economics and Food Marketing, University of Newcastle upon Tyne.

The Delphi Technique

Cooper, J., Daly, P. and Headicar, P. (1979) 'West Yorkshire Transportation Studies': 2, 'Accessibility analysis', *Traffic Engineering & Control* 20: 27–31.

Green, H. and Hunter, C. (1992) 'The environmental impact assessment of tourism development', in Johnson, P. and Barry, T. (eds) *Perspectives on Tourism Policy*, London: Mansell.

Linstone, H. and Turoff, M. (eds) (1975) *The Delphi Method: Techniques and Applications*, Reading, Mass: Addison-Wesley.

Ludlow, J. (1971) *The Delphi Method: A Systems Approach to the Utilization of Experts in Technological and Environmental Forecasting*, Godstone: University Microfilms.

Taylor, R.E. and Judd, L.L. (1994) 'Delphi forecasting', in Witt, S. and Moutinho, L. (eds) *Tourism Marketing and Management Handbook*, 2nd edn, Hemel Hempstead: Prentice-Hall.

Wytconsult (1976) 'West Yorkshire Transportation Studies: issues, objectives, measures and standards', Wytconsult, document 104.

Sustainability indicators

Andersen, V. (1991) *Alternative Economic Indicators*, London: New Economics Foundation (NEF), Routledge.

Environmental Challenge Group (1994) *Environmental Measures: Indicators for the UK Environment*, London: RSPB, WWF-UK, NEF.

Jacksonville Community Council (annual) *Quality Indicators of Progress*, Jacksonville, Fla.: JCC.

Local Government Management Board (LGMB) (1995) *Indicators for Local Agenda 21: A Summary*, Luton: LGMB.

London Borough of Merton (1995) *Indicators for a Sustainable Future*, London: LBM.

McGillivray, A. and Zadek, S. (eds) (1995) *Accounting for Change: Indicators for Sustainable Development*, London: New Economics Foundation.

Pearce, D., Markandya, A. and Barbier, E. (1994) *Blueprint 3: Measuring Sustainable Development*, London: Earthscan.

Sustainable Seattle (1993) *Indicators for Sustainable Community 1993*, Seattle, Wash.

APPENDIX 3
A SELECTION OF ORGANISATIONS CONCERNED WITH THE IMPACTS OF TOURISM

Annapurna Conservation Area Project (ACAP) – Nepal
Arbeitskreis Tourismus und Entwicklung (ATE) – Switzerland
Association Sénégalaise de Développement Rural Intègre (ASDRI) – Senegal
Campaign for Environmentally Responsible Tourism (CERT) – UK
Center for Responsible Tourism – USA
Center for Solidarity Tourism – Philippines
Centro de Attenzione al Turismo (CAT) – Italy
Earth Preservation Fund – USA
Ecumenical Coalition on Third World Tourism (ECTWT) – Thailand
Ecumenical Travel Office – Cyprus
Equations – India
Hawaii Ecumenical Coalition on Tourism (HECOT) – USA
Jagrut Goenkaranchi Fouz (JGF) – India
Justice in Tourism Network – Aotearoa/NZ
KOLAM Alternate Tours and Soft Travel – India
Retour – Netherlands
Rural Development by Tourism – Thailand
Stichting Toerisme & Derde Wereld – Netherlands
Studienkreis für Tourismus e.V. (SfT) – Germany
Third World Tourism European Ecumenical Network (TEN) – Germany
Tourism Concern – UK
Tourism Task Force (World Council of Churches) – Switzerland
Tourism Watch – Germany
Transverses – France

REFERENCES

Adams, K. (1984) 'Come to Tana Toraja, "Land of the Heavenly Kings": travel agents as brokers of ethnicity', *Annals of Tourism Research* 11, 3: 469–85.
—— (1991) 'Distant encounters: travel literature and shifting images of the Toraja of Sulawesi, Indonesia', *Terrae Incognitae* 16: 84–92.
Adams, W. (1990) *Green Development: Environment and Sustainability in the Third World*, London: Routledge.
Aditjondro, G. (1995) 'Focus on Bali', *Contours* 7, 3: 15–21.
Adler, J. (1989) 'Origins of sightseeing', *Annals of Tourism Research* 16: 7–29.
Adventure Travel Society (1996a) 'The Sixth Annual World Congress on Adventure Travel and Ecotourism: a step to wilderness and nature's magic', Englewood, Col.: ATS, 14.
—— (1996b) 'Adventure travel business', Englewood, Col.: ATS.
Albers, P. and James, W. (1988) 'Travel photography: a methodological approach', *Annals of Tourism Research* 15: 134–58.
Allan, N. (1988) 'Highways to the sky: the impact of tourism on South Asian mountain culture', *Tourism Recreation Research* 13: 11–16.
Allen, J. (1992) 'Post-industrialisation and Post-fordism', in S. Hall, D. Held and T. McGrew (eds) *Modernity and Its Future*, Oxford: Polity Press.
Allen, J. and Hamnett, C. (eds) (1995) *A Shrinking World? Global Unevenness and Inequality*, Milton Keynes: Open University Press.
Allen, J. and Massey, D. (eds) (1995) *Geographical Worlds*, Milton Keynes: Open University Press.
Amandala (1992) 'Ecology, economics, tourism and terrorism', editorial, *Amandala* 14 August: 6.
Arden-Clarke, C. (1992) 'Presentation to trade and environment policies after UNCED seminar' in A. Taylor and J. Gordon (eds) *Trade and Environment Policies After UNCED: Reconciling the Irreconcilable?*, London: South–North Centre for Environmental Policy and Global Environment Research Centre.
Arnold, M. (1992) 'Mundo Maya update: Mayan world . . . or illusion', *Tourism Link*, Belize City: Belize Tourism Industry Association.
Asociación de Talamanca para el Ecoturismo y la Conservación (ATEC) (Talamanca Association for Ecotourism and Conservation) (1991) 'What is eco-tourism?', Limón, Costa Rica.
Association of Independent Tour Operators (1996a) 'An introduction to the Association of Independent Tour Operators', London: AITO.
—— (1996b) 'The AITO Directory of Real Holidays', London: AITO.
Attenborough, D. (1986) *State of the Ark*, London: Routledge.
Aylward, B., Allen, K., Echeverría, J. and Tosi, J. (1996) 'Sustainable ecotourism in

Costa Rica: the Monteverde Cloud Forest', *Biodiversity and Conservation* 5, 3: 315–44.

Barnwell, R. (1991) 'Reserved for the future', *WWF 1991 Review*, London: World Wildlife Fund, 14–15.

Barrett, F. (1989) *The Independent Guide to Real Holidays Abroad*, London: *Independent*.

—— (1990) *The Independent Guide to Real Holidays Abroad*, London: *Independent*.

—— (1992) 'Tourists seek alternative to sun and sand', *Independent* 26 August: 34.

—— (1994) 'Slow boats to China replace lager louts', *Independent* 10 January: 10.

Barthes, R. (1981) *Camera Lucida*, London: Fontana.

Baudrillard, J. (1988) *America*, London: Verso.

Belize, Government of (1989) *Integrated Tourism Policy and Strategy Statement*, Belmopan: Government of Belize.

Belize Centre for Environmental Studies (1994) 'The complexity of carrying capacity', *The Centre Forum* January/February: 1.

Betz, H. (1992) 'Postmodernism and the new middle class', *Theory, Culture and Society* 9: 93–114.

Bird, C. (1995) 'Communal lands, communal problems', *In Focus* 16: 7–8, 15.

Bishop, M. (1983) *Selected Speeches 1979–1981*, Casa de las Americas.

Blake, B. and Becher, A. (1991) 'The new key to Costa Rica', San José: Publications in English.

Blunt, A. (1994) *Travel, Gender, and Imperialism*, London: Guildford Press.

Boggan, S. and Williams, F. (1991) 'WWF bankrolled rhino mercenaries', *Independent on Sunday* 7 November: 6.

Boo, E. (1990) *Ecotourism: The Potential and Pitfalls*, vols I and II. Baltimore, Md: World Wildlife Fund.

Boorstin, D. (1961) *The Image: A Guide to Pseudo Events in America*, New York: Harper Row.

Borzello, A. (1991) 'Postcard from a truck', *In Focus* 2: 20.

—— (1994) 'The myth of the traveller', *In Focus* 14: 7.

Bourdieu, P. (1984) *Distinction: A Critique of the Judgement of Taste*, London: Routledge & Kegan Paul.

—— (1986) 'From rules to strategies', *Cultural Anthropology* 1: 110–20.

—— (1987) 'What makes a social class? On the theoretical and practical existence of groups', *Berkeley Journal of Sociology* 22: 1–17.

Bourdieu, P. and Eagleton, T. (1994) 'Doxa and common life: an interview', in S. Zizek (ed.) *Mapping Ideology*, London: Verso.

Bradt, H. (1989) *Backpacker's Africa*, Chalfont St Peter, Bucks: Bradt Publications.

Brandon, K. (1993) 'Basic steps toward encouraging local participation in nature tourism projects', in K. Lindberg and D. Hawkins (eds) *Ecotourism: A Guide for Planners and Managers*, North Bennington, Vt: Ecotourism Society.

Briassoulis, H. (1992) 'Environmental impacts of tourism: a framework for analysis and evaluation', in H. Briassoulis and J. van der Straaten (eds) *Tourism and the Environment*, London: Kluwer Academic.

British Airways (1995) *Annual Environment Report 1995*, Hounslow: British Airways.

Britton, R. (1979) 'The ubiquitous tourist brochure', *Annals of Tourism Research* 6, 3: 318–29.

Britton, S. (1981a) 'Tourism and economic vulnerability in small Pacific states: the case of Fiji', monograph 23, Canberra: Development Studies Centre, Australian National University.

—— (1981b) 'Tourism, dependency and development: a mode of analysis', occasional paper 23, Canberra: Development Studies Centre, Australian National University.

—— (1981c) 'The spatial organisation of tourism in a neo-colonial economy: a Fiji case study', *Pacific Viewpoint* 21, 2: 144–65.

—— (1982) 'The political economy of tourism in the Third World', *Annals of Tourism Research* 9: 331–58.

—— (1991) 'Tourism, capital and place: towards a critical geography of tourism', *Environment & Planning D: Society & Space* 9, 4: 451–78.

Britton, S. and Clarke, W. (1987) *Ambiguous Alternative: Tourism in Small Developing Countries*, Fiji: University of the South Pacific.

Brockelman, W. and Dearden, P. (1990) 'The role of nature trekking in conservation: a case study in Thailand', *Environmental Conservation* 17, 2: 141–8.

Brooks, E. (1990) 'The hidden Kingdom of Mustang', *Geographical Magazine* September: 15.

Bruner, E. (1989) 'Of cannibals, tourists, and ethnographers', *Cultural Anthropology* 4, 4: 438–45.

Bryden, J. (1973) *Tourism and Development: A Case Study of the Commonwealth Caribbean*, London: Cambridge University Press.

Budowski, G. (1995) 'Responsible tourism: new trends', address to Conference on Sustainable Tourism, San José, Costa Rica.

Bugnicourt, J. (1977) *Tourism with No Return*, Development Forum, Geneva: United Nations.

Burns, P. and Holden, A. (1995) *Tourism: A New Perspective*, London: Prentice-Hall.

Butler, R. (1975) 'Tourism as an agent of social change', occasional paper 4, Department of Geography, Trent University, Ontario.

—— (1980) 'The concept of a tourist area cycle of evolution: implications for management of resources', *Canadian Geographer* 24, 11: 5–12.

—— (1991) 'Tourism, environment, and sustainable development', *Environmental Conservation* 18, 3: 201–9.

Calder, S. (1994a) 'Make sure you live to tell the tale', *Independent* 5 November: 35.

—— (1994b) 'The myth makers' guide to the world', *In Focus* 14: 6.

Cardenal, L. (1991) 'Río San Juan: El repoblamiento amenaza la vida del bosque', *Barricada* 17 January: 4.

Caribbean Conservation Corporation (1991) *Proyecto Paseo Pantera: Preserving Biological Diversity in Central America*, New York: Caribbean Conservation Corporation.

Carothers, A. (1993) 'The green machine', *New Internationalist* 246: 14–16.

Carriere, J. (1991) 'The crisis in Costa Rica: an ecological perspective', in D. Goodman and M. Redclift (eds) *Environment and Development in Latin America: The Politics of Sustainability*, Manchester: Manchester University Press.

Cater, E. (1992) 'Profits from paradise', *Geographical Magazine* March: 17–20.

—— (1994) 'Ecotourism in the Third World: problems and prospects for sustainability', in E. Cater and G. Lowman (eds) *Ecotourism: A Sustainable Option?*, Chichester: Wiley.

—— (1995) 'Consuming spaces: global tourism', in J. Allen and C. Hamnett (eds) *A Shrinking World? Global Unevenness and Inequality*, Milton Keynes: Open University Press.

Chambers, R. (1994a) 'The origins and practice of participatory rural appraisal', *World Development* 22, 7: 953–69.

—— (1994b) 'Participatory rural appraisal (PRA): analysis of experience', *World Development* 22, 9: 1253–68.

—— (1994c) 'Participatory rural appraisal (PRA): challenges, potentials and paradigm', *World Development* 22, 10: 1437–54.

Chatterjee, P. (1995) 'Mexico: World Bank to bail out banks by cutting environment, other loans', Washington, DC: Inter-Press Third World News Agency.

Chatterjee, P. and Finger, M. (1994) *The Earth Brokers: Power, Politics and World Development*, London: Routledge.

Child, B. (1996) 'The practice and principles of community-based wildlife management in Zimbabwe: the CAMPFIRE programme', *Biodiversity and Conservation* 5, 3: 369–98.

Child, G. (1996) 'The role of community-based wild resources management in Zimbabwe', *Biodiversity and Conservation* 5, 3: 355–68.

Chomsky, N. (1989) *The Culture of Terrorism*, London: Pluto Press.

Chung, H. (1994) 'People's spirituality and tourism', *Contours* 6, 7/8: 19–24.

Clark, J. (1990) 'Carrying capacity: the limits to tourism', paper presented to the Congress on Marine Tourism, Hawaii.

Clark, S. (1983) 'Introduction', in B. Marcus and M. Taber (eds) *Maurice Bishop Speaks: The Grenada Revolution 1979–1983*, New York: Pathfinder Press.

Cohen, E. (1972) 'Toward a sociology of international tourism', *Social Research* 39, 1: 164–82.

—— (1974) 'Who is a tourist?: a conceptual clarification', *Sociological Review* 22, 4: 527–55.

—— (1979a) 'A phenomenology of tourist experiences', *Sociology* 13: 179–201.

—— (1979b) 'The impact of tourism on the hill tribes of Northern Thailand', *Internationales Asienforum* 10, 1–2: 5–38.

—— (1979c) 'Rethinking the sociology of tourism', *Annals of Tourism Research* 6, 1: 18–35.

—— (1985) 'Tourism as play', *Religion* 15: 291–304.

—— (1989) '"Primitive and remote" hill tribe trekking in Thailand', *Annals of Tourism Research* 16, 1: 30–61.

Colchester, M. (1994) 'Salvaging nature: indigenous peoples, protected areas and biodiversity conservation', discussion paper, Geneva: United Nations Research Institute for Social Development.

Coward, R. (1996) 'Sun, sand and encounters with otherness', *Guardian* 27 May: 11.

Crick, M. (1989) 'Representations of international tourism in the social sciences: sun, sex, sights, savings, and servility', *Annual Review of Anthropology* 18: 307–44.

Croall, J. (1995) *Preserve or Destroy: Tourism and the Environment*, London: Calouste Gulbenkian Foundation.

—— (1996) 'On the road to disaster', *Guardian* 14 August: 5.

Crompton, R. (1993) *Class and Stratification*, Oxford: Polity Press.

Culler, J. (1988) 'The semiotics of tourism', in J. Culler (ed.) *Framing the Sign: Criticism and Its Institutions*, Oxford: Basil Blackwell.

Cultural Survival Quarterly (1982) 'The tourist trap: who's getting caught?', *Cultural Survival Quarterly* 6, 3.

—— (1990a) 'Breaking out of the tourist trap: part one', *Cultural Survival Quarterly* 14, 1.

—— (1990b) 'Breaking out of the tourist trap: part two', *Cultural Survival Quarterly* 14, 2.

Curran, J., Ecclestone, J., Oakley, G. and Richardson, A. (eds) (1986) *Bending Reality: The State of the Media*, London: Pluto Press.

Daltabuit, M. and Pi-Sunyer, O. (1990) 'Tourism development in Quintana Roo, Mexico', *Cultural Survival Quarterly* 14, 1: 9–13.

Daly, H. (1993) 'From adjustment to sustainable development: the obstacle of free trade', in Earth Island Press (ed.) *The Case Against Free Trade: GATT, NAFTA and the Globalisation of Corporate Power*, San Francisco, Cal.: Earth Island Press.

de Kadt, E. (ed.) (1979) *Tourism: Passport to Development?*, Oxford: Oxford University Press.

Dearden, P. (1988) 'Tourism in developing societies: some observations on trekking in highlands of North Thailand', in L. D'Amore and J. Jafari (eds) *Tourism: A Vital Force for Peace*, Montreal: First Global Conference: 207–16.

—— (1991) 'Tourism and sustainable development in Northern Thailand', *Geographical Review* 81: 400–13.

Dearden, P. and Harron, S. (1992) 'Tourism and the hilltribes of Thailand', in B. Weiler and C. Hall (eds) *Special Interest Tourism*, London: Belhaven Press.

—— (1993) 'Alternative tourism and adaptive change', *Annals of Tourism Research* 21: 81–119.

Deihl, C. (1985) 'Wildlife and the Maasai', *Cultural Survival Quarterly* 9, 1: 37–40.

Dobson, A. (1995) *Green Political Thought*, London: Routledge.

Doxey, G. (1975) 'A causation theory of visitor–resident irritants: methodology and research inferences', in *The Impact of Tourism Proceedings of the Travel Research Association 6th Annual Conference*, San Diego, Cal.

—— (1976) 'When enough's enough: the natives are restless in old Niagara', *Heritage Canada* 2: 26–7.

Drake, S. (1991) 'Local participation in ecotourism projects', in T. Whelan (ed.) *Nature Tourism*, Washington, DC: Island Press.

Eagleton, T. (1991) *Ideology*, London: Verso.

Eames, A. (1994) 'Train cruises', *High Life* September: 86–96.

Eckersley, R. (1986) 'The environment movement as middle-class elitism: a critical analysis', *Regional Journal of Social Issues* 18: 24–36.

—— (1989) 'Green politics and the new class: selfishness or virtue?', *Political Studies* xxxvii: 205–23.

—— (1992) *Environmentalism and Political Theory: Towards an Ecocentric Approach*, London: University College London Press.

Economist (1995) 'Asia goes on holiday', *Economist* 20 May: 57–8.

Economist Intelligence Unit (1995) 'The Gambia', in *EIU Country Report*, 3rd quarter, London: EIU.

Edwards, R. (1992) 'Rape of the Himalayas', *Guardian Weekly* 19 and 21 June: 25.

Elkington, J. and Hailes, J. (1992) *Holidays that Don't Cost the Earth*, London: Gollancz.

Epler Wood, M. (1991) 'Global solutions: an ecotourism society', in T. Whelan (ed.) *Nature Tourism*, Washington, DC: Island Press.

—— (1994) 'Membership Directory background', *ecotourism society 1994 International Membership Directory*, Bennington, Vt.: Ecotourism Society.

Errington, F. and Gewertz, D. (1989) 'Tourism and anthropology in a post-modern world', *Oceania* 60: 37–54.

Ethical Consumer (1994) 'The EC Ecolabel', *Ethical Consumer* 32, November.

Evans, K., O'Hare, G. and Thompson, C. (1994) 'Soft tourism facing hard choices in The Gambia', *In Focus* 13: 10 and 17.

Evans-Pritchard, D. (1993) 'Mobilisation of tourism in Costa Rica', *Annals of Tourism Research* 20, 4: 778–9.

Evans-Pritchard, D. and Salazar, S. (1992) 'What is ecotourism?', San José: Eco Institute of Costa Rica and ULACIT (Universidad Latino americana de Ciencia y Tecnología).

Everitt, J. (1987) 'The torch is passed: neocolonialism in Belize', *Caribbean Quarterly* 33, 3/4: 42–59.

Fanon, F. (1967) *The Wretched of the Earth*, London: Penguin.

Featherstone, M. (1987) 'Lifestyle and consumer culture', *Theory, Culture and Society* 4, 1: 55–70.

—— (1988) 'In pursuit of the postmodern: an introduction', *Theory, Culture and Society* 5, 2/3: 195–217.

—— (1991) *Consumer Culture and Postmodernism*, London: Sage.

Feifer, M. (1985) *Going Places*, London: Macmillan.

Ferguson, J. (1990) *Grenada: Revolution in Reverse*, London: Latin America Bureau.

Fernandes, D. (1994) 'The shaky ground of sustainable tourism', *TEI Quarterly Environment Journal* 2, 4: 4–35.

Fletcher, J. (1989) 'Input-output analysis and tourism impact studies', *Annals of Tourism Research* 16, 4: 514–29.

Flynn, M. (1996) 'Report on Guatemala', *Mesoamerica* 15, 8: 3–4.

Forbes, R. and Forbes, M. (1993) 'Special interest travel: creating today's market driven experiences', in J. Ritchie and D. Hawkins (eds) *World Travel and Tourism Review*, Wallingford: CAB International.

Forsyth, T. (1996) *Sustainable Tourism: Moving From Theory to Practice*, London: Tourism Concern.

Forum for the Future (1996) 'Securing the future', pamphlet, Cheltenham, Forum for the Future.

Foucault, M. (1980) *Power and Knowledge*, Hemel Hempstead: Harvester & Wheatsheaf.

Frank, A. (1991) 'For a sociology of the body: an analytical review', in M. Featherstone, M. Hepworth and B. Turner (eds) *The Body: Social Process and Cultural Theory*, London: Sage.

Friends of the Earth (1992) 'The eleven days that tried to change the world', letter to members and supporters, London, Friends of the Earth.

Frueh, S. (1988) *Tourism to Protected Areas*, Baltimore, Md: World Wildlife Fund.

Fukuyama, F. (1989) 'The end of history?', *The National Interest* summer: 1–18.

—— (1992) *The End of History and the Last Man*, London: Penguin.

Fussell, P. (1980) *Abroad: British Literary Travelling between the Wars*, Oxford: Oxford University Press.

Galand, P. (1994) 'I don't want to be an accomplice', *Envío* 12, 153: 12–13.

Gehrels, B. and Rankin, A. (1995) 'Conservation: pandas before people?', *Orbit* winter: 8–9.

Gellhorn, M. (1990) 'Too good for tourists', *Independent Magazine* 3 November: 70–74.

Ghai, D. (ed.) (1994) *Development and Environment: Sustaining People and Nature*, Oxford: Blackwell.

Giddens, A. (1989) *Sociology*, Cambridge: Polity Press.

Gonsalves, P. (1993) 'Divergent views: convergent paths: towards a Third World critique of tourism', *Contours* 6, 3/4: 8–14.

—— (1995) 'Structural adjustment and the political economy of the Third World', *Contours* 7, 1: 33–9.

Goodall, B. (1992) 'Environmental auditing for tourism', in C. Cooper (ed.) *Progress in Tourism, Recreation and Hospitality Management*, vol. 4, *Environmental Issues*, London: Belhaven.

Gordon, R. (1990) 'The prospects for anthropological tourism in Bushmanland', *Cultural Survival Quarterly* 14, 1: 6–8.

—— (1994) 'A powerhouse of revenue', *Financial Times* 30 September: xvi.

Gordon-Walker, R. (1993) 'A real adventure (and so cheap!)', *Independent* 13 August: 19.

Gouldner, A. (1979) *The Future of Intellectuals and the Rise of the New Class*, London: Macmillan.

Grant, M. (1996) 'Influencing the "other"', *In Focus* 19: 16.

Green, D. (1995) *Silent Revolution: The Rise of Market Economics in Latin America*, London: Cassell.

Green, H. and Hunter, C. (1992) 'The environmental impact assessment of tourism development', in P. Johnson and B. Thomas (eds) *Perspectives on Tourism Policy*, London: Mansell.

344

Green, W. (1991) 'Rio San Juan Exploratory', informal report for Twickers World and Costa Rica Expeditions.

Green Flag International (1990) press release, 18 December.

Green Horizons Travel (1995) 'Holidays that don't cost the earth', information pack, Green Horizons Travel.

Gregory, D. (1994) *Geographical Imaginations*, Oxford: Blackwell.

Guardian (1993) 'Pass Notes: package holidays', *Guardian* 20 August: 3.

Guerrin, M. (1991) 'For Cartier-Bresson the focus is now firmly on drawing', *Guardian Weekly* 8 December: 14.

Gunn, C. (1994) *Tourism Planning*, London: Taylor & Francis.

Gunson, P. (1996) 'Marketing men put curse of tourism industry on Mayas', *Guardian* 28 September: 14.

Habermas, J. (1981) 'New social movements', *Telos* 49: 33–7.

Hall, C. and Jenkins, J. (1995) *Tourism and Public Policy*, London: Routledge.

Hall, D. and Kinnaird, V. (1994) 'Ecotourism in Eastern Europe', in E. Cater and G. Lowman (eds) *Ecotourism: A Sustainable Option?*, Chichester: Wiley.

Hall, S. (1992a) 'The question of cultural identity', in S. Hall, D. Held and T. McGrew (eds) *Modernity and Its Future*, Oxford: Polity Press.

—— (1992b) 'The West and the Rest: discourse and power', in S. Hall and B. Gieben (eds) *Formations of Modernity*, Oxford: Polity Press.

—— (1995) 'New cultures for old', in D. Massey and P. Jess (eds) *A Place in the World? Places, Cultures and Globalization*, Milton Keynes: Open University Press.

Ham, M. (1995) 'Cashing in on the silver-backed gorilla', *Features: Ecotourism Special*, London: Panos Institute.

Hamilton, A. (1995) 'Ecocentrics', *Guardian Weekend* 17 June: 54–6, 59.

Harrison, D. (1992) *Tourism in Less Developed Countries*, London: Belhaven.

Harrison, P. (1979) *Inside the Third World*, London: Penguin.

Harvey, D. (1973) *Social Justice and the City*, London: Edward Arnold.

—— (1989a) *The Urban Experience*, Oxford: Blackwell.

—— (1989b) *The Condition of Postmodernity*, Oxford: Blackwell.

Hawkins, R. and Middleton, V. (1993) 'The environmental practices and programmes of travel and tourism companies', in J. Brent Richie and D. Hawkins (eds) *World Travel and Tourism Review: Indicators, Trends and Issues*, Wallingford: CAB International, 163–71.

—— (1994) 'International environmental regulation and control', in S. Witt and L. Moutinho (eds) *Tourism Marketing and Management Handbook*, Hemel Hempstead: Prentice-Hall.

Hayward, T. (1994) *Ecological Thought: An Introduction*, Oxford: Polity Press.

Herman, E. (1992) *Beyond Hypocrisy: Decoding the News in an Age of Propaganda*, Boston, Mass.: South End Press.

Higinio, E. and Munt, I. (1993) 'Eco-tourism gone awry', *NACLA Report on the Americas* xxvi, 4: 8–10.

Hills, T. and Lundgren, T. (1977) 'The impact of tourism in the Caribbean', *Annals of Tourism Research* 4: 248–57.

Hirsch, F. (1976) *The Social Limits to Growth*, Cambridge, Mass.: Harvard University Press.

Hitchcock, M., King, V. and Parnwell, M. (eds) (1993) *Tourism in South-East Asia*, London: Routledge.

Holder, J. (1990) 'The Caribbean: far greater dependency on tourism likely', *Courier* 122 July/August: 74–9.

Holmberg, J., Thomson, K. and Timberlake, L. (eds) (1993) *Facing the Future: Beyond the Earth Summit*, London: International Institute of Environment and Development and Earthscan.

Hong, E. (1985) *See the Third World While it Lasts: The Social and Environmental Impact of Tourism with Special Reference to Malaysia*, Penang: Consumers' Association of Penang.

Hough, J. (1988) 'Obstacles to effective management of conflicts between national parks and surrounding human communities in developing countries', *Environmental Conservation* 15, 2: 129–36.

Huyssen, A. (1984) 'Mapping the postmodern', *New German Critique* 33: 5–52.

Imam, A. (1994) 'SAP is really sapping us', *New Internationalist* 257: 12–13.

Inglehart, R. (1977) *The Silent Revolution: Changing Values and Political Styles among Western Publics*, Princeton, NJ: Princeton University Press.

—— (1981) 'Post-materialism in an environment of insecurity', *American Political Science Review* 75: 880–900.

Ingram, C. and Durst, P. (1989) 'Nature oriented tour operators: travel to developing countries', *Journal of Travel Research* 28, 2: 11–15.

Inskeep, E. (1987) 'Environmental planning for tourism', *Annals of Tourism Research* 14: 118–35.

—— (1991) *Tourism Planning: An Integrated and Sustainable Development Approach*, New York: Van Nostrand Reinhold.

International Institute for Environment and Development (1994) *Whose Eden? An Overview of Community Approaches to Wildlife Management*, London: IIED.

International Union for the Conservation of Nature (1985) 'United Nations list of national parks and protected areas', Gland: IUCN.

—— (1989) *Recursos: Annual Report IUCN Programme for Central America*, San José: IUCN.

Jackson, P. (1991) 'Mapping of meanings: a cultural critique of locality studies', *Environment & Planning A* 23, 2: 215–28.

—— (1992) *Maps of Meaning*, London: Routledge.

Jacobs, M. and Stott, M. (1992) 'Sustainable development and the local economy', *Local Economy* 7, 3: 212–31.

Jager, M. (1986) 'Class definition and the esthetics of gentrification: Victoriana in Melbourne', in N. Smith and P. Williams (eds) *Gentrification of the City*, London: Allen & Unwin.

Jameson, F. (1984) 'Postmodernism, or the cultural logic of late capitalism', *New Left Review* 146: 53–92.

—— (1991) *Postmodernism, or the Cultural Logic of Late Capitalism*, London: Verso.

Jardine, C. (1994) 'Beware: rough road ahead', *Daily Telegraph* 3 November: 13.

Johnston, B. (1990) 'Introduction: breaking out of the tourist trap', *Cultural Survival Quarterly* 14, 1: 2–5.

Josephides, N. (1994) 'Tour operators and the myth of self-regulation', *In Focus* 14: 10–11.

Kallen, C. (1990) 'Ecotourism: the light at the end of the terminal', *E Magazine* July/August.

Kay, C. (1989) *Latin American Theories of Development and Underdevelopment*, London: Routledge.

Kaye, M. (1994) 'Costa Rica at the crossroads: mass development or nature based tourism?', in 'A call for an international dialogue on the future of Costa Rica tourism', San José: mailing list on Internet.

Kincaid, J. (1988) *A Small Place*, London: Virago.

King, A. (1995) 'Migrations, globalisation and place', in D. Massey and P. Jess (eds) *A Place in the World?*, Milton Keynes: Open University.

Klatchko, J. (1991) 'Sherpa trek', *New Internationalist* 222: 27.

Kohl, J. (1996) 'Recruiting a PRA team for Ecuador', email communication, 21 March.

Korten, D. (1996) 'Development is a sham', *New Internationalist* 278: 12–13.

Krippendorf, J. (1987) *The Holidaymakers: Understanding the Impact of Leisure and Travel*, London: Heinemann.

Kuhn, T. (1962) *The Structure of Scientific Revolutions*, Chicago: University of Chicago Press.

Kutay, K. (1989) 'The new ethic in adventure travel', *Buzzworm: The Environmental Journal* 1, 4: 31–6.

Kvalöy, S. (1993) 'A tale of two countries', *Resurgence* 159: 14–17.

Lascelles, D. (1992) 'Conflict and dilemmas', in J. Quarrie (ed.) *Earth Summit '92*, London: Regency Press.

Lash, S. (1991) *Sociology of Postmodernism*, London: Routledge.

Lash, S. and Urry, J. (1987) *The End of Organized Capitalism*, Cambridge: Polity Press.

—— (1994) *Economies of Signs and Space*, London: Sage.

Lavery, P. (1971) *Recreational Geography*, London: David & Charles.

Lea, J. (1988) *Tourism and Development in the Third World*, London: Routledge.

—— (1993) 'Tourism development ethics in the Third World', *Annals of Tourism Research* 20: 701–15.

Leiss, W. (1983) 'The icons of the marketplace', *Theory, Culture and Society* 1, 3: 10–21.

Lewis, D. (1990) 'Conflict of interests', *Geographical Magazine*, December: 18–22.

Lindberg, K. and Hawkins, D. (eds) (1993) *Ecotourism: A Guide for Planners and Managers*, North Bennington, Vt: Ecotourism Society.

Lindberg, K. and Huber, R. (1993) 'Economic issues in ecotourism management', in K. Lindberg and D. Hawkins (eds) *Ecotourism: A Guide for Planners and Managers*, North Bennington, Vt: Ecotourism Society.

Linden, I. (1993) 'Free lunches and free markets', *CIIR News*, December: 3.

Lipietz, A. (1995) *Green Hopes: The Future of Political Ecology*, Oxford: Polity Press.

Lipman, G. (1992) 'The role of travel and tourism industry – promoting sustainable ecotourism', paper presented at Royal Geographical Society Conference, 'Ecotourism – a sustainable option?', London.

Lohmann, L. (1991) 'Who defends biological diversity?', in V. Shiva, P. Anderson, H. Schuching, A. Gray, L. Lohmann and D. Cooper (eds) *Biodiversity: Social and Ecological Perspectives*, Penang: World Rainforest Movement.

McAfee, K. (1991) *Storm Signals: Structural Adjustment and Development Alternatives in the Caribbean*, London: Zed Books.

MacCannell, D. (1973) 'Staged authenticity: arrangements of social space in tourist settings', *American Journal of Sociology* 79, 3: 589–603.

—— (1976) *The Tourist: A New Theory of the Leisure Class*, New York: Sulouken.

—— (1992) *Empty Meeting Grounds: The Tourist Papers*, London: Routledge.

McCellan, D. (1986) *Ideology*, Milton Keynes: Open University Press.

McClarence, S. (1995) 'A discrete charm', *Independent on Sunday* 3 September: 55.

McClintock, A. (1994) 'Soft-soaping empire: commodity racism and imperial advertising', in G. Robertson, M. Mash, L. Tickner, J. Bird, B. Curtis and T. Putnam (eds) *Travellers' Tales*, London: Routledge.

McGrew, A. (1992) 'The state in advanced capitalist countries', in J. Allen, P. Braham and P. Lewis (eds) *Political and Economic Forms of Modernity*, Oxford: Polity Press.

—— (1995) 'World order and political space', in J. Anderson, C. Brook and A. Cochrane (eds) *A Global World? Re-ordering Political Space*, Milton Keynes: Open University Press.

Machlis, G. and Tichnell, D. (1985) *The State of the World's Parks: An International Assessment for Resource Management, Policy and Research*, Boulder, Col.: Westview Press.

McIvor, C. (1994) 'Management of wildlife, tourism and local communities in Zimbabwe', discussion paper, Geneva: United Nations Research Institute for Social Development.

—— (1995) 'Clash on environment', *New African*, December: 35.

McKercher, B. (1993) 'Some fundamental truths about tourism: understanding tourism's social and environmental impacts', *Journal of Sustainable Tourism* 1, 1: 6–16.

McNeeley, J., Miller, K., Reid, W., Mittermeir, R. and Werner, T. (1990) *Conserving the World's Biological Diversity*, Washington: IUCN, World Resources Institute, Conservation International, WWF-US and World Bank.

McNeely, J., Thorsell, J. and Ceballos-Lascurain (undated) *Guidelines: Development of National Parks and Protected Areas for Tourism*, Madrid and Paris: World Tourism Organisation and United Nations Environment Programme.

McNeil, J. (1996) *Costa Rica: The Rough Guide*, London: Rough Guides.

Madeley, J. (1996) 'Foreign exploits: transnationals and tourism', briefing paper, London: Catholic Institute for International Relations.

Mader, R. (1996) 'Honduras notes', email communication, 8 October.

Mahr, J. and Sutcliffe, S. (1996) 'Come to Burma', *New Internationalist* 280: 28–30.

Maldonado, T., Hurtado De Mendoza, L. and Saborio, O. (1992) *Análisis de Capacidad de Carga para Visitación en las Áreas Silvestres de Costa Rica*, San José: Fundación Neotrópica.

Marcus, B. and Taber, M. (eds) (1983) *Maurice Bishop Speaks: The Grenada Revolution 1979–1983*, New York: Pathfinder Press.

Marshall, J. (1994) 'Papagayo isn't eco-tourism', *Tico Times* 8 April: 2 and 36.

Marx, K. (1965) *Capital*, London: Lawrence & Wishart.

Mason, P. (1995) 'A rationale for codes of conduct in the Arctic', paper presented to conference on 'Shaping Tomorrow's North', Thunder Bay, Canada.

Mason, P. and Mowforth, M. (1995) 'Codes of conduct in tourism', occasional paper, University of Plymouth.

—— (1996) 'Codes of conduct in tourism', *Progress in Tourism and Hospitality Research*, 2: 151–67.

Massey, D. (1995) 'Imaging the world', in J. Allen and D. Massey (eds) *Geographical Worlds*, Milton Keynes: Open University Press.

Mathieson, A. and Wall, G. (1982) *Tourism: Economic, Physical and Social Impacts*, London: Longman.

Mesoamerica (1996) 'Costa Rica', *Mesoamerica* 15, 4: 9.

Miller, D. (1995) *Acknowledging Consumption*, London: Routledge.

Monbiot, G. (1994) *No Man's Land*, London: Macmillan.

—— (1995) 'No man's land . . . ', *In Focus* 15: 10.

Morris, F. (1995) 'Ecopaz travel/study seminars', email communication, 15 October, Terespolis, Brazil.

Morrison, P. (1995) 'No more heroes', *Wanderlust* January: 68.

—— (1996) 'Tales of the unexpected', *Wanderlust* February/March: 84.

Morrow, L. (1995) 'I came, I saw, I spoiled everything', *Time* 10 July.

Mortlock, E. (1988) *Postguide to Indonesia*, Ashbourne: Moorland.

Mowforth, M. (1996) 'Co-operativa Longo Maï, Costa Rica', *Newsletter of the Environmental Network for Central America* 19: 6–7.

Mulberg, J. (1993) 'Economics and the impossibility of environmental evaluation', paper presented to conference on 'Values and the Environment', Guildford: University of Surrey.

Munt, I. (1992) 'A great escape?', *Town and Country Planning* 61, 7/8: 212–14.

—— (1994a) 'Eco-tourism or ego-tourism?', *Race and Class* 36, 1: 49–60.

—— (1994b) 'The Other postmodern tourism: travel, culture and the new middle class', *Theory, Culture and Society* 11, 3: 101–24.

—— (1995) 'The travel virtuosos', *Contours* 7, 2: 29–34.

Murphree, M. (1996) 'Wildlife in sustainable development: approaches to community participation', in *African Wildlife Policy Consultation*, London: Overseas Development Administration.

Murphy, P. (1983) 'Perceptions and attitudes of decision-making groups in tourism centres', *Journal of Travel Research* 21, 3: 8–12.

—— (1985) *Tourism: The Community Approach*, London: Routledge.

Naipaul, V. (1962) *The Middle Passage*, London: Picador.

Nash, D. (1989) 'Tourism as a form of imperialism', in V. Smith (ed.) *Hosts and Guests: The Anthropology of Tourism*, Oxford: Blackwell.

Nelson, J. (1993) 'The great global greenwash: BM vs the environment', *CAQ* 45, spring.

New Internationalist (1984) 'Visions of poverty, visions of wealth: tourism in the Third World', *New Internationalist* 142, December.

—— (1988) 'Photography: how we see the world', *New Internationalist* 185, July.

—— (1993) 'Wish you were here', *New Internationalist* 245, July.

—— (1994) 'Simply . . . how Bretton Woods re-ordered the world', *New Internationalist* 257: 14–15.

Noronha, R. (1979) 'Paradise reviewed: tourism in Bali', in E. de Kadt (ed.) *Tourism: Passport to Development?*, Oxford: Oxford University Press.

Norris, S. (1994) 'Credit on a high road with no bank', *Independent* 23 July: 38.

O'Connor, J. (1984) *Accumulation Crisis*, Oxford: Blackwell.

Offe, C. (1985) 'New social movements: challenging the boundaries of institutional politics', *Political Science Review* 6, 4: 483–99.

Olerokonga, T. (1992) 'What about the Maasai?', *In Focus* 4: 6–7.

Olindo, P. (1991) 'The old man of nature: Kenya', in T. Whelan (ed.) *Nature Tourism*, Washington, DC: Island Press.

Open University (1995) *The Shape of the World: Study Guide 3*, Milton Keynes: Open University.

O'Riordan, T. (1978) 'Participation through objection: some thoughts on the UK experience', in B. Sadler (ed.) *Involvement and Environment: Proceedings of the Canadian Conference on Public Participation*, Edmonton: Environment Council of Alberta.

Otis, J. (1992) 'Belize counting the cost of pact with tourism developers', *The Nation* 7 September: 33.

Overseas Private Investment Corporation (1995) 'Program Handbook', Washington, DC: OPIC.

Oxfam (1995) *The Oxfam Poverty Report: A Summary*, Oxford: Oxfam.

Panos Institute (1995) 'Ecotourism: paradise gained or paradise lost?', Panos media briefing 14, London: Panos Institute.

Patterson, K. (1992) 'Aloha for sale', *In Focus* 4: 4–5.

Pattullo, P. (1996) *Last Resorts: The Cost of Tourism in the Caribbean*, London: Latin America Bureau.

Pearce, D. (1989) *Tourist Development*, London: Longman.

—— (1995) *Tourism Today: A Geographical Analysis*, London: Longman.

Pearce, D. and Moran, D. (1994) *The Economic Value of Biodiversity*, London: Earthscan.

Pearce, F. (1990) 'Exchanging chances', *Guardian* 13 April: 10.

Pearce, P. and Moscardo, G. (1986) 'The concept of authenticity in tourist experiences', *Australia and New Zealand Journal of Sociology* 22, 1: 121–32.

349

Pels, D. and Crebas, A. (1991) 'Carmen or the invention of a new feminine myth', in M. Featherstone, M. Hepworth and B. Turner (eds) *The Body: Social Process and Cultural Theory*, London: Sage.

People's United Party (1989) *Belizeans First*, Belize: PUP.

Perez, L. (1974) 'Aspects of underdevelopment in the West Indies', *Science and Society*, 37: 473–80.

—— (1975) 'Tourism in the West Indies', *Journal of Communications* 25: 136–43.

Permanent Peoples' Tribunal (1994) 'Verdict on IMF and World Bank policies with respect to international law and the right to self determination', *Envio* 12, 161: 32–9.

Picard, M. (1991) 'Cultural tourism in Bali', in M. Hitchcock, V. King and M. Parnwell (eds) *Tourism in South-East Asia*, London: Routledge.

Pleumarom, A. (1990) 'Alternative tourism: a viable solution?', *Contours* 4, 8: 12–15.

—— (1994) 'The political economy of tourism', *Ecologist* 24, 4: 142–7.

Plog, S. (1972) 'Why destination areas rise and fall in popularity', *Cornell HRA Quarterly* November: 13–16.

Poon, A. (1989a) 'Competitive strategies for Caribbean tourism: the new versus the old', *Caribbean Affairs* 2, 2: 74–91.

—— (1989b) 'Competitive strategies for a "New Tourism"', in C. Cooper (ed.) *Progress in Tourism, Recreation and Hospitality Management*, vol. 1, London: Belhaven.

—— (1993) *Tourism, Technology and Competitive Strategies*, Wallingford: CAB International.

Porritt, J. (1984) *Seeing Green*, Oxford: Blackwell.

Postman, N. (1985) *Amusing Ourselves to Death: Public Discourse in the Age of Show Business*, London: Methuen.

Pratt, M. (1992) *Imperial Eyes: Travel Writing and Transculturation*, London: Routledge.

Pretty, J. (1995) 'The many interpretations of participation', *In Focus* 16: 4–5.

Price, C. (1993) 'The irrelevance of discounted cash flow to environmental values', in *Proceedings of the Conference on Values and the Environment*, Faculty of Human Studies, University of Surrey.

Programme for Belize (1989) 'Goal and objectives of Programme', Belize City: Programme for Belize.

Prosser, R. (1994) 'Societal change and the growth in alternative tourism', in E. Cater and G. Lowman (eds) *Ecotourism: A Sustainable Option?*, Chichester: Wiley.

Rao, N. (1991) 'Tourist as a pilgrim: a critique', conference on 'Postmodernism and the Search for the Other', Delhi: Delhi University.

Richter, L. (1983) 'Tourism politics and political science: a case of not so benign neglect', *Annals of Tourism Research* 10: 313–35.

—— (1994) 'The political fragility of tourism in developing nations', *Contours* 6, 7/8: 32–8.

Richter, L. and Richter, W. (1985) 'Policy choices in South Asian tourism development', *Annals of Tourism Research* 12: 201–17.

Ritzer, G. (1993) *The McDonaldization of Society*, Newbury Park, Cal.: Pine Forge Press.

Robins, K. (1991) 'Tradition and translation: national culture in its global context', in J. Corner and S. Harvey (eds) *Enterprise and Heritage: Crosscurrents of National Culture*, London: Routledge.

Rojek, C. (1993) *Ways of Escape: Modern Transformations in Leisure and Travel*, London: Macmillan.

Rosenthal, R. (1991) 'Sustainable tourism: optimism v pessimism', *In Focus* 1: 2–3.

Rovinski, J. (1991) 'Private reserves, parks and ecotourism in Costa Rica', in T. Whelan (ed.) *Nature Tourism: Managing for the Environment*, Washington, DC: Island Press.

Rughani, P. (1993) 'From tourist to target', *New Internationalist* 245: 7–12.

Ryel, R. and Grasse, T. (1991) 'Marketing ecotourism: attracting the elusive ecotourist', in T. Whelan (ed.) *Nature Tourism: Managing for the Environment*, Washington, DC: Island Press.

Sachs, W. (1992a) 'Whose environment?', *New Internationalist* 232, June: 4–6.

—— (ed.) (1992b) *The Development Dictionary: A Guide to Knowledge as Power*, London: Zed Books.

—— (ed.) (1993) *Global Ecology*, London: Zed Books.

Said, E. (1991) *Orientalism*, London: Penguin.

San José Audubon Society (1992) 'An invitation to ecotourism', San José: SJAS.

Savage, M., Barlow, J., Dickens, P. and Fielding, T. (1992) *Property, Bureaucracy and Culture: Middle-class Formation in Contemporary Britain*, London: Routledge.

Schmidt, C. (1995) Personal communication, 'interview in Belize', 12 October, Punta Gorda, Belize.

Scott, A. (1992) 'Political culture and social movements', in J. Allen, P. Braham and P. Lewis (eds) *Political and Economic Forms of Modernity*, Milton Keynes: Open University Press.

Seabrook, J. (1995) 'Far horizons', *New Statesman and Society* 11 August: 22–3.

Selwyn, T. (1993) 'Peter Pan in South East Asia: the brochures', in M. Hitchcock, V. King and M. Parnwell (eds) *Tourism in South-East Asia*, London: Routledge.

—— (1994) 'Tourism and myth', *In Focus* 14: 4–5.

—— (1995) *The Tourist Image: Myth and Myth Making in Tourism*, London: Wiley.

Shaw, G. and Williams, A. (1994) *Critical Issues in Tourism: A Geographical Perspective*, Oxford: Blackwell.

Sherman, P and Dixon, J. (1991) 'The economics of nature tourism: determining if it pays', in T. Whelan (ed.) *Nature Tourism*, Washington, DC: Island Press.

Shiva, V. (1988) *Staying Alive: Women, Ecology and Development*, London: Zed Books.

—— (1993) 'The greening of the global reach', in W. Sachs (ed.) *Global Ecology*, London: Zed Books.

Shivji, I. (1973) *Tourism and Socialist Development*, Dar-es-Salaam: Tanzania Publishing House.

Short, L. (1991) 'Crafts: bridge or melting pot?', *Orbit* 42: 6.

Shurmer-Smith, P. and Hannam, K. (1994) *Worlds of Desire, Realms of Power*, London: Edward Arnold.

Sidaway, R. (1994) 'The limits of acceptable change', report prepared for Countryside Commission, Edinburgh.

Sills, M. (1991) 'On the development trail', *Geographical Magazine* September: 4–7.

Simons, P. (1988) 'Belize at the crossroads', *New Scientist* 29 October: 61.

Sklar, H. and Everdell, R. (1980) 'Who's who on the Trilateral Commission', in H. Sklar (ed.) *Trilateralism: The Trilateral Commission and Elite Planning for World Management*, Boston, Mass.: South End Press.

Sloan, E. (1993) 'Eco-tourists threatening to trample tiny Belize to death', *The Nation* 28 January: 32.

Smith, V. (ed.) (1989) *Hosts and Guests: The Anthropology of Tourism*, Philadelphia, Penn.: University of Pennsylvania Press.

Sniffen, J. (1995) 'UNEP impact assessment meetings', email communication, 20 June.

Soares, M. (1992) *Debt Swaps, Development and Environment*, Rio de Janeiro: Brazilian Institute for Economic and Social Analysis.

Sontag, S. (1979) *On Photography*, London: Penguin.

Spinrad, B. (1982) 'St. Lucia', in S. Seward and B. Spinrad, *Tourism in the Caribbean: The Economic Impact*, Ottawa: International Development Research Centre.

Srisang, K. (1992) 'Third World tourism: the new colonialism', *In Focus* 4: 2–3.

Stancliffe, A. (1995) 'Agenda 21 and tourism: an introductory guide', paper available from Tourism Concern, London.

Stauth, G. and Turner, B. (1988) 'Nostalgia, postmodernism and the critique of mass culture', *Theory, Culture and Society* 5: 509–26.

Stewart, J. and Hams, T. (1991) *Local Government for Sustainability*, Luton: Local Government Management Board.

Stonich, S., Sorenson, J. and Hundt, A. (1995) 'Ethnicity, class and gender in tourism developments: the case of the Bay Islands, Honduras', *Journal of Sustainable Tourism* 3, 1: 1–28.

Stott, D. (1989) 'Deforestation in Nepal', *Geographical Review* 3, 1: 32–5.

Survival International (1991) 'Tourism: special issue', *Survival* 28.

—— (1995) 'Tourism and tribal peoples', background sheet, London: Survival International.

—— (1996) 'Parks or people?', *Survival* 35.

—— (undated) 'Survival: a unique organisation for tribal peoples', booklet, London: Survival International.

Sutcliffe, S. (1995) 'Burma: the alternative guide', London: Burma Action Group.

Tapper, R. (1993) 'Tourism: is it a non-consumptive use?', Godalming: World Wildlife Fund-UK.

Teye, V. (1986) 'Liberation wars and tourism development in Africa: the case of Zambia', *Annals of Tourism Research* 13: 589–608.

—— (1988) 'Coups d'etat and African tourism: a study of Ghana', *Annals of Tourism Research* 15: 329–56.

Thomas, C. (1988) *The Poor and the Powerless: Economic Policy and Change in the Caribbean*, London: Latin America Bureau.

Thomas, N. (1994) *Colonialism's Culture: Anthropology, Travel and Government*, Oxford: Polity Press.

Thompson, C., O'Hare, G. and Evans, K. (1995) 'Tourism in The Gambia: problems and proposals', *Tourism Management* 16, 8: 571–81.

Thrift, N. (1989) 'Images of social change', in C. Hamnett, L. McDowell and P. Sarre (eds) *The Changing Social Structure*, London: Sage.

Thrupp, L. (1990) 'Environmental initiatives in Costa Rica: a political ecology perspective', *Society and Natural Resources* 3: 243–56.

Thurot, J. and Thurot, G. (1983) 'The ideology of class and tourism: confronting the discourse of advertising', *Annals of Tourism Research* 10: 173–89.

Tickell, O. (1992) 'After the Summit', *Green Line* July, 98: 3.

Tinker, J. (1992) 'Endpiece', *United Nations Conference on Environment and Development: A User's Guide*, special bulletin 1, United Nations Environment Programme-UK and International Institute for Environment and Development, London.

Tourism Concern (1992) *Beyond the Green Horizon: Principles for Sustainable Tourism*, Godalming: World Wildlife Fund.

—— (1994a) 'Greenwash contradictions', *In Focus* 14: 13.

—— (1994b) 'Rainforest SOS', *In Focus* 13: 8.

—— (1996a) *Trading Places: Tourism as Trade*, London: Tourism Concern.

—— (1996b) 'Our holidays, their homes', campaign update, spring, London: Tourism Concern.

—— (1996c) *Writing Your Dissertation on Sustainable Tourism: Tips for Student Projects*, London: Tourism Concern.

—— (no date) *Trekking in the Himalayas*, information sheet, London: Tourism Concern.

Truong, D. (1991) *Sex, Money and Morality: Prostitution and Tourism in South East Asia*, London: Zed Books.

Turner, C. and Manning, P. (1988) 'Placing authenticity – on being a tourist: a reply to Pearce and Moscardo', *Australia and New Zealand Journal of Sociology* 24, 1: 136–9.

Turner, L. (1974) *Multinational Companies and the Third World*, London: Allen Lane.

—— (1976) 'The international division of leisure: tourism and the Third World', *World Development* 4, 3: 253–60.

Turner, L. and Ash, J. (1975) *The Golden Hordes: International Tourism and the Pleasure Periphery*, London: Constable.

Ultra-Unlimited (1991) 'Eco-tours out of Cobán', brochure, Cobán: Guatemala.

UNCED (1992) 'Resolution 44/228', New York: United Nations.

United Nations Commission on Human Rights (1993) 'Report on the situation of human rights in Myanmar', New York: UNCHR.

United Nations Development Programme (1994) *Human Development Report*, New York and Oxford: UNDP/Oxford University Press.

United Nations Environment and Development-UK (1993) 'Convention update: Biodiversity Convention, 30 ratify', *Connections* autumn: 9.

—— (1994) 'Travel and tourism – the environment is our business', *Connections* autumn: 7.

Urry, J. (1990a) *The Tourist Gaze*, London: Sage.

—— (1990b) 'The consumption of "tourism"', *Sociology* 24: 23–35.

van den Abbeele, G. (1980) 'Sightseers: the tourist as theorist', *Diacritics* 10: 3–14.

Veblen, T. (1925) *The Theory of the Leisure Class*, London: Allen & Unwin.

Wall, G. (1993) 'International collaboration in the search for sustainable tourism in Bali', *Journal of Sustainable Tourism*, 1, 1: 38–47.

Wall, G. and Dibnah, S. (1992) 'The changing status of tourism in Bali, Indonesia', in C. Cooper (ed.) *Progress in Tourism, Recreation and Hospitality Management*, vol. 4: 120–30, London: Belhaven.

Walter, J. (1982) 'Social limits to tourism', *Leisure Studies* 1: 295–304.

Warde, M. (1992) 'Where thieves prosper', *Guardian Weekend* 14 November: 51–2.

Warford, J. (1987) 'Personal communication to M. Mowforth from Acting Director of the World Bank's Environment Department', 4 August, Washington, DC.

Watson, G. and Kopachevsky, J. (1996) 'Tourist carrying capacity: a critical look at the discursive dimension', in C. Cooper and A. Lockwood (eds) *Progress in Tourism and Hospitality Research*, vol. 2 no. 2, Chichester: Wiley.

The Way Ahead (1996) 'Forum for the short term?', *The Way Ahead* 26: 4–5.

Weaver, D. (1991) 'Alternatives to mass tourism in Dominica', *Annals of Tourism Research* 18, 3: 414–32.

Weber, W. (1993) 'Primate conservation and ecotourism in Africa', in C. Potter, J. Cohen and D. Janczewski (eds) *Perspectives on Biodiversity: Case Studies of Genetic Resource*, Washington, DC: American Association for the Advancement of Science, 129–50.

Wells, M. and Brandon, K. (1992) *People and Parks: Linking Protected Area Management with Local Communities*, Washington, DC: World Bank, World Wildlife Fund and US Agency for International Development.

West, P. and Brechin, S. (eds) (1991) *Resident Peoples and National Parks*, Tucson, Ariz.: University of Arizona Press.

Wheat, S. (1994) 'Tourism Concern interview', *In Focus* 14: 8–9.

Wheeler, T. and Lyon, J. (1994) *Guide to Bali and Lombok*, Hawthorn, Victoria: Lonely Planet.

Wheeller, B. (1992) 'Alternative tourism: a deceptive ploy', in C. Cooper and A. Lockwood (eds) *Progress in Tourism, Recreation and Hospitality Management*, vol. 4, London: Belhaven.

—— (1993a) 'Sustaining the ego', *Journal of Sustainable Tourism* 1, 2: 129–39.

—— (1993b) 'Willing victims of the ego trap', *In Focus* 9: 14.

—— (1996) 'In whose interest?', *In Focus* 19: 14–15.

Whelan, T. (ed.) (1991) *Nature Tourism: Managing for the Environment*, Washington, DC: Island Press.

Whitehorn, K. (1963) quoted in TIPA (Truth and Integrity in Public Affairs) occasional paper 7: 9.

Wight, P. (1994) 'Environmentally responsible marketing of tourism', in E. Cater and G. Lowman (eds) *Ecotourism: A Sustainable Option?*, Chichester: Wiley.

Williams, R. (1988) *Keywords*, London: Fontana.

Witt, S. and Moutinho, L. (eds) (1994) *Tourism Marketing and Management Handbook*, Hemel Hempstead: Prentice-Hall.

Wolff, I. (1991) 'Weekending in El Salvador', *Independent on Sunday* 27 October: 59–61.

Wood, K. (1991) 'Belize cleans up in eco-tourism stakes', *Observer* 6 October: 65.

Wood, K. and House, S. (1991) *The Good Tourist*, London: Mandarin.

World Commission on Environment and Development (1987) *Our Common Future*, New York: United Nations.

World Conference on Sustainable Tourism (1995) 'Charter for sustainable tourism', Lanzarote, Canary Islands, 27–28 April.

World Tourism Organisation (1990) *Compendium of Tourism Statistics 1985–1989*, Madrid: WTO.

—— (1991) *What is WTO?*, Madrid: WTO.

—— (1994a) *Tourism Market Trends: Americas 1985–1994*, Madrid: WTO.

—— (1994b) *Tourism Market Trends: Africa 1985–1994*, Madrid: WTO.

—— (1994c) *Tourism Market Trends: East Asia and the Pacific 1985–1994*, Madrid: WTO.

—— (1994d) *National and Regional Tourism Planning: Methodologies and Case Studies*, London: Routledge.

—— (1995a) *Compendium of Statistics 1985–1989*, Madrid: WTO.

—— (1995b) *Global Tourism Forecasts to the Year 2000 and Beyond*, vol. 1, Madrid: WTO.

—— (1997) *Yearbook of Tourism Statistics*, vol. 1, Madrid: WTO.

World Travel and Tourism Council (1991) *Travel and Tourism in the World Economy*, Brussels: WTTC.

WTTC, WTO and Earth Council (1995) *Agenda 21 for the Travel and Tourism Industry: Towards Environmentally Sustainable Development*, London: WTTC.

Wright, E. (1985) *Classes*, London: Verso.

Wright, M. (1996) 'Have mouse, will travel', *Independent on Sunday* 10 March: 56.

Yearley, S. (1995) 'The transnational politics of the environment', in J. Anderson, C. Brook and A. Cochrane (eds) *A Global World? Re-ordering Political Space*, Milton Keynes: Open University Press.

Young, H. (1994) 'Community-based tourism development in Belize: government policies and plans in support of community initiatives', in *Guide to Community-based Ecotourism in Belize*, Belize: Ministry of Tourism and Environment and Belize Enterprise for Sustained Technology.

Zaba, B. and Scoones, I. (1994) 'Is carrying capacity a useful concept to apply to human populations?', in B. Zaba and J. Clarke (eds) *Environment and Population Change*, Liège: Ordina Editions.

Zukin, S. (1987) 'Gentrification: culture and capital in the urban core', *Annual Review of Sociology* 3: 129–47.

INDEX

Note: Page numbers in **bold** type refer to **figures**
Page numbers in *italic* type refer to *tables*
Page numbers followed by 'N' refer to notes

355

Picard, M. 313
Pleumaron, A. 173
Plog, S. 109
poaching 176, 267
politics 281–8; global 34, 35; tourism
    230, 281–8
pollution 128, 271
Poon, Auliana 52, 54, 55
Porritt, Jonathon 161, 214
positive local sanctions 87
post-Fordism 26, 30, 101; consumption
    56; and tourism *27*
Postmodernism 31–3, 126
power 37, 38–43, 44–83;
    environmentalism 181–4; geography
    43; relationships 4, 278, *319*
Pratt, M. 109, 270
preservation 172–9
Pretty, J. 113, 238; typology of
    participation 240, *241*, 245, 247
Price, C. 222–3
professionalisation 138
profit maximisation 199
Programme for Sustained Development
    (PSD): Gambia 283
Prosser, R. 91
protected areas: categories *178*; global
    growth 177
psychological capacity 250
Puerto Rico 228
Puerto Viejo de Talamanca 226
Pululahua Geobotanical Reserve 246

Quintana Roo, Mexico 52, 181, 182

racism 69, 80, 173; commodity 76
rail travel: colonial 69
rainforest 66–7, 174, 186, 233; Belize
    183
Rao, N. 38, 70
Rara Avis 227
Rara Lake: National Park 48
record of achievement (ROA) 138
Recursos 184
refugees 243
regions: peripheral 91
research tourism 232
resettlement 262–9
Richter, L. 153, 281, 282
Rio Summit (1992) 22, 23, 169, 295
Ritzer, G. 16
Robins, K. 39
Rojek, C. 135, 146

Romero: Archbishop Oscar 81
Rosenthal, R. 128
Royal Geographical Society (RGS) 137
Rughani, Pratap 200
Rwanda 266–9
Ryel, R. and Grasse, T. 90, 201

Sachs, W. 42, 51, 78, 185
safari 258; Kenya 95; walking 73
Said, E. 59, 78, 270
St Lucia 194
Samuel, Robert 132
San Blas Archipelago 259
San George, Robert 176
San José 310
San José Audubon Society (SJAS) 211
Sarawak 274, 275
satellite linkage 274
Savignac, Antonio Enriquez 197
Schmidt, C. 255
Schoorl, Jaap 268, 269
Scones, I. and Zaba, B. 40
Seabrook, J. 53
self-mobilisation 244
Selwyn, T. 84, 200
Serengeti National Park 263, 265
services: consumption 32; sector industry
    25
sex tourism 330N
Shaba reserve: Kenya 48
Shanghai 15
Sherman, P. and Dixon, J. 267
Sherpa 272; Expeditions 128, 135
Shining Path guerilla movement 75
Shiva, V. 51, 169, 181
Shivji, I. 50
Short, L. 59
Shurmer-Smith, P. and Hannam, K. 60
Sideway, R. 251
Sloan, E. 310
Smith, V. 86, 89, 109, 153
Soares, M. 170
social class 125–8, 133; reproduction
    145; taste 130
social costs 120
social tourism 53
social-perceptual capacity 250
socio-environmental organisations
    156–87
Sontag, S. 78, 79
South Africa 174
South America 60, 76, 80
Spinrad, B. 194